Animal Bioethics: Principles and Teaching Methods

This book was partly developed within the AFANet Socrates Thematic Network for Agriculture, Forestry, Aquaculture and the Environment (2000-2004). This project has been carried out with the support of the European Community in the framework of Socrates programme. The content of this book does not necessarily reflect the position of the European Community, nor does it involve any responsibility on the part of European Community.

Animal Bioethics

Principles and Teaching Methods

Edited by:

M. Marie
S. Edwards
G. Gandini
M. Reiss
E. von Borell

Wageningen Academic
P u b l i s h e r s

Subject headings:
Moral philosophy
Education
Communication

Photo cover by Jean-Pierre Avois

ISBN 9076998582

First published, 2005

Wageningen Academic Publishers
The Netherlands, 2005

Preface

It is a matter of fact that, nowadays, animal bioethics is assuming a growing importance in our societies. Issues such as, for example, the conditions of farm animals' breeding and housing, transport and slaughter, the introduction of genetic engineering into animal production, the use of animals in experimentation, or wildlife in captivity for public demonstration, entertainment or even in a domestic environment as 'new' companion animals, are subjects of discussion and question our conceptions of the human-animal relationship. This situation can be attributed to multiple causes, including the evolution of practices and uses of animals, the abuses and misuses which have been brought to the attention of the public, the higher sensitivity of the general public to these issues, the growing influence of pressure groups and the consideration of such societal demands at the political level, and the evolution of the moral framework in a post-modern, multicultural, global society. As a consequence, there has been a reappraisal of the legal framework, first at different paces in individual European countries, then in a co-ordinated way in Europe from 1978 onwards. The subject is now under consideration at the international level under the auspices of the OIE (the World Organisation for Animal Health).

This situation emphasises the increasing moral responsibility of professionals in their actions involving animals, and the necessity for them to communicate with the wider society in this respect. In order to respond to these concerns, many higher education institutions are engaged in a process of setting up new courses and curricula devoted to societal demands and the moral aspects of human action, and, specifically in veterinary, agronomy and science studies, to animal bioethics. This trend is rather new: as revealed by a survey performed in 2003 in 17 European countries, 43% of such courses were created during the previous five years. It appears also that the pedagogic practice in such courses is quite variable between institutions in terms of content, volume, background of the teachers, and teaching methods. However, these courses represent only a small share of the global curricula, and the topic is still largely excluded from biology textbooks. Accordingly, there is a need to share experience regarding this subject in terms of materials and teaching methods in order to help teachers to develop new courses and to spread good practice. This book, aimed at teachers, students, professionals, and every individual wanting to develop skills for communication and moral reasoning, is designed to fill this gap.

In the first chapters, this book addresses fundamental aspects of animal bioethics: concepts relating to the human-animal relationship and the corresponding moral positions are examined from different perspectives, according to the historical, philosophical, religious, cultural and economic points of view. The contributions of antic, virtue, deontologic, utilitarianist, contractualist or post-modern ethics are reviewed. Is humankind unique and different from animals (dualism), or do we all belong to the same community (monism), and what consequences should be inferred from the answer about ethical conduct towards animals? What are the implications linked to theo-, anthropo-, zoo- (or patho-), bio- or eco-centred world views? These questions are also tackled. Whilst animal welfare is often considered as the first, if not the only, entry to bioethics, broader approaches such as the respect for animal integrity or intrinsic value (based on cases such as blind or featherless hens obtained by selective breeding, or pigs genetically engineered in order to be used for xeno-transplantation) have also to be taken into account. Through the different contributions, the points of view of different parties and driving forces such as stockpersons, scientists, consumers, animal protectionist associations, lawyers, and economic forces, are explored. It is hoped that this in-depth analysis will contribute to the enhancement of the content of the teaching and give an understanding of the dynamics of the concept of bioethics, in the past as well as in the future.

The second part of this book is about objectives, strategies and methods of teaching and communication. Beyond developing knowledge on fundamental aspects of ethics, with special reference to animal bioethics, the objectives are to increase sensitivity and awareness about this subject, to develop skills in moral reasoning and judgement for problem solving, and, also, to shape the moral psychology and moral behaviour of individuals.

The proposed methods are mainly student-centred, facilitating an active participation of the learner, with intensive use of the problem-based learning approach. They present a large range of complexity, from simple games, to exercises of variable length, case studies (either constructed or real-life dilemmas), role playing, and structured tools such as the ethical matrix and the reflexive equilibrium method. They are adapted to the different stages of the learning process or to the final evaluation, and help in supporting the different levels of attainment, from specific phases (such as motivational, or analytical steps) to elaborated handling of complex cases.

Description of existing courses, with reviews of experience, emphasizing interdisciplinary approaches and the link between ethical theory and concrete cases,

Animal Bioethics: Principles and Teaching Methods

on one hand, and the in-depth analysis of the content of a syllabus such as that devoted to animal welfare, on the other hand, will help in developing curricula, identifying relevant material and shaping 'ideal libraries'.

Three different indexes (i: philosophical concepts, schools, authors; ii: cases, situations; iii: teaching methods and objectives) have been designed to help locate relevant information and references within the book.

The concept of this book arose during the course of the AFANet project, the European Socrates Thematic Network for Agriculture, Forestry, Aquaculture and the Environment (2000-2004), within the Aristoteles activity devoted to animal bioethics teaching. This activity incorporated the above-mentioned survey of contemporary practice in bioethics teaching, two workshops (Nancy 2002: "Teaching animal bioethics in agricultural and veterinary higher education in Europe"; Dublin 2004: "Design and implementation of case studies in animal ethics teaching"), and the development of an on-line course database (http://www.ensaia.inpl-nancy.fr/bioethics/). We hope that it will help the sharing of experience and the spreading of good practice in this still evolving domain.

M. Marie
S. Edwards
G. Gandini
M. Reiss
E. von Borell

Contents

Contents

Contents

Contents

A mirror of myself? Monist and dualist views of animals

Monica Libell
Dept of History of Sciences and Ideas, Lund University, Sweden

Abstract

Mankind has always used animals as the Other, to compare himself with them. Animals have provided the mirror in which mankind has defined himself[1]. In their perceived similarities and dissimilarities with humans, he has distinguished the boundaries which separates him from the Other. Since antiquity, this relationship has been lively debated between at least two groups of interpreters. The "dualists" have contended that humans are unique and decidedly different from animals, whereas "monists" have argued that we are essentially the same, metaphysically and/or physiologically. Along with this ambiguous perception of animals, discussions concerning our moral relationship have followed. Dualists have often more or less dismissed human obligations to animals. Monists, on the other hand, have usually stressed our responsibilities. During the 19th century, the boundaries between these groups grew increasingly blurred as physiological evidence pointed to evolutionary similarities, indeed kinship. Monists - for instance, natural scientists - convinced about the evolutionary kinship between humans and animals, could emphasize a dualist ethical outlook. Similarly, dualists who based their beliefs on anthropocentric Christian thinking, nevertheless found themselves at times showing an unusual concern for animals due to civilizational or Christian demands. Only believers in a metaphysical unity, seem to have retained a monist worldview and ethics. Today, the issue still creates discussion. How similar are we to animals and how should these similarities inform our ethical conduct toward them?

1. Introduction

Humankind has an ancient relationship to other species. Since our species first evolved, we have been surrounded by other living beings. We have used other

---

[1] I am well aware of the ethical problems involved in using terms such as *mankind, he/him* when referring to humans, or *it* when referring to animals and similarly "*the* animal" or "*the* human being" as if they were a collective unit. However, as this is an historical text, I have chosen this usage to avoid anachronism.

species for a number of purposes. We have lived of them, eaten them, used their fur and skin; we have made them work for us; we have cut open their bodies to satisfy our curiosity and to search for physiological truths; we have developed strong emotional ties with some of them, and they have become intimate friends and life companions. Animals have constantly been a source for human satisfaction, wishes, and needs. From an existential perspective, animals have provided the mirror image of ourselves; we have seen ourselves in them and compared ourselves to them in order to understand ourselves. Animals have in this sense become the Other, the one who is different, yet familiar, who is threatening, yet sympathetic. This dialectic has led to an ambivalent relationship which has persevered through time and remains so still today.

Commonly people believe that we today have a better understanding of what animals constitute and therefore what the right relationship ought to be. But, looking at history, one can only conclude that although we today seem to have a much clearer idea of what different species need, their biological make-ups and mental states, we have not made any significant changes in our ethical conduct. Nor have our ethical arguments changed much. Both reasons for and against animal ethical concerns have stayed surprisingly stable throughout history. The arguments proposed by eminent moral philosophers today were already advanced by intellectuals 2000 years ago. A reason for this might be that most issues debated today were discussed then as well, such as vegetarianism, hunting, vivisection, recreational activities such as gladiator spectacles, and most other practices in which people exploited animals for human purposes. And not surprisingly, since human practices using animals have remained relatively stable over time, arguments have done the same. Although we today may think that animal ethics has decisively moved forwards, it has in fact made little progress if we compare the situation of animals with other groups oppressed through history, such as women and slaves. Whereas women secured legal rights for themselves in the early 20th century and slaves in the 19th century, when slavery was officially abolished, and in the 1960s with the civil rights movement, animal welfare advocates have not been able to outlaw any "field" of practice in the Western world in which animals are exploited, not even recreational cruelties, such as the modern gladiator spectacle, the Spanish bullfight[2].

---

[2] Individual practices such as bear baiting, cock fighting etc have of course been outlawed, but no entire field of animal use has been banned, for instance, hunting (including recreational hunting), animal experimentation, slaughtering. Only some moderations have been achieved in these fields.

Animal Bioethics: Principles and Teaching Methods

2. Monists and dualists

Historically, one can find three time periods when animal welfare topics were lively debated. The first phase was, as already mentioned, Antiquity. The second phase started in the late 18th century with the Enlightenment and continued through the 19th century up to the outbreak of the First World War. The last period begun in the revolutionary years in the late 1960s and 1970s when demands for emancipation also were extended to animals. Between these periods, one could argue, an anthropocentric worldview reigned, which gave primacy to human needs.

If one studies history more closely a specific issue emerges which seems to have permeated the entire debate and remains the key matter to this day. Indeed, it might unlock the mystery of human ambivalence and help explain why the debate over time has remained so surprisingly stable. It concerns the issue of human similarity or dissimilarity with animals. Are people and animals essentially the same or are there unique differences between them? If so, are these differences in degree or in kind?

Those who have opted for the position that humans and animals are essentially the same and have contended that only smaller variations separate the species, we can call monists. Monists generally have assumed a theory of unity, in which everything is considered related to each other. Those who, on the other hand, have viewed the human species as endowed with unique features, something it does not share with any other species, we can call dualists. The dualists divide up life in two categories: humans in the first and animals and plant life in the second. The features they typically single out as uniquely human are a highly developed intelligence and complex linguistic ability[3].

These two conflicting ideological perspectives are usually accompanied by specific ethical views. With few exceptions, the monist approach has been a greater advocate for animal welfare, whereas the dualist perspective has focused on human welfare. This division holds sway through the 18th century, but the distinctions are blurred in the 19th and the first half of the 20th century. It should, however, be pointed out that the question of similarity or dissimilarity is partly located on a mental and spiritual level rather than on a purely physiological. It is the *feeling* of being kindred or not being kindred with the animal world which has often decided the proponent's ethical position in the matter. Although natural scientific discoveries may have

---

[3] The terms "monist" and "dualist", are here used as simply meaning a perception of "unity" or "division". As basically philosophical concepts they are of course not limited to discussions concerning animal-human relationships.

further supported convictions of connectedness, they have seldom sufficed in themselves. In fact, as we will see, scientific studies into the physiological similarities have also been used to underscore the differences.

The monist and dualist positions will be the main focus of attention in this chapter. I will discuss how these views through history have formed our understanding of animal nature and informed our ethical conduct towards them.

The monist and dualist outlooks on life seem to have been founded already in Antiquity. Many ancient thinkers believed that humans and animals were related through an all-encompassing world-soul, and a belief in reincarnation was common. Whether one had been born into a person or another living being was chance. Hence, ancient thinkers urged people to be kind to animals; some groups, such as the Pythagoreans, even practiced vegetarianism. But in 4th century BC, a physician by the name of Alcmaeon decidedly contended that man was unique in his rational and linguistic abilities. These traits could not be found among any other species. In this way, Alcmaeon started the dualist tradition, which we commonly associate with Aristotle, namely that humankind should be distinguished from other living beings and situated at the top of the hierarchy of earthly life (Dierauer, 1998, p. 45)[4].

The more benevolent monist view of animals, as compared to the dualist view can also be found in the Bible. They can be seen as being represented in the two different myths of creation found in Genesis. The first myth tells us of a world of unity and equality in which nature is animated. Man and woman are created simultaneously, they are vegetarians and live of plants, fruits and berries. Animals are their companions and co-workers. In the second, which can be found directly following the first[5], nature is perceived as something that must be subjugated and tamed by hard and steady labor. The equal order that we could find in the first myth has been changed to a hierarchical structure. The man is created first, while animals and the woman are created later, in that very order. Man's control over animals and nature is established as Adam is assigned the task of naming them. He designates the beasts "by their own names", which denotes that their names in some way reflect their true natures. In this process, human language became linked to knowledge and ultimately to power over the Earth. Animals became subjugated to the rule of man and could be used in whatever way he wished. But not only animals were named by man; Eve also received her name "woman" from him. And as she lured him to eat the forbidden apple, one of God's punishments was to condemn women to a life in submission to men's rule. As in the case of animals, men acquired control over

---

[4] Dierauer contends that this tradition was not initiated by Aristotle, to whom it is usually credited.
[5] To note however, is that the second myth is considered to have been written before the first.

women. Western societies have chosen to follow the second creational myth, which has led Western society in a hierarchical, dualist, and patriarchal direction.

3. The victory of the dualist position

The strict hierarchical structure was most pertinently exhibited during the Middle Ages. The dualists argued that God had created man in his own image and that animals and nature were created solely as men's physical resources. Mankind therefore had no moral obligations to them. The monists, on the other hand, contended that creation was unified and that God had created animals with the same loving and benevolent hand with which he had created man. Humankind therefore had a pressing responsibility towards animals - as God's creatures.

Another monist idea trying to explain animals' presence on Earth was that they were sent as moral messengers, in that they embodied symbolic meanings. Or possibly man was a microcosm in whom all animals were represented. These anthropocentric apprehensions perceived animals only as significant for their roles in human physical, moral, and theological development. However, they also pointed to a similitude, even kinship, that transgressed the dualist boundary erected between animal and man. Some even suggested that man, enmeshed in this web of life, should follow nature in his actions (Harrison, 1998).

During the 17th century, the dualist approach solidified. The French philosopher René Descartes argued that mankind was peremptorily different from the animal kingdom. Created in the image of God, man must be unique; thus, there has to be a clear difference between man and animal, not one of degree but of kind. Whereas humans were endowed with rationality, Descartes argued, animals were only matter. In fact, animals could be compared to machines, without awareness and feelings. They even lacked the ability to experience pain. His theory gained great approval among the clergy, not because people necessarily believed it, actually he wasn't really convinced himself, but because it secured man's sacred position in God's creational plan. The case for dualism was strengthened.

Contesting animals' capacity to suffer can possibly be seen as an even more radical and powerful line of reasoning against the ethical consideration of animals than Alcmaeon's argument that they cannot think. Even if animals are not rational, man still might consider them sensate beings. But Descartes' theory of "animal automata" effectively destroyed this possibility for moral consideration.

However, at the end of the 18th century, this dualist approach was increasingly challenged. In an attack on the Cartesian theory of animals, Voltaire complained about the poor judgment shown by scientists of his day clinging to this superseded view, which portrayed animals as machines without consciousness and feeling. He wrote:

"Barbaric people take the dog, which so immensely excels man regarding friendship, and tie him down while alive to show you his veins. You discover in him all of the senses that you find in yourself. So answer me, you 'mechanist:' why has nature endowed him with of all these prerequisites for emotion, if he cannot feel? Has the animal received nerves in order to be insensitive to pain? Do not presume such a brazen contradiction in nature!" (Voltaire, 1764)

Voltaire's view foreshadowed a change in the moral perception of both animals and humans. During the latter part of the 17th century, a substantial civilizational and humanitarian drive took form, which aimed to educate and reform society. Charity work for poor people spread, opposition against slavery grew, demands for a more humane treatment of prisoners and old and sick people were pronounced, better care for children was demanded as well as for women's rights. In this company one also finds animal welfare concerns. Together these movements strove for a better, more moral and civilized society, in which no animal or human suffered evil.

Initially, the animal ethical idea underlying these reform efforts was dualist and anthropocentric, in that it stemmed from human needs. Kindness to animals was part of the agenda of educating people into good citizens and raising the standard of society. People who were pitiless toward animals, some argued, were at risk of becoming callous to their fellow humans, possibly even develop into criminals.

4. Animal similarities

Along with these civilizational demands, scientific studies also contributed. Through anatomical and physiological examinations, the physical body became the focus of increasing attention, and similarities between the human and the animal body became apparent. Continuing Voltaire's thoughts, many wondered whether animals were not indeed equipped with the same senses as humans. And did they not feel pain as intensely as we do? Since animals seemed to be endowed with they same organs as humans and exhibited a similar structure, it seemed reasonable to expect that they in many ways resembled us and shared many of our traits. But no one could say *how* similar they were and *how* these similarities should inform our moral conduct. German newspaper editor and enlightener F. H. Eichholz suggested that since animals probably are endowed with the same five senses as we are and therefore feel pain just as intensively as we do, we owe them direct duties. The moral treatment of animals, he added, may even be a decisive criterion for our humanity (Eichholz, 1805, p. 12).

With these concerns, it is not surprising that the ancient spiritual notion of kinship was revived. Was not reality essentially united, with all of its parts ultimately connected with each other? Vegetarian societies arose, which rejected bloody diets as well as traditional hierarchical thinking. Among these ideas, Indian and Buddhist notions flourished, in which vegetarianism along with a belief in reincarnation and compassion with other life was pronounced.

Possibly the only academics, aside from certain Christian theologians, who uncompromisingly promoted traditional dualist arguments, were natural law proponents such as Pufendorf, Thomasius, and Grotius, who contended that rights and law only apply to human interests and therefore exclusively regulate intrahuman relationships. But the larger trend of sympathy for animals, which had started in the latter part of the 18th century grew stronger throughout the 19th, a trend that even touched the über-rationalist Kant, who described the relationship between animals and humans in terms of (mutual) gratitude and love (Kant, 1924). Throughout the 19th century, animal welfare societies mushroomed all over Europe. Their primary focus lay on what they considered *cruelty*. Cruelty was somewhat vaguely defined as *unnecessary suffering*. The distinction between unnecessary and necessary suffering was just as nebulous as it is today, but often it boiled down to *public* abuse. Hence, to beat a dog in the street was perceived as offensive, whereas the same treatment at home was considered a private matter. It is worth noting that in many countries the same view governed husbands' treatment of their wives.

5. Natural scientific monism - ethical dualism

With the arrival of Charles Darwin's book in 1859, *On the Origin of Species*, the case for a biological kinship with animals was strengthened. Even if the Swedish botanist Carl von Linné already in the 18th century had classified the human species in the same order as the apes, there was much uncertainty as to what this kinship really entailed. Darwin asserted that nature in a constantly on-going evolutionary process had developed all existing species and that mankind was the result of this process, rather than of divine creation. This purely biological connection not only challenged the belief in a creational God but also the idea that humans differed from animals in an absolute fashion. How should these new insights inform our moral relationship with animals? Monists argued that if we are kindred with the lower animals through evolution, it seems reasonable to think that we know something about them and that animal welfare can no longer be solely derived from personal feelings but is a scientifically anchored issue. Hence, if we know that animals suffer in our hands, this insight has moral significance. In this way, the scientific perspective more or less willingly promoted a monist view of animals. But a common biological ancestry hardly changed the position of the dualists. Animals were still not rational beings and they were still unable to speak. Hence, our moral obligations need not be extended to them. Natural scientists kept this dualist ethical outlook while adopting a natural scientific monist perspective, footed on the Darwinian theory of evolutionary similarities. With the contention that animals and humans are closely related, the life sciences, including medicine, would benefit enormously. And by detaching ethical responsibilities from biological similarities - as was done in, for instance, experimental physiology, a science that spearheaded scientific medicine - physiology rocketed as a science in the latter part of the 19th century. But in society in general, the perception of similarity and kinship received an even greater cultural impetus, which substantially contributed to a benevolent attitude toward animals; the ever more popular keeping of pets. With diminishing contact with the countryside and without being financially dependent on animals for their livelihood, the growing urban middle and upper-class citizenry acquired pets, at an unforeseen rate. Primarily dogs became popular and they were readily turned into members of the family. Through the study of their behavior and psychological traits, people obtained both first-hand insight into the animal kingdom and recognized distinct traces of themselves in the nun-human Other. And with pets sharing home and hearth, the animals easily developed into individual personalities, to which humans developed strong emotional ties. The urban citizenry, which was decidedly dualist, ran into problems as their emotions were extended to individuals who were not humans. The solution was to distinguish dogs (and at times cats) from the rest of the animal kingdom. An unabashed anthropomorphism arose, which tried to show that dogs were more similar to humans than to other animal species. (Libell, 2001, 230 ff). See Figure 1.

Although the dualist line now was culturally drawn between humans and pets on one side, and animals in general on the other, the traditional dualist boundary was challenged. It could be argued that pets unknowingly acted as ambassadors for the whole animal kingdom. Through the knowledge people acquired from their interactions with their pets and the emotions they developed, human understanding of the living conditions and needs of other species increased and facilitated

Figure 1. Reproduced from Ernst von Weber, Die Folterkammern der Wissenschaft (Berlin und Leipzig 1879), p. 29. Image from a German antivivisection booklet, picturing a dog by the bar with a pint of beer and pipe. The message is to question the use of animals in experimentation, who to such a degree resemble humans.

sympathy for their plight. Animal welfare was slowly losing its former anthropocentric direction and more attention was given to the needs of the animals themselves. The concept of animal ethics and animal rights arose. The questions that appeared on the agenda of the animal welfare societies shifted. Not only was physical pain discussed but also mental and emotional distress. Issues that remind us of the animal welfare arguments of today.

What, however, one needs to remember is that these 19th-century humanitarian organizations remained dualist in their approach. They had their foothold in the conservative, religious, and upper strata of society and they generally kept an anthropocentric and Christian outlook, which emphasized the dignity of man. But through their emphasis on civilized and gentle conduct and the introduction of pet dogs as civilized newcomers in society, the cruel treatment of sensate beings was increasingly shunned.

6. Monist ethicists

There were however exceptions to the dualist perspective among the animal welfare proponents, thinkers who entertained a monist worldview that combined natural scientific findings with metaphysical ethics. A representative of this perspective was German philosopher Arthur Schopenhauer. He found that contemporary scientific findings in physiology revealed amazing similarities between humans and animals, and he even developed a pre-Darwinian evolutionary theory. Only the human brain seemed to present a significant difference. But the ethical significance lay, according to Schopenhauer, not primarily in the brain function but in the physiological and metaphysical unity. He believed in the Hindu notion of a primordial "Will," which permeates all entities in the universe and ties them together into one unified coherent reality. On the surface, every being may appear different and gives the impression that the world is made up of billions of unique individuals. As long as we remain on this superficial level of reality, dualism will reign and we perceive other individuals as peremptorily different from ourselves. This dualist illusion is according to Schopenhauer the root of all cruelty. As long as we see others as different from ourselves, only egoism can arise. But beneath the surface, Schopenhauer maintains, every creature is cast by the same primordial force. This uniting force is disclosed to us through compassion. The feeling of compassion cuts through the veil of superficial differences between individuals and reveals the sameness of the Other. We identify with the other being as we can feel her pain as ours. And since the force permeates both humans and animals, the process of identification and compassion expresses itself in both groups. Schopenhauer's monist worldview led him to an "ethics of compassion" and he rejected all cruelty and demanded animal rights (Schopenhauer, 1837, §19.7; Schopenhauer, 1851, §177).

Another thinker, who like Schopenhauer created a spiritual framework around the fascination for life, was Albert Schweitzer, a doctor in medicine, philosophy, and theology (awarded the Nobel peace prize in 1952). The foundation of his ethics is one's own feeling of being alive. Out of this intuitive insight, an existential context is disclosed: "I am life that wants to live, in the midst of life that wants to live." This biological wonder, which includes all life, rather than just human life, should be the concern of Christian love. Hence, he infused his ethics of "reverence for life" with a Christian dimension. As the creation of the universe is God's work, we shall show respect for all of its parts. Just as you revere your own life and your own will to live, he contended, you shall marvel at and respect other individuals' lives and will to live, regardless whether it is an ant, flower, pig or a human being. To Schweitzer, all life has equal worth and we may only harm other life in order to sustain our own. (Schweitzer, 1923, sections XVII, XXI) His monist ethics became a cornerstone in several animal ethical and environmental organizations up to the 1970s.

Schweitzer developed his ethics in conscious opposition to the threatening takeover of technology and the natural sciences in the 19th century. Life was increasingly seen as a source of resources for human exploitation. But he also differed from the usual metaphysical and humanist currents that often remained dualist, separating human life from other life forms. And eventually his and Schopenhauer's monist ethical systems, that tried to merge natural scientific findings with metaphysics, lost out to more natural scientifically minded thinkers, while humanists became increasingly marginalized.

Concomitant with Schweitzer's ethical efforts, a group of people appeared who based their claims on an ideology they labeled ecology, and they named themselves Monists. The league was organized under the guidance of the coiner of ecology, the German zoologist, Ernst Haeckel. He promoted a holistic anti-mechanistic approach to life and reality. A fierce enemy of Christianity, he attacked the dualist approach as anthropocentric, unsustainable, and artificial. But although the monists entertained an ethics which included animals, they were usually less concerned with animal ethical aspects than with, for instance, animals' interactions with their environment and "their place in the cycle of energy use" (Bramwell, 1989, 39 ff). And though retaining an ecologically spiritual approach, their perspective nevertheless entailed strong natural scientific elements that probably paved the way for even more purified natural scientific thinking.

7. The defeat of humanist ethics

As the century drew to a close, technology and the natural sciences were the pronounced victors. The success of natural scientific thinking and its methodology led to the attempt at extrapolating its thinking to other academic fields (primarily to medicine and the social sciences). The humanities (including the humanitarian dualists), with their emphasis on human culture, religion, and ethics, fitted poorly with the natural scientific perspective. As the first half of the 20th century saw the spreading and culmination of the monist natural scientific approach, the humanist dualists saw their former fields of domination disintegrating. The humanist values as guidelines for cultural behavior in society were replaced by natural scientific reasoning.

With the Nazi takeover in Germany a seemingly monist perspective was presented. The Darwinian theories of the late 19th century, which had successfully been used in the European conquering of, for instance, African territory and the establishment of colonies, were by the Nazis transferred to intra-European culture. Although Darwin himself had made no ethical claims in his theories, his notion of the survival of the fittest was by the European colonialists easily taken as evidence of the inferiority of the "African race," to which the European race stood out as evolutionary more evolved and fit. The same idea of inequality among races permeated the Nazi ideology, in which the Aryan people was singled out as the superior race. To this vision, they attached the Nietzschean ideal of predator mentality. This mindset cut through the dualist boundary between humans and animals. Nazis wished to foster a culture of strong, healthy, unsentimental Aryan people, which favored wild animal predators over pets and animal prey (see also Arluke and Sanders, 1989, 132-166)[6]. For instance, pet-keeping was found revolting, since it is based on the cultivations of emotions and a need for social security as it is linked to "an emphasis on individual autonomy," which of course fitted poorly with the Nazi demand for obedience (Sax, 1997). Hence, although the traditional dualism between humans and animals was challenged, a new dualism arose between, on the one side, the ideal of the purified, yet primitive Aryan man and animal predators and, on the other hand, groups of animals and humans deemed inadequate, weak, defective or in any other way not conforming to the ideal.

This quasi-monistic reign was terminated with the end of World War II and an austere anthropocentric era commenced. With the murderous consequences of the perceived monistic Nazi regime as a backdrop, an uncompromising division between human and animal value seemed warranted. Hence, the 1950s and 1960s showed little understanding for animal ethical concerns. Instead a busy era of mechanization and industrialization was begun, in which animals played little more than the role of physical resources for human needs.

However, burgeoning attempts were made within the humanities at recapturing some of the fields lost to the natural scientific monists. And in the 1970s, animal ethics again became a theme within the humanities. Moral philosophers were unsurprisingly first at the scene. In 1975, Peter Singer wrote the book *Animal Liberation,* which would play a momentous role in the modern animal ethical debate. In a monistic fashion, Singer drew parallels between racism, sexism, and speciesism, i.e. discrimination of other species. He kept the natural scientific belief in logic and applied it to ethics. By examining ethical criteria, he noted that traditional criteria such as language and rationality are valuable human traits, but they are not relevant ethical criteria. Whether one can communicate one's pain or not has little to do with the severity of the pain experienced. Instead, rather than language, he singled out the ability to experience pain as a decisive basic ethical factor. And with ethics based on this physiological criterion, he concluded that speciesism cannot be defended on logical grounds any more than racism can (Singer, 1993, 16-82).

Since Peter Singer's book, animal ethical debates have spread in Western society and for the last 20 years an ever-growing stream of books fill the shelves of the

---

[6] Hitler as well as other high-ranking Nazi officials held pets, but it was the image of wild animals that was conjured up as moral ideals. They held that beneath the veil of the domesticated tame pet, lured the true nature of the wild animal.

bookstores. Proponents from different academic and cultural fields seem to take an interest in the issue. One can find philosophers, feminists, sociologists, anthropologists, ethologists, historians, and also theologians and Christian advocates commenting. While theologians and Christian advocates earlier focused on human welfare, many today perceive no conflict by encompassing animals in their ethical thinking. They focus on the details of how animals can be integrated into the moral sphere, rather than on what grounds they can be excluded.

8. Concluding remarks

In this chapter, I have argued that throughout history, two currents of thought seem to have run parallel. The monist perspective has asserted that humans and animals are essentially the same - that they are related, metaphysically or biologically. The dualist approach, on the other hand, has maintained that there are fundamental differences between humans and other species, and it has specifically singled out rationality as a unique human feature. In this race for recognition, the dualist approach has continuously kept the upper hand. In the 19th century, the two perspectives got intertwined as Darwin's theory of evolution, made a case for biological kinship. Many dualists, among others, the natural scientists, although accepting the monist conclusions of the theory, retained a dualist ethics. Animals were our next of kin but they still lacked rationality and could therefore be used for human needs and wishes.

As the process of industrialization released dogs from their work-duties in society, they were easily turned into beloved pets by the urban citizenry. With the intimate interaction between human and pet, humans caught a glimpse of the animal mind and behavior. The status of dogs increased dramatically, but they tacitly also became ambassadors for other animals whose plight in society was increasingly recognized.

The Nazi regime in Germany brought an end to humanitarian ethics but it also cut across the usual dualist human/animal division. This division, which to many scholars in the post-war era has appeared as monist, still today remains a hindrance for the progress of animal ethics in the intellectual debate. Apart from bold attempts from (primarily) a number of moral philosophers (and a few American jurists), other academics and intellectuals favorably disposed to animal welfare arguments have maintained a dualist and anthropocentric attitude. The threatening image of Nazi Germany, necessitates a strict division between humans and other species. Any compromising stance, they fear, might relativize human value and result in a repetition of the murderous consequences of the Nazi regime (see also Noske, 1989,

80ff). However, the most fundamental challenge to animal ethics is probably the economic dependence on animal exploitation in human society. As in the case of dogs, history evinces that as long as animals are profoundly integrated into the economic web of society, as producers of meat, fur, skin, etc, substantial ethical concerns are difficult to advance. Notwithstanding the financial and ideological benefits of a dualist ethics, Peter Singer has argued that in order to avoid unjust discrimination, an impartial view and a monist value system is the only valid ethical basis[7]. Considering that dualist views throughout history have secured oppressive boundaries between men *and* women, white *and* black people etc, while today's monist approach has created a more equal society for, for instance, black people and women, it seems reasonable to assume that the adoption of a monist ethics might produce similar results to the benefit of animals.

References

Arluke, A. and Sanders, C., 1996. Regarding Animals, Philadelphia.

Bramwell, A., 1989. Ecology in the 20th century, New Haven and London.

Dierauer, U., 1998. "Das Verhältnis von Mensch und Tier im griechisch-römischen Denken" in Paul Münch (ed.): Tiere und Menschen. Geschichte und Aktualität eines prekären Verhältnisses, Paderborn, p. 37-84.

Eichholz, F.H., 1805, Aufklärung und Humanität, Mannheim.

Harrison, P., 1998. „The Virtues of Animals in Seveneteenth Century Thought" in Journal of the History of Ideas, 59.3, p. 463-484.

Kant, I., 1924. Eine Vorlesung Kants über Ethik (ed.) Paul Menzer, Berlin; See the section: "Von den Pflichten gegen Tiere und Geister."

Libell, M., 2001. Morality Beyond Humanity: Schopenhauer, Grysanowski and Schweitzer on Animal Ethics, Lund.

Noske, B., 1989. Humans and Other Animals, London and Winchester.

Sax, B., 1997. "What is a 'Jewish dog'? Konrad Lorenz and the Cult of Wilderness" in Society and Animals nr. 1, vol. 5, p. 3-21.

Schopenhauer, A., orig. publ. 1837. Über die Grundlage der Moral.

Schopenhauer, A., orig. publ. 1851. Parerga und Paralipomena.

Schweitzer, A., 1923. Kultur und Ethik, München.

Singer, P., 1993. Practical Ethics, Cambridge U.P.; see his two chapters on equality.

Voltaire, F.M.A. de, orig. publ. 1764. Dictionaire philosophique portatif ; keyword: "Bête".

---

[7] I classify Singer's ethics as monist. He does not, however, use this term to characterize his ethics.

Religious resources for animal ethics?

Benjamin Taubald
Institute for Social Ethics, Vienna University, Schottenring 21, A-1010 Vienna, Austria

"The idea of man in European history is expressed in the way in which he is distinguished from the animal. Animal irrationality is adduced as proof of human dignity. This contrast has been reiterated with such persistence and unanimity by all the predecessors of bourgeois thought - by the ancient Jews, Stoics, Fathers of the Church, and then throughout the Middle Ages down to modern times - that few ideas have taken such a hold on Western anthropolog (Horkheimer and Adorno, 1972).

Abstract

The article explores whether there are religious resources for the recent debates on animal ethics. It is shown that within the Judeo-Christian traditions, human behaviour towards animals, is embedded in a general sense of respect to divine creation - an understanding of the man-animal relationship that is found in most religions. When these traditions amalgamated with the Hellenistic thought in the first centuries of our era, only a small part of them could be transformed into a new mental paradigm. The philosophic heritage of Hellenism was bisected, too, by reducing it to Aristotle's plead that animals would lack reason and therefore could be treated arbitrarily. A very similar foreshortening took place in the transition from premodern to enlightened societies. The Kantian revolution in philosophy, meant to overcome the national and historical confinements of moral universalism, excluded at the same time all practical relations to non-human beings from the realm of morality. Given the fact that the predominant positions within the animal ethics discourse still rely on the normative figures and categories of classical modern thinking, it appears to be essential to integrate the moral intuitions, formerly incorporated in our comprehensive worldview, into this discourse. Religions, as original bearers of these intuitions, could make a valuable contribution to this task.

1. Introduction

When our Stone Age ancestors created the astonishing cave paintings in France and Spain 15.000 and more years ago, they did not only establish one of the earliest testimonies of human ability, but also provided mankind with a problem that has bothered it in various degrees up to this very day. We can, to some extent, reconstruct the techniques and circumstances under which the cave paintings were produced, but we can hardly apprehend what those 'artists' wanted to 'express', what motivated them and what their 'works' meant for them and their tribal groups (and we should doubt whether any of these words accurately describe what happened in the caves). Yet we see there was a problem. Man has stepped out of the unconscious unity with nature. He lived by hunting, and intensively felt the killing of his prey. He was struggling for an interpretation of his relation to nature, that simultaneously threatened and nourished him and thus became an existential discomfort. And we understand that he thought about this relation in "religious" terms (in a broad sense), including the basic idea of a "sacred", i.e. separated, area, and some concepts that have been preserved in shamanistic and totemistic beliefs.

Within the rampantly globalizing western culture, the question about man's relation to non-human nature and especially to animals is obviously not being asked as a religious question - if it is asked at all, albeit much engagement in the field of environmental and animal ethics is still driven by some kind of religious motivation or conviction. But even if one does not share these convictions, it seems that the history of man-animal relationship cannot be understood without referring to religions or particular forms of religion. So two questions arise. Firstly, what actually is the religious heritage of our modern Western culture, that spreads all over the planet, and what attitudes towards non-human nature in general, and animals specifically, did it bring forward? And secondly, we shall have to consider the transformation processes that the interpretation of the man-animal relationship is undergoing in the transition to a secular culture. My thesis will be that our secular culture is still tied to its religious roots in some way, not in terms of religious beliefs or comprehensive doctrines, but in the way of a conceptual formation of its moral foundations. To understand the role of animals in Western thought, these foundations have to be clarified.

2. Western culture, religious traditions, and the destruction of the environment

In 1967, historian Lynn White published his classical article on „The Historical Roots of Our Ecological Crisis". White's answer to the relation between the special, technical-instrumental handling of nature on the one hand and the Christian religion that has formed our culture on the other is very unambiguous: in his view, Christianity is directly responsible for the former. "Especially in its Western form,

Christianity is the most anthropocentric religion the world has seen. (...) Man shares, in great measure, God's transcendence of nature."[8] White is referring here to the creation of man in God's image, i.e. the patristic teaching of Adam foreshadowing the incarnation of Christ and the deification of human nature. This is tearing apart the connection between man and nature, it has demolished the dignity of nature and made it subject to man's command. "Christianity, in absolute contrast to ancient paganism... not only established a dualism of man and nature but also insisted that it is God's will that man exploit nature for his proper ends. (...) By destroying pagan animism, Christianity made it possible to exploit nature in a mood of indifference to the feelings of natural objects."[9]

In the debate on animal ethics that has arisen since the 1970s, White's point of view has become a commonly accepted and often repeated truism - notwithstanding the criticism it had to take from historians. White's article itself was „almost a sacred text for modern ecologists" (Thomas, 1984). It is surprising that Judaism and Christianity (and one would have to add Islam in consequence) should differ so essentially from virtually every other known religion (Watens, 1987). Religions in general tend to back modes of behaviour that favour continuance and sustainability; they endorse a non-evasive relation to nature and make violent interventions a matter of justification. Asiatic religions in India and China, professing cosmic unity and harmony, make it their basic principle not to infringe upon the integrity of nature - up to the extreme practice of Jainism, where man is obliged to wear a face mask and to purify drinking water to prevent the swallowing of tiny animals. Is the Judeo-Christian religion the one exception here? Or is the evidence of White's argument just the other way round: because our culture shows such a unique obliviousness when it comes to its relationship to non-human nature, it seems plausible that our formative religious tradition invalidates nature?

---

[8] L. White jr., "The Historical Roots of Our Ecological Crisis," Science, Vol. 155. (1967) 1203-1207; repeatedly reprinted. I quote from: D. VanDeVeer/C. Pierce (eds.), The Environmental Ethics and Policy Book. Philosophy, Ecology, Economics, Belmont 1994, 45-51, here p.49. For the debate on White's theses cf. E. Whitney, Lynn White, Ecotheology, and History, in: Environmental Ethics 15 (1993) 151-169; for theology, cf. H. Baranzke/H. Lamberty-Zielinsky, Lynn White und das dominium terrae (Gen 1,28b). Ein Beitrag zu einer doppelten Wirkungsgeschichte, in: Biblische Notizen H. 76 (1995) 32-61. An important collection is H. Halter/W. Lochbühler (eds.), Ökologische Theologie und Ethik, Graz - Wien - Köln 2000 (Cf. Vol. I, 21-187).

[9] White loc.cit. 49. White stresses that he is not dealing with an inherent interpretation of Judeo-christian tradition, but with a description on a historical form of Christianity, and he sees other strands of tradition in the religious heritage (cf. L. White, Continuing the Conversation, in: I.G. Barbour (ed.), Western Man and Environmental Ethics, Reading 1973, 60ff.). Many of his successors abandoned this distinction and understood White's theses as a standard description of Christianity. - A hundred years before White, Arthur Schopenhauer named the Jewish religious heritage the "source of the rough and ruthless treatment of animals in Europe." (A. Schopenhauer, Preisschrift über Grundprobleme der Ethik, § 19) Cf. his Parerga und Paralipomena, Vol. II, §177, a text with a striking anti-semite flavor.

3. The Judeo-Christian heritage

In trying to establish the perspective of the Judeo-Christian religion to decide this question, the answer seems to be simple. One takes the Bible and just has to read the first page of the so-called Old Testament:

"Then God said, 'Let us make man in our image, after our likeness; and let them have dominion over the fish of the sea, and over the birds of the air, and over the cattle, and over all the earth, and over every creeping thing that creeps upon the earth.'
So God created man in his own image, in the image of God he created him; male and female he created them.
And God blessed them, and God said to them, 'Be fruitful and multiply, and fill the earth and subdue it; and have dominion over the fish of the sea and over the birds of the air and over every living thing that moves upon the earth.'"[10]

„Subdue the earth" - it is hard to see anything different from the "mood of indifference to the feelings of natural objects" in this text. White is completely right that this was the way the creation account has been understood, handed down and interpreted in the Christian church. Only within recent decades, the debate on environmental ethics has given a push to biblical studies that has generated important insights into the biblical understanding of the man-animal-relationship (cf. the following groundbreaking works: Lohfink, 1988; Koch, 1991; Zenger, 1987; Janowski, 1993; Janowski, 1993a; Schmitz-Kahmen, 1997). It has been known for a long time that the first chapter of the Book of Genesis belongs to the so-called Priestly Source, which is a rather young layer within the Old Testament. Composed in the sixth century BC., it responds to the experiences of the Exile to Babylon and the confrontation with the civilizations of the Ancient Orient. Especially the wording "have dominion", which was commonly translated "rule" (*rdh* in Hebrew), in this context has a meaning that is very different from what we would expect. G.W.F. Hegel thought the 'rule' of the oriental despot in the Ancient World to be the best example for the arbitrary rule of a single person, who alone was free, while all others were in chains. But in fact in the Ancient Orient the concept of 'rule' meant almost the opposite. The ruler was everything but free, he was strictly bound by divine law, and his rule was definitely not arbitrary. It was his task to maintain the order of creation and to care for every limb - i.e., to protect the weak, to uphold them. His dominion is frequently compared to the life of a shepherd who cares for the goods with which he has been entrusted.

---

[10] Genesis 1,26-28.

This preservation of the divine order of creation is also the actual meaning of the biblical texts and the "dominion" in Genesis 1; and this order does not know a hostility or rivalry between man and animal. They both inhabit a world carefully prepared for them; "every plant yielding seed which is upon the face of all the earth, and every tree with seed in its fruit" is given to man for food - so they live as vegetarians - while "every green plant" is for land animals and birds[11]. Man is in a field of tension between unity with nature, especially kinship with animals (an essential similarity which the creation account expresses by the fact that man and animals are created on the same day, the sixth[12]) and his unique position symbolized by his creation in God's image. But what distinguishes man from nature, is just his responsibility for nature; only in this relation the distinction exists.

Close to this mentality are the Psalms, which - as prayers - express the religious self-image of the Old Israel very directly. The great 104th Psalm describes creation as a house of life that gives residence to man and animals alike. The Creator cares for man, he lets the earth put forth bread and wine; and he cares for the animals as well:

"The stork has her home in the fir trees.
The high mountains are for the wild goats;
the rocks are a refuge for the badgers...
O Lord, how manifold are thy works!
In wisdom hast thou made them all;
the earth is full of thy creatures."[13]

It is not said here that man is the end of creation, or that the creatures were made for man - up to the mythical monster Leviathan, which God did "form to sport in it"[14]. Living nature, or nature as a whole, remains inaccessible and incomprehensible to man, as a realm of its own. This point is also made in the two orations of God at the end of the Book of Job, where the claim is explicitly rejected that man could understand the inner principles of the cosmos. The wild animals "laugh" at him[15].

---

[11] Genesis 1,29sq.

[12] This is about the "beasts of the earth according to their kinds and the cattle according to their kinds, and everything that creeps upon the ground according to its kind" (Genesis 1,25) More distant to man are the fish and the birds ("the great sea monsters and every living creature that moves, with which the waters swarm, according to their kinds, and every winged bird according to its kind", Genesis 1,21) that were created on the fifth day.

[13] Psalm 104, 17b-18.24.

[14] Psalm 104,26. - In the Psalms, there are also words like Psalm 8,7: "Thou hast given him dominion over the works of thy hands; thou hast put all things under his feet". These text is closely related to the Priestly Source, the "dominion" must be understood as above.

[15] Job 39,7.18.

But only in its full integrity, the cosmos is the way it was meant to be, a many-voiced praise of the Lord:

"Praise the Lord from the earth,
you sea monsters and all deeps,
fire and hail, snow and frost,
stormy wind fulfilling his command!
Mountains and all hills,
fruit trees and all cedars!
Beasts and all cattle,
creeping things and flying birds!
Kings of the earth and all peoples,
princes and all rulers of the earth!
Young men and maidens together,
old men and children!
Let them praise the name of the Lord..."[16]

So the religious consciousness of the Old Israel shows, with regard to the relation of man to animal, numerous elements that it shares with many other religions; man's relation to the environment is primarily unfolded in the concept of creation; this already implies a certain order that is binding for all beings, but especially for man according to his greater ability of action compared to non-human nature; and there is, in some sense, a rank order where man stands out, but beyond this, nonetheless there is an inherent value of non-human nature that stems directly from the act of creation.

Yet this order did not persist. The world - as it was intended to be - would have known no alienation of man from nature, nor a divisiveness within nature; according to Genesis, man did not only live without killing animals, animals themselves were not hostile to each other. The "green plants" were assigned to them for their nourishment. Only the loss of the original harmony brought forward the deadly enmity of man and animals, and animals among each other. "Now the earth was corrupt in God's sight, and the earth was filled with violence"[17], the Priestly Source comments the consequences of what is called the Fall of Man. It is the unrestricted rule of violence that causes the Deluge, and instead of the non-violent order of creation, a system of limitation of violence is erected afterwards. The killing of men as well as the killing of animals is accepted as a reality that cannot be overcome

---

[16] Psalm 148,7-13a.
[17] Genesis 6,11.

but merely confined[18]. Among man, the institution of blood revenge prevents an uncontrolled explosion of violence, as it limits the exertion of vengeance to the principle of "one for one". The killing of animals, primarily for food purposes, is basically justified: "Every moving thing that lives shall be food for you; and as I gave you the green plants, I give you everything."[19]. An increasingly complex system of rules for clean and unclean food regulates this violence. A man who, beyond this, treats his animals in a bad manner, is still a bad man[20]. Even the animals themselves are included in these regulations and bound to forego violence against men[21]. Religious commandments, like the Sabbath, apply partly to animals who live in community with men,[22] partly those animals are protected by specific religious prescriptions, such as the well-known "You shall not muzzle an ox when it treads out the grain."[23]. Animals are under divine protection. As far as predatory relationships between animals are concerned, this texts tell us nothing about them; there might be some kind of law for this, as some passages suggest[24], but it remains unknown to man.

But in this situation, the memory of the original order of creation remains alive and bears the utopian hope for restoration. This hope is not only for the Jewish people, but a universal hope for all people, even for the whole non-human world. "Man and beast thou savest, O Lord".[25] The ripest expression of the eschatological hope is found in the words and visions of Isaiah:

"The wolf shall dwell with the lamb,
and the leopard shall lie down with the kid,
and the calf and the lion and the fatling together,
and a little child shall lead them.

---

[18] René Girard has investigated this dimension of the Judeo-Christian tradition, cf. R. Girard, Je vois Satan tomber comme l'éclair, Paris, 1999.

[19] Genesis 9,3.

[20] This has become proverbial in Ancient Israel: cf. Proverbs 12,10.

[21] Cf. Genesis 9,5. Those animals who break the law shall be subjected to capital punishment, cf. Exodus 21,28.

[22] Cf. Exodus 20,10; Deuteronomy 5,12.

[23] Deuteronomy 25,4. - To give another example, the Old Testament repeats three times: "You shall not boil a kid [i.e., of goat or sheep] in its mother's milk" (Exodus 23,19; 34,26; Deuteronomy 14,21). The meaning of this law is that it would pervert the order of creation to use the mother's milk, that should nourish the young, for the purpose of cooking it instead. Now it could be argued that these prescriptions do not aim at the animals themselves, but at something 'behind' them. But for the mind of the Ancient World, this distinction would simply make no sense. The mother nourishing the kid is not an arbitrary sign for the order of creation, it is itself this order.

[24] The 104th Psalm says, "The young lions roar for their prey, seeking their food from God" (V.21).

[25] Psalm 36,6b.

The cow and the bear shall feed;
their young shall lie down together;
and the lion shall eat straw like the ox.
The sucking child shall play over the hole of the asp,
and the weaned child shall put his hand on the adder's den.
They shall not hurt or destroy
in all my holy mountain;
for the earth shall be full of the knowledge of the Lord
as the waters cover the sea."[26]

Every thing lost when the order of creation was gone, is reestablished here; friendship of man with man, man with animal, and animal with animal. It is all part of the "knowledge of the Lord". Man remains in an exceptional position within this harmony; his is the responsibility for creation, but this stewardship is of such a non-violent and danger-free kind that a child can perform it. This is a hope and a vision. But at the same time it is the true essence of nature, albeit hidden beneath the surface and awaiting its perfection.

According to the Christian faith, these times have already begun in Jesus, the son of Mary, the Christ. Through him, and with him, the eschatological visions became reality and were even surpassed by the incarnation of the Lord. Jesus' disciples and the early community in Jerusalem literally believed that they would see the Day of the Lord and the end of this world within their own lifetime. Subsequently, the apocalyptic dimension of faith dominated their testimonies. Jesus had already anticipated peace with nature when he lived in the desert for forty days before his public appearance; "he was with the wild beasts; and the angels ministered to him"[27]. Paul professes the hope for redemption of the world at a very prominent place in his theology: "the creation itself will be set free from its bondage to decay and obtain the glorious liberty of the children of God. We know that the whole creation has been groaning in travail together until now"[28]. St. John's magnificent vision of the new heaven and the new earth concludes the New Testament.

---

[26] Isaiah 11,6-9. Cf. Ebach, 1986

[27] Mark 1,13b. In his imitation, many of the saints who have lived in the wilderness since the days of St. Paul the Hermit lived in peace with nature. In the legends of the Saints and Desert Fathers, wild animals frequently protect and nourish the saints: the raven which brings bread, the lions who accompany them and, at last, dig the grave etc.

[28] Romans 8,21sq. - Admittedly, it is also Paul who, dealing with the prohibition to muzzle an ox when it treads out the grain, asks sarcastically: "Is it for oxen that God is concerned?" (1 Corinthians 9,9) Considering the context of this chapter, one may not attach too much importance to this polemic question.

What does this mean to the interpretation of non-human nature? At the same time, nature is regarded as the good creation of God, bearing the dignity of its creator; and as something that has been corrupted by the Fall of Man, and needs to be surmounted to reestablish the original order. In the first place, this applies to man, who is in need for redemption as much as nature is. This is a delicate equilibrium. To the Christian believer, redemption has become a reality in the resurrection of Christ. Only the eschatological hope, which is an act of faith, and not of reason, can extend this redemption to non-human nature. Within the small communities of the early church, these disparate elements could be kept in balance. But the bonds ripped, when Christianity became a state and world religion in the Hellenistic-roman world[29].

4. Christian transformation of Hellenistic thinking

It was a very important decision for early Christianity to engage the Hellenistic world of thought and to encounter the philosophic tradition. Actually, what we call "theology" is the result of this development - a phenomenon that is not found to the same extent in other religions, including the Jewish. However, one might judge the consequences[30], we should expect that Christianity did not remain unchanged by the assimilation of the content of Greek philosophy. The apocalyptic dimensions of the religious traditions, however, proved to be resistant to their translation into the metaphysical language of Hellenistic philosophy. They remained strange to the Greek mind for several reasons. The disappointment about the imminent expectation of the end of the present age had already provoked a serious crisis in the belief of the early church and weakened this strand of tradition. Its disquieting and subversive content came even more under pressure, when Christianity climbed from a persecuted and marginalized sect, in some small district on the outskirts of the Roman Empire, to the official state religion and only too quickly devoted itself to the support of political power. The result was that Christianity did not recognize (and still does not) the "apocalyptic" as a part of its own heritage, but considers it almost as an insulting word, half on its way to heresy.

This way, the religious and secular traditions that emphasized the role of man alone, and neglected non-human nature, reinforced each other. This led to a selective

---

[29] The broader background for this discussion is the debate about the hellenization of Christianity that has bothered theology since the beginning of the 20th century. For an introduction, cf. C. Andresen, Antike und Christentum, in: Theologische Realenzyklopädie, Bd.3, Berlin -New York 1978, 50-99.

[30] This has been traditionally a quarrel among catholic and protestant theologians; Adolf v. Harnack in his "Essence of Christianity" (1900) interpreted the development of dogma, that was inspired by Hellenism, as a decay from original Christian teaching, that was initially overcome by Luther.

appropriation of the Greek traditions of interpretation of the man-animal relationship. Greek philosophy (Dierauer, 1977; Sorabji, 1993; Münch, 1998) had contained within it an extraordinarily rich tradition that stressed the kinship of man and animal. Pythagoras and Empedocles taught transmigration of the soul and believed that it was possible to be reincarnated as an animal, or even as a plant, and vice versa. So the difference between man and animal, is only one of degree. This is why they advocated vegetarianism. Even Plato thought it to be possible for the souls to migrate between human and animal beings, although he did not reject the eating of meat. Aristotle was the first of the great Greek philosophers to deny reason (logos), thought (dianoia) and intellect (nous) to animals. Now whatever exists, exists for an end - Aristotle, as a natural scientist, knows that nature does not give birth to anything that is arbitrary or meaningless. Man's being is perfected in a life according to reason, in the life of the philosopher who leads a divine life of pure contemplation or theoria. What is the point of the existence of animals? Obviously, their real meaning is to exist for the sake of man - as food and labour[31]. The lack of reason also causes that men and animals are completely different; but, for Aristotle, where there is no commonality, there is by definition no relation of justice[32]. Man simply cannot treat an animal "just" or "unjust", in the same way as he cannot treat a plant or a stone in a just or unjust manner. In his behaviour towards animals, there is no moral point of view for man.

Aristotle's positions did not remain unchallenged; in fact they could not even gain acceptance in his own school. His successor as head of the peripatos, Theophrastus, judged it as wrong to kill those animals who would not harm man, reintroducing 'injustice' to the man-animal relationship; he also spoke against animal sacrifice. It is controversial whether he rejected the killing of animals for food purposes as well. His successor, Strato, abandoned the distinction of perception and thinking, and therefore again ascribed reason to all animals. Yet a different position was taken by the Stoics, whose philosophy was motivated by ethical concern. While for classical Greek philosophy, for Plato and Aristotle, the concrete political community (polis) is the unquestioned horizon of ethical reflection, the Stoics were struggling to overcome this particularity and to endorse a world-wide community of all human beings. For the first time, they brought up the idea of an inherent equality of all men, based on the belonging (oikeiôsis) of all reasonable beings, men and gods. But the humanitarian idea has its price; by extending the borders of the community, they

---

[31] Cf. Politics 1,8 (1256b15-26). It has been noted that this passage is somewhat singular in Aristotle, but it is a consequence of his teleological thinking.

[32] Ethica Nicomachea 8,11 (1161b2-3). As early as in the 7th century, Hesiod marked the difference between men and animals by saying that Zeus gave justice (dikê) to man, while animals would kill each other. Cf. also Aristotle's idea that we would wage a "just war" against animals as they were trying to harm us: Politics 1,8 (1256b25-30).

are simultaneously locked up. Animals, which lack reason by definition, are again and definitely excluded from the moral community. So man can use them arbitrarily.

At the time when early Christianity spread in the antique world, stoicism was just one of several philosophical strands in Hellenism. The question whether animals had soul and reason was undecided, and so were the consequences one would have to draw concerning the treatment of animals. The Middle and Neo-Platonists, in particular, made astonishing statements with respect to this issue. Plutarch's work "On the Eating of Meat" (*De esu carnium*), which comes down from the first century AD, pleads for respectful treatment of animals and a vegetarian lifestyle. "What meal", Plutarch asks, "is not expensive? One for which no animal is put to death. Shall we reckon a soul to be a small expense?... But for the sake of some little mouthful of flesh we deprive a soul of the sun and light, and of that proportion of life and time it had been born into the world to enjoy." (Plutarch; cf. Newmyer, 1995, 19ff). We find the most elaborate defense of reason and moral status of animals in Porphyry in the third century, especially in his work "On Abstinence from Animal Food" (*De abstinentia*). Above all, he delivers the thought (probably again following a suggestion by Plutarch) that for our treatment of animals it is not so important whether they have reason or not, although he would argue that they are rational. But what really counts is that they are sentient beings, that they feel horror and pain.

Christianity developed a social and mental affinity to the Stoics and made their thoughts prevail, because both shared a universalist approach. So neither Origen in the third, Gregory of Nyssa in the fourth, Nor Augustine in the fifth century had any doubts that animals are irrational and were created for the sake of man exclusively. Their harsh judgements should probably be seen in connection with the fact that different positions were (quite accidently) taken by adversaries of Christianity: Porphyry, the most eminent critic of Christianity in antiquity, Celsus, against whom Augustine wrote, and the heretic Manicheans who favoured vegetarianism. In particular, Augustine would become a dominant figure within Christianity. He was the one who finally and conclusively asserted Aristotle's point of view that there can be no community in justice (*societas iuris*) and community of law (*societas legis*) with animals (cf. Augustinus). In his "City of God" he explains: "if we read 'Thou shalt not kill', we do not... accept it is said about irrational living things, whether flying, swimming, walking, or crawling, because they are not associated in a community with us by reason, since it is not given them to have reason in common with us. Hence it is by a very just ordinance of the Creator that their life and death is subordinated to our use."[33]

---

[33] De civitate Dei 1,20.

Augustine's authority carried this position up to modern times. But at least on the margins of the Christian tradition, other strands remained alive. The Desert Fathers had always acknowledged the dignity of non-human creation; eminent theologians, like Basil of Caesarea and John Chrysostom, who lived as a hermit for years, followed them. Both were rooted in the monastic tradition that reaches up to St. Francis. Also, the antique knowledge of zoology was preserved, from the second-century "Physiologus" (cf. Heiser, 1997, for some motives of the Physiologus) to medieval Albertus Magnus who wrote a commentary on Aristotle's zoological works. We find the spiritual traditions of Judaism in Maimonides, the great Jewish philosopher of the Middle Ages, who points out that man could very well get along without the greater part of living creation - not without all plants, but virtually without all animals - and that therefore the principle of creation is not utility, but plentitude[34].

A short view on Islam shows, contrary to widespread prejudice, that these elements are also obvious here. It is true that in the Quran an anthropocentric worldview can be found as well, where the creation of animals for the sake of man is promoted. The 16th Surah, named "The bees", reads:

"And cattle He has created for you (men): from them ye derive warmth, and numerous benefits, and of their (meat) ye eat.
And ye have a sense of pride and beauty in them as ye drive them home in the evening, and as ye lead them forth to pasture in the morning.
And they carry your heavy loads to lands that ye could not (otherwise) reach except with souls distressed: for your Lord is indeed Most Kind, Most Merciful,
And (He has created) horses, mules, and donkeys, for you to ride and use for show; and He has created (other) things of which ye have no knowledge."[35]

But as in the Judeo-Christian tradition, animals' participation in salvation is emphasized. God's loving care for animals is reported repeatedly[36]. The 16th Surah also tells that animals, as the angels do, "revere their Lord, high above them, and they do all that they are commanded"[37]. The eschatological destination of animals is also explicit: "they (all) shall be gathered to their Lord in the end"[38].

---

[34] Cf. his "Guide for the Perplexed" from the end of the 12th century, 3,13.25.
[35] XVI,5-8.
[36] E.g. in the 11th Surah: "There is no moving creature on earth but its sustenance dependency on Allah. He known the time and place of its definite abode and its temporary deposit: All is in a clear Record." (XI,6(8) u.ö.)
[37] XVI,49(51)f.; cf. XXIV,41 u.ö.
[38] VI,38.

Beside these texts, love of animals is essential in many stories about the Prophet, together with exhortations not to torment animals. Umar, the second Caliph, is in the center of a great number of similar narratives. In the Middle Ages, European travelers were astonished to find animal asylums in the Islamic world. There are also pieces of evidence of vegetarianism based on ethical-theological considerations (cf. Giese, 2001).

Coming back to our culture and our religious traditions at the end of the Middle Ages, the authoritative theologian of Western history, Thomas Aquinas, Doctor of the church and doctor communis, teaches and passes on exclusively the Aristotelean and Augustinean tradition. His interpretation of the order of creation includes the brief and unambiguous statement: "All animals are naturally subject to man."[39] Their existence as such has no meaning, apart from their utility to man. "According to the Divine ordinance the life of animals and plants is preserved not for themselves but for man."[40] The chapter on justice reaffirms Augustine's principle[41] that there is no community of men and animals and therefore every kind of use is legitimate, especially the use for food. "There is no sin in using a thing for the purpose for which it is. Now the order of things is such that... the plants, which merely have life, are all alike for animals, and all animals are for man. Wherefore it is not unlawful if man use plants for the good of animals, and animals for the good of man... Now the most necessary use would seem to consist in the fact that animals use plants, and men use animals, for food, and this cannot be done unless these be deprived of life: wherefore it is lawful both to take life from plants for the use of animals, and from animals for the use of men."[42] Generally it is evident that "it matters not how man behaves to animals, because God has subjected all things to man's power."[43] After the Reformation, Thomas' views were dominant primarily in the Catholic church. But things were not very different for the protestants. Luther, also orientating himself

---

[39] Summa theologica I 96,1 corp. („Omnia autem animalia sunt hominem naturaliter subjecta.")

[40] Loc.cit. II-II 64,1 ad 1. ("... non propter seipsam, sed propter hominem." Cf. Summa contra Gentiles III,22: "Plantae sunt propter animalia, animalia vero propter hominem.") There is a strange ambiguity between this tenet and the fact that Thomas knows and endorses a spirituality that sees nature within the context of creation and redemption. As for the whole tradition, redemption itself supposes and completes nature (gratia supponit et perficit naturam). But when it comes to the foundations of human agency, these elements seem to be gone.

[41] Ibid. II-II 64,1 sed contra.

[42] Ibid. II-II 64,1 corp.

[43] Loc.cit. I-II 102,6 ad 8. - This applies for rationality. But Thomas has to admit that men's acts are not exclusively driven by reason, and therefore continues: "But if man's affection be one of passion, then it is moved also in regard to other animals: for since the passion of pity is caused by the afflictions of others; and since it happens that even irrational animals are sensible to pain, it is possible for the affection of pity to arise in a man with regard to the sufferings of animals. Now it is evident that if a man practice a pitiful affection for animals, he is all the more disposed to take pity on his fellow-men..." (ibid.) We shall come across this argument later with Kant.

towards Augustine, stresses in the first place man's unique and outstanding position when dealing with his relation to non-human nature, in the theology of grace as well as in the theology of creation. While in his sermons opposing tendencies of respect for and care of animals can be found repeatedly, these elements are not integrated in his systematic theology (cf. on this issue Peters,1991).

5. The modern transformation of philosophy

At the threshold of modern times, the religion-shaped communities of the Western world were undergoing an epochal transformation. Modern, secular, enlightened societies arose. A comprehensive self-interpretation based on historical experiences and a redefinition of societal moral values was at stake - and is, in fact, still at stake today. These processes were different in Germany, France and in England and America, but were united in the acknowledgement of the task to overcome the religious determination of society that had led to the bloody chaos of religious war.

It was René Descartes who established the paradigm, as in many other fields, for the modern interpretation of animals. The parts of his work where he does not only deny reason, but also conscious perception, to animals are well known. The "reasonable soul", that "could by no means be educed from the power of matter"[44], inhabits man exclusively and is the foundation for his consciousness, which is radically and substantially separated from the body. Animal life, however, is to be thought as a mere mechanical link of stimulus and reaction as in automata or machines. So "this proves not only that the brutes have less reason than man, but that they have none at all"[45]. Interestingly, this makes it much easier for Descartes to explain the seemingly intentional behaviour of animals, because "it is nature which acts in them according to the disposition of their organs: thus it is seen, that a clock composed only of wheels and weights can number the hours and measure time more exactly than we with all our skin"[46].

Why is this point so important to Descartes, and why does he feel repeatedly forced to take this opinion against all appearances, while the question about animal reason is a side topic at its best? Descartes gives an answer that sounds anything but

---

[44] Descartes, Discours de la méthode (1637), 5,12 („l'âme raisonnable [qui] ne peut aucunement être tirée de la puissance de la matèrie").

[45] Loc.cit 5,11 („Et ceci ne témoigne pas seulement que les bêtes ont moins de raison que les hommes, mais qu'elles n'en ont point de tout").

[46] Ibid. („c'est la nature qui agit en eux, selon la disposition de leur organes: ainsi qu'on voit qu'un horloge, qui n'est composé que des roues et des ressorts, peut compter les heures, et mesurer le temps, plus justement que nous avec toute notre prudence").

modern: "because it is of the greatest moment: for after the error of those who deny the existence of God,... there is none that is more powerful in leading feeble minds astray from the straight path of virtue than the supposition that the soul of the brutes is of the same nature with our own..."[47] It is disputed whether here and elsewhere (e.g., in the famous dedication of his Meditationes de prima philosophia to the theologians at the Sorbonne) he is driven by a genuine theological quest for knowledge. Anyway, this shows the formative effects of medieval theological-philosophical thought in modern times. And, conversely, the scheme of Descartes allowed rational theology to get rid of the problem of animal suffering. The solution could be a very simple one - animals, according to this theory, don't suffer at all. Nicolas Malebranche, philosopher and theologian, drew this conclusion as early as the 17th century.[48] While not everybody came to share this view explicitly, it had devastating practical effects. Nicholas Fontaine, a contemporary visitor to the jansenistic seminary of Port-Royal, reported: "They administered beatings to dogs with perfect indifference, and made fun of those who pitied the creatures as if they felt pain. They said the animals were clocks; that the cries they emitted when struck were only the noise of a little spring that had been touched, but that the whole body was without feeling" (Quoted from Singer, 1990).

Descartes' counterintuitive judgements did not find much response in the English-speaking countries. John Locke in the 17th and David Hume in the 18th century were willing to attribute a certain degree of reason to animals, based on pragmatic sensualism. Admittedly, there were no practical consequences, because the absolute superiority of man remained yet unchallenged;[49] but following them Descartes' positions were harshly questioned in the French debate, inter alia by Voltaire and Rousseau.[50] So it is not by accident that within these traditions Jeremy Bentham was able to find his famous words in 1789, "The day may come when the rest of the animal creation may acquire those rights which never could have been

---

[47] Loc.cit 5,12 („a cause que [ce sujet] est des plus importantes; car, après l'erreur de ceux qui nient Dieu... il n'y a point qui éloigne plutôt les esprits faibles du droit chemin de la vertu, que d'imaginer que l'âme des bêtes soit de même nature que nôtre...").

[48] Cf. N. Malebranche, De la recherche de la vérité (1678), IV,2,7: "Ils mangent sans plaisir, ils crient sans douleur, ils croissent sans le scavoir ; ils ne désirent rien, ils ne craignent rien, ils ne conissent rien." Malebranche also provided the argument that all suffering was the consequence of Adam's sin, and, as animals are not descendend from Adam, they cannot feel pain.

[49] Thomas Hobbes, on the other hand, again excludes animals from the sphere of justice as they lack the ability of making contracts, and there is no justice without covenant. Hobbes is repeating a position that Epicurus had taken in anitquity.

[50] Michel de Montaigne had promoted a philosophically rooted respect for animals even before Descartes, up to that he advocated the equality ["parité"] of man and animal. His Essais were rather popular for a while, but they never had an impact that would compare to their rank. Cf. St. Toulmin, Cosmopolis. The Hidden Agenda of Modernity, New York 1990.

withheld from them but by the hand of tyranny... The question is not, Can they reason?, nor Can they talk? but, Can they suffer?"(Bentham, 1789, ch. 17, §1 note1).

In Germany, the rationalist paradigm was dominant in the debate on the status of animals,[51] until Immanuel Kant's anthropological foundation of ethics affected a historical break. Kant's philosophy had the practical intention of setting the moral subject free from its determination by nation or confession that had proven to be so fateful, and constituting it in pure autonomy on the basis of rationality. What matters for our topic is, that man's relation to animals has no place in his theory. Man's rationality "raises him infinitely above all other creatures living on earth. Because of this, he is a person;... he is a being who, by reason of his preeminence and dignity, is wholly different from things, such as the irrational animals whom he can master and rule at will." (Kant, 1798, I §1 [AA VII 127]) In much the same way as in Aristotle and the Christian tradition, in the original sphere of morality nothing exists to set any limits on our behaviour towards animals. At least there remains a side argument against cruelty to "the animate, but nonrational part of creation": "for it dulls his shared feeling of their pain and so weakens and gradually uproots a natural predisposition that is very serviceable to morality in one's relations with other men." (I. Kant, 1797, §17 [AA VI 443]). As far as it is a duty to uphold the means of humanity, there is sort of an indirect duty to animals; "considered as a direct duty, however, it is always a duty of man to himself."[52]

Again the exclusion of animals was the price for a universal moral community of all men. As the philosophical concepts and categories became more and more rigid, certainly to a very high degree in Kant, it seemed to be impossible to give moral reasons for a better treatment of animals. At least, there was a difference between theory and practice - in the secular context as well as in the religious. When the Society for the Prevention of Cruelty to Animals [53] was founded in 1824, an Anglican priest was also among its 22 founders. In Germany, where the first such societies were founded in 1830s, Pietism and enlightenment theology were the main pillars of the protection of animals. Rev. Christian Adam Dann (d. 1837) is regarded as the "Father of the German Tierschutzbewegung" (cf. Ingensiep, 2001). This is also the tradition of Albert Schweitzer, whose slogan "reverence for life" (Schweitzer, 1923, Kap. 21) became so influential. Schweitzer himself always regarded his thoughts as a synthesis of the philosophical and theological heritage and would never disregard the spiritual dimension of the relationship to nature. By the end of the 19th century,

---

[51] Among others, Leibniz assumed a graduated substantiality of animal soul, and Wolff even upheld its individual immortality.

[52] Ibid.

[53] Since 1840 Royal Society for the Prevention of Cruelty to Animals, and still existing.

and based on these practical experiences, the first antecedents of the modern philosophical debate on animal ethics can be found, such as Henry Salt in England and Ignaz Bregenzer in Germany (cf. Salt, 1892; Bregenzer, 1894). Theology, though, still had its difficulties integrating the moral principles that motivated these actions - especially the magisterium of the Catholic church. It was only in 1987 that a papal document acknowledged that "when it comes to the natural world, we are subject not only to biological laws but also to moral ones, which cannot be violated with impunity."[54] 1989 the First European Ecumenical Asssembly in Basel made the long overdue confession for the Christian churches: "We have failed to witness to God's care for all and every creature and to develop a life-style expressing our self-understanding as part of God's creation." (European Ecumenical Assembly, 1989). The churches express the "commitment to seeking ways out of the divisions between humanity and the rest of creation; out of the dominance of human beings over nature... into a community of human beings with all creatures where their rights and their integrity are respected."[55]

6. Integrating the tradition?

The recent debate in moral philosophy is, as far as its normative foundations are concerned, much the same as it was in the 18th and 19th century. In the field of animal ethics, the most eminent approaches tie themselves more or less directly to utilitarian (Peter Singer) or deontologic (Tom Regan) traditions. Authors like Mary Midgley, who explicitly reflect the historic and hermeneutic dimension of the concepts they use, are the exception (cf. Midlgey, 1983). This is why I want to go back to Kant's reconstruction of the foundations of morality.

Enlightened philosophy was rooted in historical experiences - from the Thirty Years War to the French Revolution - that had one thing in common: the decay of institutions that were the traditional bearers of morality. To the Enlightened mind, religion and all institutions that were legitimatized by religion, were deeply discredited. The whole system of morality, not only in a philosophical sense, but as a social and political phenomenon, had to undergo a profound transformation; a

---

[54] Ioannes Paulus PP. II, Sollicitudo Rei Socialis (1987), Nr.34. - Five years later, when the "Catechism of the Catholic Church" was issued, the authors were obviously not aware that passages like the following were frustrating for people concerned with the protection of animals: "It is contrary to human dignity to cause animals to suffer or die needlessly. It is likewise unworthy to spend money on them that should as a priority go to the relief of human misery. One can love animals; one should not direct to them the affection due only to persons" (Catechism of the Catholic Church [1993], Nr.2418). It is not just by accident that this is from the chapter "Respect for persons and their goods".
[55] Loc.cit 45.

translation that not only would have to regain the persuasive force that the old institutions had lost, but assign it to a new system of convictions and plausibilities. Hence the need to reformulate ethical convictions as the moral law of the autonomous subject that still reflects the "starry heaven above me".[56] In this sense, as Jürgen Habermas stated recently in his reflections on "Faith and Knowledge" (Habermas, 2001), Kant "gave the first great example for what is indeed a secularising but at the same time a saving deconstruction of religious truths. In Kant, the authority of God's commandments finds an unmistakable echo in the unconditional worth of moral duties. With his concept of autonomy it is true that he destroys the traditional idea of being children of God. But he forestalls the banal consequences of an emptying deflationing by a critical relativising of the religious content."

What we call today a "human rights culture" is, at least partly, the result of this effort. It is more than a legalistic or philosophical construction, but became a part of our cultural identity and a natural benchmark for moral agency. So while this is a success story of modern thinking, as far as interpersonal relations are concerned, the realm of non-human nature did not find a vocabulary that would allow to reword the moral intuitions in this sphere. Even if the consciousness of this problem is growing, the attempt to establish a corresponding social practice must be related to a theoretical discourse of justification to provide an enduring foundation for this practice. A mere copy of the discourse that was designed to fit to the interpersonal sphere will not be sufficient, since it does not meet the moral intuitions of the majority of moral agents.

It is this process of transformation that is at stake today. A concept of rationality that is aware of its own historical condition can only try to overcome this aporia in an attempt to retrace its roots. This attempt shows, in the first place, the polyvalent position of the animal in Western thinking. What we use to identify as 'the' tradition is the outcome of a process that was influenced by an assemblage of historical contingencies that buried the richness of our entire heritage. It is a major task for religion today to rediscover the strands it has lost in history, and this has already begun in Christianity, as well as in Judaism and Islam. A theoretical attempt to do so, however, is not enough. According to the practical paradigm of philosophy, actors and institutions must prove that they are credible enough to be the bearers of this tradition; and given the long history of betrayal to the dignity of non-human creation, this requirement will not be easy to fulfill. Then and only then can the communicative context be established that would allow an authentic expression of how we interpret our existence within the natural world.

---

[56] This is, of course, the famous "Conclusion" of Kant's "Critique of Practical Reason" (1788): "Two things fill the mind with ever new and increasing wonder and awe, the more often and the more seriously reflection concentrates upon them: the starry heaven above me and the moral law within me"." (AA V 161f.)

Perhaps, it is more than pious wishful thinking to expect this task to be important for the ethical debate. The moral traditions of Judeo-Christianity and Hellenism alike were bisected when they amalgamated. Later, the transformation of the comprehensive moral content of religious traditions into modern thinking failed. It would be a great mistake to think that ethical practices, and moral convictions that give guidance to this practices, could be imposed at one's choice; wittingly or unwittingly, they remain tied to their original contexts. They can be altered, reinterpreted, willfully destroyed or simply forgotten, but they can hardly be invented. I do not see so many resources for a renewed relationship towards nature apart from the religious traditions that would allow to recreate a "community of human beings with all creatures where their rights and their integrity are respected"[57]. In our specific historical situation, and according to our understanding of historicity, there is no alternative to engage the specific traditions that are so deeply incorporated in our society, even in the ways of protest and dismissal. I give the final word to Jürgen Habermas: „Moral feelings which hitherto can be expressed in an adequately differentiated way only in the language of religion can find a general resonance as soon as a saving formulation takes hold for something that is already almost forgotten but implicitly is missed. This succeeds very rarely, but sometimes. A secularisation that does not destroy is accomplished by way of translation. This is what the West, as the secularising power worldwide, can learn from its own history." (Habermas, 2001, cf. Bentham, 1789). And this is what the debate on animal ethics, in an interdisciplinary dialogue, should look for.

References

Augustinus, De moribus manichaeorum 2,17.

Bentham, J., 1789. An introduction to the principles of moral and legislation.

Bregenzer, I., 1984. Thier-Ethik.

Dierauer, U., 1977. Tier und Mensch im Denken der Antike. Studien zur Tierpsychologie, Anthropologie und Ethik, Amsterdam

Ebach, J., 1986. Ursprung und Ziel. Erinnerte Zukunft und erhoffte Vergangenheit. Biblische Exegesen, Reflexionen, Geschichten, Neukirchen-Vluyn, 75 sqq.

European Ecumenical Assembly, Peace with Justice for the Whole Creation. Basel, 15.-21.Mai 1989, Nr.43

Giese, A., 2001. "Vier Tieren auch verheißen war, ins Paradies zu kommen" - Betrachtungen zur Seele der Tiere im islamischen Mittelalter, in: F. Niewöhner/J.-L. Seban (eds.), Die Seele der Tiere, Wolfenbüttel, p. 111-131.

---

[57] Cf. fn. 55.

Habermas, J., 2001. Glauben und Wissen. Friedenspreis des Deutschen Buchhandels 2001, Frankfurt.

Heiser, L., 1997. Natur und Tiere in frühchristlicher Deutung, Köln.

Horkheimer, M. and Adorno, Th.W., 1972. Dialectic of Enlightenment, transl. J. Cumming, New York, p. 245.

Ingensiep, H.W., 2001. Zur Lage der Tierseele und Tierethik im Deutschland des 19. Jahrhunderts, in F. Niewöhner/J.-L. Seban (eds.), Die Seele der Tiere, Wolfenbüttel, p. 283-331.

Janowski, B., 1993. Herrschaft über die Tiere. Gen 1,26-28 und die Semantik von rdh, in G.Braulik/S.E.McEvenue/W.Groß (eds.), Biblische Theologie und gesellschaftlicher Wandel, Freiburg/Brsg., p. 183-189.

Janowski, B., 1993a. Gefährten und Feinde des Menschen. Das Tier in der Lebenswelt des alten Israel, Neukirchen-Vluyn.

Kant, I., 1797. Metaphysics of Morals. II: Docrtine of Virtue. Doctrine of the Elements of Ethics, §17 (= Metaphysik der Sitten, II: Tugendlehre. Ethische Elementarlehre §17 [AA VI 443]).

Kant, I., 1798. Anthropology, From a Pragmatic Point of View. Part One, First Book (= Anthropologie in pragmatischer Hinsicht abgefasst, I §1 [AA VII 127]).

Koch, K., 1991. Gestaltet die Erde, doch hegt das Leben!, in K. Koch, Spuren des hebräischen Denkens. Beiträge zur alttestamentlichen Theologie, Neukirchen-Vluyn 1991, p. 223-237.

Lohfink, N., 1988. „Macht Euch die Erde untertan"?, in N. Lohfink/G. Dautzenberg (eds.), Studien zum Pentateuch, Stuttgart, p. 137-142.

Midlgey, M., 1983. Animals and why they matter, Athens.

Münch, P. (ed.), 1998. Tiere und Menschen. Geschichte und Aktualität eines prekären Verhältnisses, Paderborn - Wien.

Newmyer, S., 1995. Commenting on Plutarch on Animals: Past and Future of a Moral Debate, in: Plutarchos. Journal of the International Plutarch Society 12, p. 19.

Peters, A., 1991. Kommentar zu Luthers Katechismen. Vol. 2: Der Glaube. Göttingen.

Plutarch, De esu carnium 994D/E.

Salt, H., 1892. Animal's Rights. Considered in Relation to Social Progress.

Schmitz-Kahmen, F., 1997. Geschöpfe Gottes unter der Obhut des Menschen. Die Wertung der Tiere im Alten Testament, Neukirchen - Vluyn.

Schweitzer, A., 1923. Kulturphilosophie II: Kultur und Ethik.

Singer, P., 1990. Animal Liberation. 2nd ed., New York.

Sorabji, R., 1993. Animal Minds and Human Morals. The Origins of the Western Debate, London.

Thomas, K., 1984. Man and the Natural World. Changing Attitudes in England 1500-1800, London, p. 23.

Watens. S., 1987. Animals, in M Eliade (ed.), The Encyclopedia of Religion, Vol. 1, New York, p. 291-296.

Zenger, E., 1987. Gottes Bogen in den Wolken. Untersuchungen zu Komposition und Theologie der priesterschriftlichen Urgeschichte, Stuttgart [2].

Ethics, morals and law relative to animals

John Hodges
Lofererfeld 16, 5730 Mittersill, Austria

Abstract

Morality is central to the definition of what it means to be human. Individuals experience morality through conscience. Society acknowledges morality by expectations of acceptable and unacceptable human behaviour. The historic relationships in Western society between morals, ethics, values, community and law are examined identifying their origins in Judeo-Christian teaching. Treatment of other humans is the prime area of ethical concern but the role of morality in the exploitive or sustainable use of natural resources is growing. Good husbandry and care of livestock have been built upon seeing humanity as a steward, not the owner, of resources. Today, society faces new moral challenges in intensive animal production systems and emerging molecular science. The temptation is to ignore the sentient nature of animals and treat them as disposable resources. The central issue is whether humanity sees itself as simply the top species with freedom to use unlimited power over all other species. The critical issue of boundaries between animal and human species is explored. The key question is whether, in the future, humanity will exercise moral stewardship and draw ethical lines against more and more exploitive treatment of livestock. Who will draw the line, when and at what point? If humanity defines itself simply as the top animal species there will be no fixed ethical lines. The ethics of the top species are always selfish. Ethical behaviour is ultimately dependent upon accepting that humanity is not simply animal, but is also transcendent and thereby has a unique responsibility for stewardship of other species.

1. Introduction

Moral conscience is a human trait. The choice to behave ethically or unethically is a human privilege. Laws of society are made by humans. Therefore the relationship between moral principles, ethical behaviour and law is a study of issues which are specifically human. In this study, the focus is upon societies' attitudes, decisions and actions in the care and management of animals being reared for food. Extreme differences are evident in the design of systems for handling animals. The alternative systems, ranging from extensive to intensive, derive in part from the way we view livestock. For example, do we see animals as dumb, sentient individuals with a long history of companionship and service to human society; or do we see them as impersonal, disposable resources measurable only in gross physical terms having no individuality, to be processed with maximum economic efficiency? Or, are there alternative, middle ways? And how are we to decide? Should these decisions be made by those who care for animals, processors in the food chain, investment financiers, animal scientists, consumers, citizens, animal welfare or animal rights groups or by governments which legislate? Should economic parameters alone be the criteria for these decisions? And if there are laws, on which moral principles and on whose values should they be built?

The values that a society attaches to its livestock shape the way the animals are treated. In simple, traditional societies where each family has a few livestock providing food, clothing, fuel and for easing the rigours of manual work, each animal is valued almost as a family member - in some traditional communities also serving as wealth and currency. The majority of livestock and people in the world at the beginning of the 21st century still live in this way. By contrast, in Western society where most livestock are kept only for food, few people ever set their eyes upon a farm animal. Scientific knowledge now offers increasingly powerful methods of reshaping the way livestock are bred, fed, managed, transported and killed. The emerging sciences of molecular genetics and gene-based biotechnologies foresee a revolutionary future for remodelling the basic nature of domestic animals to suit the economic values driving Western society.

These new biotechnologies present Western society with a totally new scenario in its moral relationships with food animals resulting in diverse and extreme positions. Some groups in society seek higher standards of animal care which many business and scientific leaders feel are unnecessary and expensive intrusions into livestock production. Some extreme "animal rights" groups regrettably engage in violence. Issues concerning animals run deep in the human breast. Some people, perhaps without deep thought on these new issues, may ask whether any moral principles are involved in the concepts of animal consciousness or suffering.

Discussion of morals and ethics is relatively new in animal science. The parameters and terms needed for intelligent and meaningful discussion and for legislation are poorly understood. To facilitate better understanding and discussion on animal management these moral, ethical and legal concepts are discussed in the context of the historic development of Western society, leading to the current challenge presented by knowledge-based livestock systems in the third millennium.

2. Moral standards, ethical norms of behaviour and culture in human society

2.1 Overview of human and animal relationships

Any study of morals and ethics in relation to domestic animals must necessarily engage in a study of why and how human societies develop certain moral standards (Good and Evil) and why some of these are formalized into law (Right and Wrong). The general decision within a society on whether Good equals Right and Evil equals Wrong is a major factor in determining the moral nature of political, economic and social systems. On more specific issues, such as treatment of animals, the balance of Good and Evil also contributes to defining the character of any society. Another vital question is how and when the moral standards in a society change.

2.2 Morals and ethics

This chapter is not a volume on moral philosophy. Nevertheless an exploration of the relationship between morals and ethics in human society is a necessary prelude to understanding how we treat animals. Morals are concerned with the principles of Good and Evil. The existence and human awareness of Good and Evil were debated by the early classical philosophers. Socrates (469-399 B.C.), the founding father of Western philosophy, had some difficulty in providing abstract definitions of Good and Evil but he clearly had no doubt that transcendent and moral principles are provided to guide human behaviour. In the three monotheistic faiths of Judaism, Christianity and Islam these principles are generally considered to have been revealed to mankind from the transcendent. Socrates provided a useful definition which clarifies the relationship between morals and ethics when he said: "Ethics is knowing the difference between Good and Evil and then choosing to do the good". Unpacking his statement we learn that morals provide us with the knowledge of Good and Evil and that this knowledge requires us to make decisions which result in ethical or unethical behaviour.

Moral awareness provides an individual with knowledge of Good and Evil and places upon him or her the necessity of making choices to act ethically or unethically in specific contexts on defined issues. Ethics is concerned with behaviour. Awareness of being able to choose to act in good or evil ways either as a life style or in individual actions lies deep and early in the history of human civilizations. The acceptable moral standards have varied in different civilizations and they may change over time. There have been earlier civilizations with practices now considered unacceptable in Western societies; for example, believing that it was Good to offer human sacrifices or to burn a widow alive with her dead husband.

Ethics concerns behavioural choices in situations in daily life based upon the underlying knowledge of Good and Evil. While it may be possible to have an academic understanding of morality which is abstract and floating free of application in any specific human experience, most classical and current exponents of human behaviour insist that ethics must be embedded in human activities, decisions and experiences of daily life.

2.3 Ethics and community

In real life people do not necessarily do what they know to be morally Good. Individuals and institutions may or may not follow the moral precepts and expectations of their society. Positive ethical behaviour flows from willingness to implement the moral standards of society in daily life and in relationships. The earliest philosophers and most religious teachers recognize that human civilization is a community experience. In pondering the impact of individual and institutional moral behaviour they conclude, almost without exception, that any human society is doomed if its members and especially its leaders do not practise a positive ethical code. Races and nations may successfully make war on other communities but, without ethics, long-term survival of their own community is at risk. Here in this study, we are especially concerned with the ethics of professional animal scientists and the leaders in agriculture and livestock production.

Socrates identified the community aspects of morality by observing that "Societies are not made of sticks and stones, but of men whose individual characters, by turning the scale one way or another, determine the direction of the whole" (Socrates, in Plato's Republic: Book 8). Another similar and more recent source in the 16th century, the English poet John Donne (1994, Ed.), brings us the well-known truth that "No man is an island".

The Greek philosophers and their civilization influenced European and hence Western society. Socrates, capturing the view of many great thinkers and wise men, was convinced that the identity of a person and his or her character is shaped and reinforced by a life which embraces Good and rejects Evil. Although he acknowledged the brokenness of human nature and knew that choosing evil may enable people to satisfy their self-interest, Socrates knew that inner peace and an integrated identity or character were the products of making good moral choices. For those persons who assume that there is no moral resource in the universe,

Socrates' position and that of other sages can be ignored. Choice is, after all, one of the evidences of morality. However, human history, filled as it is with stories of both fulfilled and broken lives and social systems, supports the view that there is a moral infrastructure in the universe of which mankind is a part. Many recent examples of the impact of good morals and of the social consequences of negative ethical behaviour by leaders and institutions may be found in the terrible events of the 20th century studied by Glover (2001) who seeks to unravel the moral history of that century.

2.4 Culture, normative values and behaviour

We are each born into a culture with defined beliefs, values and assumptions common to our community of people who are linked together by historic, geographic, ethnic, linguistic or religious bonds. Such a community, often a nation or sub-set of a nation, demonstrates a visible culture which in the popular sense includes styles of clothing, architecture, food, furniture, rites of passage and folk festivals. There are, however, more important cultural traits which are less obvious until one moves into and lives in another community. The characteristics which to some extent make one feel an outsider are deeper than the visible signs. Language is a central feature of the deeper culture and is often a necessary vehicle to understand the less obvious but highly important values, assumptions and social beliefs which affect the way a community sees themselves, others and life. These values and assumptions shape socio-economic behaviour, family life, education, the care of animals, the food chain and the acceptance or rejection of the transcendent in life.

"The way we do things around here" (Newbigin, 1986) defines normative behaviour in a culture. It is often easier to observe than to describe. The reason people do things differently is because they think that is the "right" way things should be done. The visitor who has previously experienced only his or her own culture may feel and may even express superiority at the strange practices of people with another worldview.

2.5 Values

Everyone has values. Values are what we think before we start to think. Simply put, values are the objectives that matter most to a person and for which she or he will work, invest time, energy and resources and to which priority of interest is given on a consistent basis (Hodges, 2003a). At the personal level, values might (or might not) include spouse, children, parents, golf, music, travel, wealth, reputation etc.

Values drive a person's behaviour and influence her or his responses in decision-making at both the macro and micro-levels of life. We are willing to spend our resources and time, even our whole lives, serving our values. Those who say they have neither personal values nor an agenda of issues to which they are committed are either naïve or unwilling to disclose their values - even sometimes to themselves. In some cases a person may suppress and deny his or her values in the interest of maintaining respectability while acting upon hidden self-seeking values. Often values are unspoken assumptions or pre-suppositions. Values define the quality of life.

2.6 Quality of life

This term is not easily defined in the abstract. But it is experienced. Individuals, families, groups, communities and nations have clear ideas of what factors add to or diminish the quality of their life. Their expectations may vary with age and maturity. Quality of life is not identical with standard of living, which is more commonly described in gross economic terms such as a family's financial income or the average Gross National Product (GNP) of a state. Quality of Life for most people includes components with non-monetary value. Life is, after all, not a set of data but a personal and community experience. As Jesus said, quoting Moses: "Man does not live by bread alone".

3. Origin of ethics, culture and values in Western society

We now focus on the particular context of Western society, specifically Europe and its daughter cultures in North America in both of which the intensification of animal production has its origin. In addition to the influence of the Greek philosophers upon the historic, cultural roots of European society, Judeo-Christian values have also provided a deep foundation of morality for more than a thousand years - not always practised and sometimes misinterpreted in the pursuit of power - but always present. We now examine briefly these Biblical perspectives on moral knowledge and ethical behaviour in which European culture has been grounded for so long. The deep nature of morality, already discussed by Socrates as a basic feature of being human, is also spelled out in the early Biblical view of the Tree of the Knowledge of Good and Evil (Bible: Genesis, 2) showing that mankind has moral knowledge but struggles and often fails to make good choices.

Later, Jesus provided an ethical guideline which thoughtful people including many leaders of all major religious persuasions have accepted as the world-class, foundational ethical principle for the whole of life. This simple but profound ethical statement, known today as the Golden Rule, states: "Treat others the way you would like them to treat you". In applying this principle throughout his teaching Jesus illustrated it by many real life situations such as his famous story known as the Good Samaritan who found a disabled stranger lying on the highway who was a victim of violent robbery; he interrupted his own agenda, took him to a hospice and paid for his care until the victim recovered. Jesus did not take the stance of a moral philosopher. Instead, he assumed people already know the principles of Good and Evil and he taught and empowered them to behave ethically by choosing Good in daily life. Mother Teresa epitomized this type of ethical behaviour in the 20th century by institutionalizing in her Missionaries of Charity the care of abused and neglected people dying on the fringes of society and, in so doing, impacted many people throughout the world by showing the power of Good in ethical action. She and her co-workers not only helped many individuals, but also ennobled the character of human society. In 1979 Mother Teresa was awarded the Nobel Prize for Peace.

3.1 European norms of ethical behaviour

From the statements of Socrates and Jesus we may summarize the historic characteristics of European moral behaviour as follows:
As humans, we know that life has a moral basis; we can choose to act positively or negatively in specific situations; our daily decisions affect others; good ethical behaviour is guided by considering the best interests of others and of society at large; implementing evil by non-ethical behaviour depletes life; choosing to live ethically builds community of life.

For convenience we might call this summary the European Ethical Norm. These assumptions have been part of the underlying moral culture of Europe for generations and have provided a foundation for many formal laws. Where European society has had no specific laws the Golden Rule was an unwritten law of normal behaviour. Power-seeking leaders have sometimes misled European society thus interrupting the slow advance of civil society. But, in general, most people behaved fairly and justly and expected others also to do "the decent thing".

An older term for Ethical Norm is the word Virtue. It was popular in earlier centuries but went out of fashion in the 20th century. Virtue means conformity of life and conduct to the standards of recognized moral laws. Virtues refer to moral excellence. The term probably went out of favour because of the flavour of perfection - which everyone knows is not found in any person or society. However, some recent writers

(MacIntyre, 1989) feel it is a term which can again be used meaningfully to define good and positive experiences of life.

3.2 Evil in the history of Europe

The great philosophers' understanding of Good and Evil as moral principles affecting human existence recognizes that people can and do choose to do Evil. Followers of the Biblical revelation well understood that the moral infrastructure forces people, societies and governments to make ethical choices between Good and Evil. The fact that people often choose Evil affirms rather than denies the moral infrastructure within which humans live. We must not be mislead by the great evils which have occurred in Europe over the centuries and particularly the large-scale evil events of the 20th century into concluding that there is no truth in morality and ethics.

Despite some periods of evil rule in European society, the European Ethical Norms remain in the human conscience. Over the centuries legislation has slowly moved towards reinforcing ethical behaviour as defined by Judeo-Christian standards which have had a unifying effect throughout Europe, albeit in slightly different cloaks and legal forms. Despite the fact that these different religious emphases have been the cause or excuse of intolerance and war, of power seeking and persecution, these Ethical Norms have tutored most people to know when they choose to do something immoral, unethical or anti-social, even when there has been no specific law against it. This unified Judeo-Christian moral basis was a strong unifying factor in the formation of the European Union during the last 50 years of the 20th century leading to a community of 25 nation states early in the third millennium. By contrast the possible extension of the European Union to countries with different historic cultures and moral precepts would not be so easy on the EU or on the joining country.

Resulting from this flow of history, in the 19th and 20th centuries European and North American countries became recognized internationally not only for economic success but also for systems of ethical behaviour, justice and internal peace. Despite some brutality against other nations, they grew into models of national probity. These moral standards and infrastructure, as well as economic prosperity, are prime reasons why many poor and oppressed people leave their countries of birth wishing to live in Europe or North America either as economic emigrants or as refugees.

3.3 A new intellectual climate in Europe: the Enlightenment

A powerful though subtle change started to affect the public understanding of morality and ethics in Western society during the second half of the 20th century. This transformation has it roots in the Reformation, the Renaissance and the Enlightenment several centuries ago in the West which, since then, has been fermenting in the minds of intellectuals, religious leaders, philosophers and socio-economic reformers. The impact of these new ideas about Good and Evil are now coming to fruition in Western society and are affecting public morality, ethics and law.

When "Good Ethical Behaviour" was an assumption in society, there was little need to speak about the obvious. Thus ethics was not a common topic of conversation or debate. But in the last decade of the 20th century, ethics sprang into the public arena and is now a major issue in Western society. The secure ethical stability in Europe and its cultural colonies has been shaken by accelerating shifts in values and worldview over the last few centuries. The bastions of Christian religion and of traditional ways of behaving have been challenged especially by the Enlightenment which has brought many new perspectives on the nature of life. The Enlightenment process slowly and persuasively challenged the historic moral values in Europe removing the authority of the Judeo-Christian God from the public square, thus changing the moral basis of legislation. Consequently during the second half of the 20th century, laws which contradict personal traditional Biblical morality were passed by majority vote in many democratic Western parliaments.

The sequence of ideological systems governing human societies over many centuries in Europe moved through family, patriarchs, tribes, theocracy, churches, kingdoms, dictatorships and communism. The end of communism in Europe enabled the West to enter the 21st century with sovereign, democratic states where individual freedom to choose often claims precedence over earlier moral values, resulting in a plurality of personal lifestyles. By contrast, in the public market place, the strict discipline of economic values and science-based knowledge dominate decision-making and resource allocation.

3.4 Modernism in Western society

Flowing out of the Enlightenment came the modern era with science, technology, industrialization, new means of transport and communication, fossil fuel to replace animal power and new technology on the farm. Massive socio-economic restructuring, much good but some violent, accompanied these changes. Market economy capitalism encouraged the entrepreneurial spirit, promised the defeat of poverty and opened new vistas in almost all areas of life. This socio-economic revolution led to questions about the old cultures, assumptions about behaviour, moralities and authorities. As part of this huge paradigm shift, farm animals have therefore moved from being visible components of society where their companionship was part of the recognized quality of life to becoming disposable economic resources hidden away from public view in specialized units often without access to any natural environment or personal interaction with humans.

New assumptions on what is normal for education, travel, family life, wealth and employment have emerged. Although 70% of Europeans (Eurobarometer, 1998) and a higher percentage of Americans say they believe in God, recognition of the transcendent has moved from the public to the private areas of life. Some intellectuals, significantly lawyers, consider that morality has no absolute foundations and is simply a relative concept. Relative morality offers the option of anti-social behaviour without sanctions.

3.5 Morals and ethics in the 21st century

At the dawn of the 21st century, post-Christian Western society is dismantling its own traditional moral foundations. Many thinking people consider that this is bad news in the long-term. Without trust, accountability, stewardship, and transparency leadership can easily degenerate into the exploitation of privilege and the pursuit of personal gain. The enormous changes in Western society outlined briefly above find morality, ethics and law in a novel position. The traditional structures for transmitting from one generation to the next the ethical norms which promote community life are being eroded. Younger generations of people often have little understanding or sympathy for ways in which Western society sustained itself morally over the centuries. Religious education of children has been diluted or banned by law from school curricula. Fewer children now grow-up in homes with a positive family ethical culture. Many young people have no understanding of why ethics are needed to build community of life. Some liberals believe that floating moral standards offer a progressive step for humanity.

In the late 1990s I asked the President of the Ukrainian Academy of Medical Sciences in Kiev how ethics are decided in post-Soviet society. "Listen," he said, "the strictly enforced ethic of communism which subjected everything to the Party has gone. Today, in Ukraine, everyone makes up their own ethics". Today in the West too, we have arrived at that same smorgasbord of ethics, though from a different historical process.

Thoughtful citizens, politicians, and other leaders in society are disturbed by the changed public behaviour in society. A general reason is the notable increase in misbehaviour and abuse of knowledge, resources and finances in some public institutions. The increasing number of these scandals around the turn of the 21st century raises suspicions in the minds of citizens that some leaders in public institutions cannot be trusted to make good decisions in the interests of society as a whole. Personal and self-seeking financial agendas have been revealed which cause loss, suffering and exploitation to many individuals and to some developing countries. Even some watchdogs of financial propriety, public auditors, have been found in collusion with senior executives in public companies to deceive and to defraud investors, shareholders, consumers and the public generally.

3.6 Ethics in the public arena

The scandals in the market place show that ethical behaviour can no longer be simply a private matter. Today ethical behaviour is being discussed in the street and is prominent in public thinking, writing and debating. But it is difficult for lawmakers who know that the traditional moral principles are no longer available, and they do not know what new criteria to use as a basis for legislation. Here are some examples which indicate this growing concern with public bodies, the market place and professional groups. In 2002, the European Parliament held "The First European Ethics Summit"; in 2003, the European Union (EU) set up a "Reflection Group on Spiritual and Cultural Dimensions of Europe"; shortly after taking office in 1996 the President of the World Bank, James Wolfensohn, convened an international conference in Washington DC on "Moral and Spiritual Values in Development"; the European Society for Agriculture and Food Ethics (EurSafe) was established in 2001 as an independent professional society to provide a focus for ethical issues in agriculture and food. Many national Animal Science Societies have recently set up professional Ethics Groups and now invite papers at their conferences and in their journals. Courses on ethics have been introduced in universities and colleges for animal science, agriculture and business students.

4. Domestic animals in relation to morality, ethics and law

4.1 Ethics and animal production

Traditional farming families often extend the same concepts of Good and Evil to their animals as they do to each other. As recently as the early part of the 20th century in rural Europe and North America livestock lived in close proximity with the farming family. Laws on treatment of farm livestock were rare and were concerned with the protection of individual animals from neglect and cruelty by perverse individuals rather than prescribing conditions for systems of livestock production.

Following the end of the Second World War in 1945 large-scale, high-input livestock units were started of a type which would not have been ethically acceptable to earlier generations of farmers. The spread of these intensive livestock units arose from the values of market economy capitalism associated with science-based agriculture. During the 50 years in the second half of the 20th century these values drove farm livestock into large-scale, automated operations with minimum labour and capital investment per animal, maximum throughput per unit time and space, leading to higher profits and return on investment and to lower costs for the consumer. These economic values spread upstream into the research agendas of animal scientists whose targets include: higher prolificacy, lower mortality, increased genetic gain in production traits, lower nutrition costs, better conversion rate, increased biological efficiency, improved ratio of edible to waste tissue, less variation between animals, and shorter time from conception to market etc., (Hodges, 2003b). This is not a small business. In 2002, ten billion mammals and birds were produced for food in the USA. For more recent documentation of "factory farming" see: Eisnitz (1997) and Scully, (2002).

In the 1960s people began to express new ethical concerns over the treatment of farm livestock, starting with individuals and Non Governmental Organizations (NGOs). The protests were initially against lack of adequate care for animals in "factory farming". Some key early publications which address the issue are: Godlovitch, Godlovitch and Harris (1972) and Singer (1975).

In earlier times social pressures of acceptance or rejection had been very strong in traditional communities and were sufficient to maintain acceptable ethical standards over the treatment of livestock in farming communities. In Western societies from the 1960s onwards there was increasing division of labour, more advanced economic infrastructures and fewer family units. But traditional values were still present in society and it became clear that some laws were needed to regulate the treatment of animals in large-scale intensive production systems. As a result of intensification, farm animals had become more distant physically from human society and were viewed increasingly as resources to be used as part of a process rather than partners in the community of life. Scientists then began to ask if animals experienced discomfort or felt pain when kept in intensive units. In some countries, governments appointed specialist groups to study the issue of animal suffering, an early example was Brambell (1965), and a new scientific discipline of animal behaviour emerged in academia.

Legislators struggle over the issues of animal welfare. They face the conflicting interests of efficiency and the calls from sections of society for sentient creatures to be treated with care and dignity. Thus legislators find themselves on a battleground between traditional European values and the newly emerging values of post-Christian society. The relationship between morality and law in the treatment of animals is not easy. A general characteristic of laws is their ability to prescribe sanctions rather than to approve compliant behaviour. It is easier to legislate that people should not kill each other than to legislate that people should love each other. In the case of animals, some laws now exist, for example against de-beaking hens and making certain types of farrowing pens illegal for sows. But there are no laws which reward livestock owners for providing clean bedding regularly.

Proponents of animal welfare argue that livestock production now operates unethically as it fails to take account of the experience of animals as sentient beings. They point out that, although animals can respond to direct pain, they are unable to express their needs for normal body movements and lifestyles. The counter position based upon the ethics of utility and convenience considers that intensive systems do not inflict physical pain; that basic needs of food, shelter and disease control are provided; and further, that animals do not need to experience quality of life as humans define it. Detailed angles in this debate are not elaborated here but recent publications include: Scruton (2002); Cavalieri (2002); DeGrazia (2002).

Laws are commonly an expression of the moral norms of a society. The origins of these norms may not be clear to the majority of the current generation. Nevertheless many people feel most comfortable with the moral assumptions that they absorbed as children which lie deep in the unconscious and tutor the human conscience in what is morally right and wrong. These factors are relevant to an understanding of the interactions in society between traditional European morality and the increasing dominance of economic values in livestock production. It is an emotive issue as people easily identify with large mammals. Vegetarians and vegans often cite abuse of animals as the reason for their choice of diet. Unfortunately there are some violent extremists who protest for "animal rights" rather than animal welfare.

Protests about poor comfort levels and quality of life from the animal welfare movements and from scientific studies of animal behaviour have resulted in some legislative responses specifying, for example, minimum standards for housing, space, building design features, transport and slaughter processes. The legislative action has been more extensive in Europe than in the USA (Singer, 2003).

4.2 Gene-based technologies

In the context of agriculture and food there is a further component in this matrix of growing concern about ethics. People are aware that emerging gene-based biotechnologies are being implemented by sectional interests often without public knowledge, debate or informed consent. Intelligent people well understand that new ways of trying to solve intractable problems are needed to enable society to advance. Molecular biology promises to open some of these doors. However, molecular biology involves manipulating fundamental life processes and thereby is capable of reshaping the traditional food chain with implications for security, safety and quality. Fear and suspicion have arisen because the decisions to use new biotechnologies in the food chain are made largely by multi-national companies often with the partisan support of governments. While the stated aim is always improvement in the quantity, quality or cost of food, everyone knows that the key decision-makers are motivated by their mandate to maximize share-holder benefits. A major question is whether one sector in capitalist society should assume the mandate to decide such vital questions on behalf of all stakeholders. Food is not an optional extra for life; and people in the West can no longer grow their own food. This situation also raises ethical questions about applying biotechnology to livestock.

4.3 Public morality and the treatment of livestock

The dilemma about public morality spreads over all social and community activities, including the production of livestock for food. How should Western society in the 21st century decide on the treatment of livestock? Should it be based upon traditional care? Or, subject to the dominant assumptions of economic prosperity, should livestock be used as disposable resources for food production? Should farm animals automatically be subject to any and all the gene-based and other technologies which mankind can and will devise?

Mankind and livestock are now approaching another major boundary in their relationship. The remarkable techniques of molecular genetics and biotechnology offer the prospect of overcoming many intractable problems in the care of animals; but trying to solve these old problems with gene-based methods will also bring threats to the fundamental nature of domestic animals. Balancing the benefits and the threats will not depend primarily upon technology, but upon the way in which mankind now views livestock and uses the new biotechnologies. Out of this view will flow positive or negative ethical behaviour towards animals.

It is not only the fundamental nature of animals which is under threat; actually the basic moral nature of mankind will be tested by our stewardship of animals in this new era. Our custody of animals, which has extended over 12,000 years since the early days of domestication, now has the absolute power to change the fundamental genetic nature of animals.

The human dilemma is clear. How will mankind respond to this supreme power, which goes beyond the humane killing of animals and offers the ability to change their genetic constitution while keeping them and their offspring alive? Mankind has discovered methods to reshape the life functions of animals by breaking down the boundaries which define species. Where, in post-Christian Western society, will humanity find the appropriate moral standards and ethical behaviour to affirm and not to destroy the biological community of life?

5. The status of animals in the 21st century

5.1 Species boundaries and values attached to animals

Treatment of domestic animals in Western society in the 21st century will be determined, as in the past, by the way animals are perceived and valued by humans. But there is a new and fundamental principle that now underlies society's perception of animals and the values attached to them.

This major fundamental principle is the genetic boundary which separates the human species from livestock and the interpretation which humans as individuals, as groups or as society place upon this boundary. The power that mankind now has to change the genetic nature of domestic livestock and then to enable the transgenic animals to reproduce inevitably brings the nature of the human species into the moral and ethical equation. Mankind can no longer look objectively at a domestic animal species as something totally other and different. This changed relationship in the emerging era of gene-based biology becomes the hinge that will determine whether mankind treats his domestic livestock in positive or negative ethical ways. In so doing a society defines itself as well as the animal species. Humanity's relationship with domestic animals has entered a new moral and ethical era. We now examine the implications of breaching this species barrier.

5.2 Species differences

In scientific terms, species boundaries are defined by reproduction which takes place within and not between species. Thus, typically, the genome of each species, although carrying genetic variation in the form of alleles, is a confined gene pool. Cross-fertilization or transfer of genes between species is not normal. Species boundaries have existed for long eras, far longer than the length of time since mankind began to domesticate mammals and birds creating a new, shared community of life. Species boundaries are not a human invention. They pre-date human civilizations. That is the way things have been and are today. Mankind needs to think very carefully before breaching these ancient genetic boundaries.

The issue of species boundaries between people and domestic animals arouses deep concerns and shapes attitudes in individuals, groups and societies. There are two principal views:

1. Those who advocate more care for farm animals
People who oppose intensive livestock systems or who want to moderate intensification by introducing more care and welfare generally assume a close relationship between human and mammalian domestic animal species. This close identity with livestock provokes them to advocate kindness. This attitude is well demonstrated by the position accorded to household pets in affluent Western society. The new knowledge from the Human Genome Project of the close matching of base sequences in the genomes of human and higher animal species supports the view that mankind and domestic livestock are very close. People with this value system and interpretation of the species boundary feel empathy towards domestic farm animals. Their perception minimizes the species boundaries between mammals and humanity.

Some groups advance this position further by advocating "animal rights". This more extreme position promotes the concept of the "telos" of each animal species in a way which matches the dignity accorded to the human race. This perspective has been advanced by the emergence in recent decades of human rights leading to the recognition of increasing rights of ethnic groups, of minority groups in human society and of the women's liberation movement - all of which have encouraged the concept of rights in all areas of life. In an oblique way this movement has probably fostered among some groups the concept of "rights" for the mammalian animal species with whom humans identify closely. This issue of "animal rights" should not be confused with the more widely based animal welfare concerns in Western society described above which call for more animal care.

This view of the relationship between humanity and domestic animals which minimizes the species boundary between humans and domestic mammals extends to animals the sensitive feelings which humans would experience in unnatural, uncomfortable conditions with artificial restrictions and restraints. The aim is to express compassion for sentient and dumb animals by seeking to maintain higher levels of care by moderating insensitive systems of livestock production.

2. Those who advocate using animals more efficiently for food production

The species boundary is equally important, though for a different reason, to groups in Western society who are in the driving seat of further livestock intensification. This group usually includes animal scientists and business interests whose major aim is to increase the efficiency of livestock production, to increase profit and shareholder value of enterprises in the food chain and to reduce the unit cost of food to the consumer. Individuals and groups with these values reflect the dominant economic values as Western society enters the 21st century.

The assumptions underlying their view of the species boundaries between humanity and domestic animals are clear. Very simply the human species is superior. Society may use knowledge and power for its own purposes by keeping animals in conditions which offer economic benefits even when this leads to increasingly unnatural conditions. Although rarely articulated in words, this position is declared by actions in the ongoing promotion of large-scale intensification in livestock production - to the extent that in some countries legislation has been enacted in the interests of animal welfare, human health or the environment.

This second interpretation of the genetic boundary separating the human species and domestic animal species is not based solely upon the fact that livestock are "a dumb animal species". It goes far deeper. It is based upon a view that humans are not only of greater value than animal species but also that humans, as the more powerful species, are free to use livestock species for economic advantage *without any pre-defined limits which might be indicated by species boundaries.* For this position, the only limits are defined by economic values. This fundamental assumption about the different intrinsic value of the human and the animal species is shown by the fact that human society provides care for disadvantaged human individuals whose physical handicaps, including the inability to communicate, may be far greater than those of normal "dumb" livestock. This view of the species boundaries assumes that Homo sapiens is of greater intrinsic value and may use superior knowledge and power for its own advantage. In principle, this position is, of course, no different from the traditional attitude of society to domestic livestock.

However, the emergence of gene-based technologies brings a new dimension with the power to change the genetic nature of animal species. This dimension brings the moral and ethical issues into the arena in a new way and is now examined in more detail.

5.3 The impact of molecular biology upon concepts of species

Molecular biology undoubtedly challenges the way people think about themselves as humans and about animals. The Human Genome Project now offers identification of all the base sequences in the human genome; and comparisons with some mammals show a very high level (>95%) of correspondence. What then is the meaning of being human? If the molecular boundaries between Homo sapiens and domestic livestock are so minimal are they really of any consequence? Is it any longer valid to behave as though dumb and incapacitated humans are worth more and should be given unlimited care whereas farm animals may be treated ignominiously?

The break-through in molecular biology has placed unprecedented power in the hands of the human species. This knowledge-based power also challenges humanity carefully to define itself in relation to the rest of the biological world. Are we simply the top species with instincts to exploit and use other species for our own ends - or does humanity have a higher moral nature leading beyond mere instinctive responses to ethical decisions and behaviour? Wise men and women, philosophers and religious leaders have long argued that responsible ethical behaviour is the foundation and guarantee for long-term survival of the community of life and of civilized society.

The current batch of biotechnology techniques is only a small sample of the endless vista of options which now lie within the grasp of humanity for reshaping the genetics of every living thing. Somatic cloning, transgenic organisms, embryo manipulation, somatic stem cells, deletion and insertion of genes - these are but the beginning. The central dogma of molecular biology is that DNA, together with proteins, controls the genetic characteristics of every living organism. The emerging disciplines of proteomics and functional genetics offer the ability to identify functions and traits linked with specific DNA and proteins and then move them across species boundaries. This restructuring of biology is also a challenge to the ethical codes which hitherto have recognized species boundaries as essential infrastructures for an integrated community of life. Transgenic technologies for livestock involving the movement of genetic material across species boundaries redefine the community of biological life. They also change the nature of the relationships between mankind, the most powerful species, and the livestock species which have enabled mankind to rise from primitive to advanced qualities of life.

The use of these powerful biotechnologies clearly involves ethics. Decisions to change the genetic nature of livestock and thereby behave unethically automatically open another issue with deeper consequences which may not be immediately apparent. The choice will reaffirm or will deny the nature of humanity. Restructuring the genetic nature of livestock will mean that society inevitably chooses to renounce its responsibilities for the community of biological life which has characterized human civilization from its earliest days. In this way humanity will redefine itself and the qualities of human society.

5.4 Where are we taking our animals?

It is not difficult, but nevertheless horrifying, to visualize a future scenario of intensive livestock production extended without species boundaries or ethical limits. Imagining such a scenario using current scientific knowledge is likely to be faulty in detail because molecular biology is accelerating so rapidly. But, science, business and society being what they are, we know that scientists will continue to research and to develop more powerful techniques for molecular manipulation. To enable us to face this issue, here is one possible scenario of the future.

Uneconomic parts of the anatomy, such as legs, have been deleted reducing the amount of space needed per animal and also making better use of nutrient input. Genes for meat tissues with higher commercial value, such as beef, are inserted in pigs which, having a more efficient digestive system than ruminants, now produce beef muscle. Genes for feathers and wool have been deleted from food animals and animal skin has been thinned to minimal levels needed to contain functioning tissues. Genes for beaks and some parts of the digestive system have been deleted as nutrients are fed into the body as a drip. Small birds like quail with low body maintenance and without legs have genes from domestic hens and produce large eggs. All animals are linked permanently to a flow-line of concentrated feed laced with appropriate hormones and enzymes. Growth hormone genes have been inserted to speed growth of meat animals. Blood supply to each batch of animals is produced by special animals devoted to that purpose. A blood circuit flows around each batch of cloned animals with links to each animal thus providing control of all physiological functions and bringing the batch to market with perfect economic timing. Some animals, now designed to consist mainly of a rumen, are adapted to processing animal waste which is itself far less than in natural conditions and which is then recycled in the production line.

In this speculative scenario, genes which can improve the efficiency of food producing animals are taken from any species of animal, fish, insect, plant or microbe and inserted into the genomes of livestock. For example, maintaining body temperature is expensive. So, intensive livestock have reduced body temperature or in some cases have been turned into cold-blooded creatures using reptile genes. A major breakthrough has at last been achieved in making the slow and uncertain process of mammalian reproduction more efficient. Animals are gender neutral. Cloned embryos carrying the most highly productive genes for desired traits are produced in bulk using duplicate genes from the old male or female lines. The gender neutral animals are shaped by hormonal supply into having the former sex-linked traits according to the product desired. These embryos are stored in deep freeze ready for a shift in market demand. Embryos are now grown for their early life in an artificial uterus.

A basic animal framework with required genes for the life functions has been developed and can be used as a template for producing a variety of animal products. This basic animal framework needs a brain for neural control of tissues but the head no longer needs mouth, eyes, or ears. A production line of these animal frameworks is maintained and somatic stem cells are added to produce the meat tissue currently required. After harvesting, more stem cells are added to regenerate a further crop of animal products. The animal frame is also maintained over long periods using somatic stem cells to replace worn out parts.

Is this all a fantasy? Such scenarios are an extension of scientific and economic reductionism without ethical limit. This value system fails to understand the integration and beauty of nature and the community of life as it now exists. It may be difficult today to grasp the immense power of molecular biology for merging animal species. But these goals and many more are natural extensions of the current reductionist research agenda and capture the mindset and excitement of research scientists. This research is very well funded by public and private sectors - perhaps only ever exceeded by nuclear physics and space programmes - in which governments control both the funds and the ethics of use. Who will decide the ethics of animal production?

"A nice little dog .....transformed (by transfer of human testicles and pituitary) into a specimen of so-called humanity so revolting that he makes one's hair stand on end. This is what happens when a researcher, instead of keeping in step with nature, tries to force the pace and lift the veil. We have made our bed and now we must lie on it." (From The Heart of a Dog: Mikhail Bulgakov, 1968)

6. The role of humanity in the community of life

6.1 Where is our treatment of domestic livestock taking humanity?

The future use of emerging gene-based and molecular biotechnologies with livestock has enormous consequences for the biological species infrastructure which has been a fundamental and unalterable fact of society since humanity moved from hunting and gathering into the domestication of animals and plants leading to settled agriculture. Whether we realize it or not, the ongoing pursuit of food which has characterized human society is now moving into a completely new phase. This new era in human civilization will be characterized not only by the creation of transgenic plants and animals - new species - but also by the redefinition of humanity in moral and ethical terms. The decisions we make will reveal not only how we view animals but also how we see ourselves.

The early philosophers and religious leaders have pointed out that moral standards and ethical behaviour are the deepest characteristics defining the nature of human communities and therefore their chances of survival. In our new era, if we blindly follow the present utilitarian ethic we may survive as an animal species but the question is whether we shall still be human. Essential human traits which have characterized civilizations built by mankind include intelligence and moral awareness of Good and Evil enabling individuals, groups and societies to choose positive or negative ethical behaviour. Human societies have been built upon recognition of the Community of Life, meaning that the lead species, Homo sapiens, has acknowledged the inter-dependence of life. Scientific knowledge has revealed the integrated nature of biological life, thus affirming that no species can live alone. The constant pre-occupation of humanity with the biological complexity of the human food chain reinforces knowledge of this inter-dependence.

6.2 A reflection on the past

Since gene-based biotechnologies are a product of Western society, it is appropriate to reflect on the moral foundations and ethical values, described earlier in this chapter, which, until recently, have defined Western civilization. These standards derived from intellectual views of humanity laid down by the Greek philosophers upon which Judeo-Christian moral values and ethical behaviour were built. As discussed earlier, this set of moral and ethical values has shaped the emergence of European and Western culture as a positive Community of Life having many characteristics which attract people in other societies.

The Biblical mandate (Bible: Genesis 1) calling upon man to practise Dominion over the natural creation has been an assumption in Western society for centuries. Although sometimes misunderstood as Domination (exploitation) rather than Dominion (husbandry), this mandate has called Western civilization to recognize that the species' boundaries place upon Homo sapiens both privileges and responsibilities to use and to care for other species. This paradigm has been an underlying assumption from the early days of domestication until today.

6.3 The question for the future

At the turn of the 21st century, Homo sapiens is now able to abuse the integrity of animal species as well as to use them. In making this choice we enter the realm of moral standards and ethical behaviour. We also redefine our role as the unique "top species" with intelligence, power and moral awareness. Either we exploit domestic livestock by reducing their status in the Community of Life to that of disposable economic resources for our privileged use; or we continue to recognize that our ongoing identity and dignity as humans requires us to behave ethically and with responsibility within the Community of Life.

In other words, do we see ourselves as "top species" defined solely by our superior molecular architecture or do we continue to embrace the transcendent in life which has always characterized human civilizations - until the dawn of post-Christian Western civilization which gives no recognition of the transcendent in the market place and in public decision-making.

We face a new question in the history of humanity. Is it possible for human society to retain knowledge of God and the transcendent in private as the majority of Western society affirm, while all public decisions are based upon economic self-interest? The answer to this question is probably No - removal from the public arena of moral standards with their associated constraints upon ethical behaviour will probably lead to a society better defined by the ethics of natural selection: "Nature - red in tooth and claw".

Is it possible for a society which decides to change the nature of domestic animals in order to increase its own economic prosperity to continue the civilized ethics of refraining from disposing of the old and unproductive individuals of the human species by euthanasia? Further, if gene-based technologies are to be used without limit for economic benefit with livestock species, why should we not use eugenics to reshape the human species into more productive performance by eliminating non-conformist individuals and by restricting reproduction to those individuals having an approved genotype? In this scenario, a new elite "Super Homo sapiens" would take over. In the 18th century Kant (1948, Ed) anticipated and deplored this type of behaviour by observing that humanity is a dignity and man cannot be used merely as a means by any man.

6.4 Professional ethics or legislation?

In modern economic society with division of labour, professional groups play a highly important role in enabling a modern economy to function and in contributing to the quality of life. A new question now emerging in Western society is whether groups of professional individuals acting together with common interests in the service of society should formalize their values into ethical codes of behaviour, thus avoiding the necessity of legislators having to place limits upon their activities. Codes of ethics serve to reassure society that professional groups are not acting solely in their own interests. The issue was addressed by Kant (1948, Ed) when writing of the morally grounded categorical imperative in which he sees professionals as having distinct moral obligations. It is a new question whether animal scientists should define their own ethical codes of practice to enable them to contribute positively to the economy and at the same time to remain objective in serving the larger goal of quality of life with its several dimensions which increasingly include non-monetary factors.

The case of medical doctors is clear - they have long had a professional ethical code derived from the Greek physician Hippocrates (460 BC). Animal scientists have never had a formal ethical code. They have operated under the general assumptions of Western society "to do the decent thing" - and to treat animals with respect. But, animal scientists in the West now face a new problem. Many are employed or funded by businesses in the food chain whose values are inevitably based upon biological and economic efficiency in food production. There is a danger that animals become solely the means to a financial end, lose their own intrinsic value and are reduced to disposable resources.

7. Conclusion

The historic European position founded upon Judeo-Christian morality, values and behaviour calls man to be accountable in the renewable use of livestock, along with all the resources of the world, for improving the quality of life (Bible: Genesis, 1). This historic position contrasts with both of the interpretations about species boundaries currently held in Western society which are described above. The historic

Biblical and therefore the earlier European view is that the human species is different from all other species. A main distinguishing trait is awareness of Good and Evil and the ability to choose and behave ethically thus having both privileges and responsibilities to care accountablility for the Community of Life.

This high view of the human species mandates us to care for and to use other species, including livestock, sustainably with a long and not a short perspective. This respect for the biological Community of Life has historically been called "Husbandry" - a term which was used earlier to describe animal science. Perhaps we need to incorporate again the concepts of husbandry into animal science thus ensuring that the values underlying the production of animals and animal products are not limited only to economic and biological efficiency - but embrace ethics and the Community of Life.

The accumulated wisdom of mankind until the last few generations in the West has always recognized transcendence as part of human nature. The majority of people in Europe and in the USA say they still do believe in God. Belief in the transcendent is related to moral awareness and ethical behaviour which are major factors in whether humanity sees itself as simply top species or endowed with both privileges and responsibilities for the Community of Life.

If, while holding private beliefs in Good and Evil, society behaves abusively in the public domain towards other species for economic gain, there is a disjunction and society becomes schizophrenic. Such duality of values will result in inconsistent laws, unethical behaviour and, in the view of the moral leaders of the past, will cause this civilization to collapse.

"It is dangerous to show man too clearly how much he resembles the beast without at the same time showing him his greatness. It is also dangerous to allow him too clear a vision of his greatness without his baseness. Its is even more dangerous to leave him in ignorance of both. But it is very profitable to show him both." Blaise Pascal, 1623-62.

References

Bible, Genesis 1. Man given Dominion.

Bible, Genesis 2. Tree of Knowledge of Good and Evil.

Brambell, R., 1965. Report of enquiry into the welfare of animals kept under intensive livestock husbandry. Her Majesty's Stationery Office, United Kingdom.

Bulgakov, M., 1968. The Heart of a Dog. Harville Press, p.108.

Cavalieri, P., 2002. The animal question: Why non-human animals deserve human rights. Translated from Italian by Catherine Woollard. Oxford University Press.

DeGrazia, D., 2002. Taking animals seriously: Mental life and moral status. Cambridge University Press.

Donne, J., (Ed.) 1994. The Works of John Donne. The Wordsworth Poetry Library, United Kingdom.

Eisnitz, G., 1997. Slaughterhouse. Prometheus.

Eurobarometer, 1998. European Union: Social Attitudes. European Union Eurobarometer.

Godlovitch, S., Godlovitch, R., and Harris, J. (Eds.), 1972. Animals, men and morals. Taplinger.

Glover, J., 2001. Humanity: A moral history of the twentieth century. Pimlico, Random House, United Kingdom.

Hodges, J. 2003a. Livestock Production Science. 82. 1-2. Pages 259-291. Elsevier.

Hodges, J. 2003b. Livestock, Ethics and Quality of Life. Journal of Animal Science, 81 (11), 2887-2894.

Kant, I.,1948, Ed. Groundwork of the Metaphysics of Morals, trans. H.J.Paton (The Moral Law), London, Hutchinson.

MacIntyre, A., 1989. Whose Justice, Which Rationality? Notre Dame, University of Notre Dame Press.

Newbigin, L., 1986. Foolishness to the Greeks. Wm. B. Erdmans Publishing Co. USA.

Scruton, R., 2002. Animal rights and wrongs. London Metro.

Scully, M., 2002. Dominion: The power of man, the suffering of animals and the call to mercy. St. Martin's.

Singer, P., 1975. Animal Liberation. New York Review/Random House. Rev. Ed. 1990 and 2001 (Ecco).

Singer, P., 2003. Animal liberation at 30. New York Review of Books, L, 8.

The animal issue: diversity in values and thoughts

Vincent Pompe
Van Hall Institute, Department Animal Management, Leeuwarden, The Netherlands

Abstract

In modern society animals are an indispensable human utility. We value animals for a series of connotations and functions, which varies from scientific research and meat production to company and conservation. Behind the diversity of values lies a similarly diverse pattern of thought and theories. Each of these perspectives consists implicitly of three sub-perspectives: a view about the key dispositions of humans; a view about the welfare dispositions of (higher) animals; and a view about our moral relation with animals.

In this paper, I present an overview of thoughts of contemporary philosophers with some historical rooting and I construct an anthropocentric-zoocentric framework, in which the three sub-perspectives will be clarified. From that framework I will analyse the welfare issue, which outcome will lead to a distinction between animal-related and human-related moral concepts. Finally, I will analyse the different strategies to determine the moral justification of animal use for human purposes.

All the overviews will show the complexity of the animal issue and will reveal the diversity of values and thoughts. A pragmatic solution will be presented to work with this diversity.

1. Introduction

Domestication of animals is a human activity that started early in the history of civilisation, in which humans catches and tames animals and controls their nutrition, care and reproduction. The early goals of domestication, that remain more or less the same throughout time, were basic, such as food, symbolism, and elementary scientific endeavour. It is only in the last century that capitalistic and technological powers have caused an increase in the functions that animals serve.

New functions are developed to match the ongoing expansion of human desire. We expand the variety of traditional meats with edible products of exotic animals. We gain leisure from animals, e.g. by sport, zoos, natural parks or as collector's items. We train them to provide us with special service, such as search and rescue, guidance and housekeeping assistance. In science we not only use animals as objects for study and experimentation, but also as standardised material to test pharmaceutical products and as entities for genetic modification. Socially, we regard pets more as a human surrogate and we search for ways to prolong the human-animal bond, e.g. wheelchairs for the canine handicapped, chemotherapy for animal cancer patients and a genetic clone to replace a lost feline (Copy Cat).

The development of domestication tends to go in two opposite directions. One is the dehumanisation of the animal in order to make it a detached instrument. The other is a further humanisation of the animal, in order to strengthen the human-animal bond.

Behind the diversity of functions lies a similar diverse pattern of values, thoughts and theories. The different ways we think about animals is just as diverse as the different ways we use them. Theories on animal ethics have their own complexity, because the philosophies consists implicitly of three sub-perspectives. One is the view about the key dispositions of humans: what makes us different from animals and what is our proper relation with animals? Some say that man is superior to the animals due to his mental powers and will take an anthropocentric stand. Others will emphasise that a human is an animal and therefore equal to animals, and advocate zoocentrism. Besides a human-oriented view, philosophy can also present an animal-oriented view. This perspective tells us something about the welfare dispositions of (higher) animals. Some think that animals have their own psychology, i.e. the mental ability to think, to experience desires and emotions, while others see animals more as behaviouristic beings that can only show stimulus-response-oriented behaviour. A third aspect of an animal philosophy is the view about our moral conduct towards animals. Are we allowed to do almost everything to animals or must we be very restricted? What ethical deliberation strategy do we use to decide whether our conduct is right or wrong, e.g. by looking at the consequences of our action or at the action itself in relation to moral principles or moral traits?

By analysing the three sub-perspectives of thoughts of contemporary philosophers on the animal issue, the complexity of the issue and therefore the broad scope of animal ethics will become clear.

2. The emergence of zoocentrism

Throughout history, the human-animal relationship has been almost exclusively anthropocentric. Directed or influenced by monotheistic religions, as Judaism, Christianity and Islam, humankind regarded itself as superior to the animal kingdom. Human dispositions as thought and rationality made the inequality obvious and therefore justified. The animal issue became a social one, and hence an ethical one, in the last quarter of the 20th century. The anthropocentric attitude towards animals was distinctly challenged by Peter Singer, Tom Regan and Bernard Rollin. Instead of scrutinising human dispositions that justify the difference between us and the animal, they concentrated on the similarities within the vertebrate domain in order to ground moral consideration for animals. Singer formed a sentientistic justice, Regan called for deontological respect for 'subjects of life' and Rollin based the moral status on the disposition 'telos'. With these theories a zoocentric approach became prominent.

Peter Singer has been very influential in the debate concerning animals and ethics. The publication of his *Animal Liberation* in 1973 marked the beginning of a growing and increasingly powerful movement in the United States, Europe and elsewhere. Singer attacks the views of those who wish to give the interests of animals less weight than the interests of human beings. His key issue is the meaning and extension of justice: to treat equal interests equally. Despite the many differences amongst human, race, religion, intelligence, age etc, one finds at least one equality among them all: the fact that they can have pain and can suffer. This common interest is the foundation of his Principle for Equality.

If the foundation of the Principle of Equality is the ability to have pain, then animals have the same interest. Singer allies at this point with Jeremy Bentham who reduced the human nature to two basic emotions: pain that causes suffering and pleasure that leads to happiness. According to Bentham these emotions belong to the nature of animals too. In order to suffer, one does not need reason neither the ability to talk.

The essence of the Principle of Equality is that we give in our moral deliberations equal weight to the similar interests of all those affected by our actions. Babies and animals are alike that they cannot reason and talk, but can suffer. If we give moral protection to babies why not to animals? We tend to ignore this injustice and favour human interests. We give more weight to human interests at the expense of animal suffering in situations in which we cannot justify it. This is what Singer calls speciesism, a form of injustice that we have to condemn in the same way as we condemn sexism and racism.

Tom Regan (1983) promotes in his book 'The Case for Animal Rights' a new line of zoocentrism by modifying Kantian anthropocentrism. Regan borrows Kant's concept of an objective end-in-itself, but expands it to animals in order to give them direct rights.

Regan rejects the narrow scope of Kant's end-in-itself concept, because it only applies to a human as a rational being that is capable of acting morally through reason, of being autonomous and of possessing personhood. According to Regan, animals have mental dispositions that make them objective ends-in-themselves too. Animals have a long array of mental dispositions that they share with humans. This collection entails, among others, beliefs, desires, perceptions and emotional lives with feelings of pleasure and pain. Animals are conscious beings and therefore subjects-of-a-life. We have to respect animals, being ends-in-themselves, and we cannot use them merely as a means to our ends. Animals have direct rights to be protected against any human intervention, and we humans have direct duties towards their well-being.

Regan differs from Singer in that he does not focus on the suffering of animals as such. Animal ethics is not just about suffering. Animal experiments remain morally wrong even if we limit the suffering by the use of anaesthetics. The idea of farm industry must be condemned even if we give the animals more space. What is morally wrong is not the fact that we cause pain and suffering, but our attitude of using animals as a source for food, science and sports. By looking at the moral attitude Regan developed a principle-oriented rights view. This view acknowledges and respects the inherent value that animals posses. This does not mean that Regan takes rights to be absolute. When the rights of different individuals conflict, then someone's rights must be overridden, with the condition that we must try to minimize the rights that are overridden. We are not permitted to override someone's rights just to make everyone better off. In that kind of case we are sacrificing rights for utility, which is never permissible in Regan's view.

The philosophy of Rollin (1983) can be placed between that of Singer and Regan. He agrees with Regan that animals have direct rights, but one must not take these right too absolutely. With Singer he agrees that interests are founded on welfare, but welfare is more than just the experience of pain or pleasure. The moral status of animals must be based on the concept of telos. A telos is a set of natural activities, physiological and behavioural, that is determined by evolution and is genetically embedded. Rollin makes use of Aristotle's concept of telos and he expands it. According to the Greek master, telos refers to a basic ontology: everything is always changing and moving, and has some aim, goal, or purpose. Everything has potential which may be actualised, e.g. an acorn may become an oak tree. The process of change and motion which the acorn undertakes is directed at the realisation of its potential. Rollin applies the telos concept to animals. By their telos, all animals are trying to develop their natural behaviour: a cat attempts to live as a cat, a pig tries to live as pig. Animals do what they have to do by nature and, as Rollin states, they are conscious of their pursuit. Animals are aware of their species-specific behaviours.

Due to their telos, animals have intrinsic value that we have to respect. This entails that we must give moral consideration to their natural behaviour, and hence must regard every restriction on their telos as a moral wrong. However, the moral right of animals is not founded on their intrinsic value as such, but on the *development* of the natural behaviour evolving from the telos. This makes, according to Rollin (1995), genetic modification of animals acceptable, because by that technique we create a new telos and therefore a new source of natural behaviour. However, there is a moral restriction. Genetic modification is not allowed when it causes suffering, pain or a hindrance of the new animal's striving for preservation.

The interests of humans and animals may get in conflict, because we have a telos too. In that case we can use animals only if the human interest is substantial and we respect with our best abilities the intrinsic value of animals, i.e. their ability to manifest their natural behaviours.

Zoocentric philosophers have in common that they regard the human species as not self-evidently superior to the rest of the animal kingdom. They consider the moral relevancy of human rationality as low, and animal's mental abilities, including their level of consciousness, as high. Their equality-based approach is to restrict the human use of animals - not in all situations, but certainly in those cases when the human interests lack necessity.

3. Anthropocentric counterattack

From the mid 1980s to the present day, the animal issue is in an in-depth debate, because the anthropocentric responses to the zoocentric movement have been elaborated. The human dispositions are re-expressed in order to restore the justification of inequality. Raymond Frey does this by explaining the fabric of interests, Peter Carruthers by emphasising the contractarian foundation of morality, Roger Scruton by advocating a 'no-rights-without-duty' theory and Mary Midgley by introducing human emotional aspects into the issue.

Animals are sentient but not sapient. That is the key claim of Ramond Frey (1980). In the debate on moral rights for animals, Frey underlines the distinction between a need and a want. Animals do have needs but no wants and can therefore not have any rights.

Frey regards a need as a condition to function well. This denotation does not only apply to living beings but also, for example to machines. In order to function properly, animals need food and water and machines need oil. On the other hand, to have a want one must have a series of mental conditions. A want is inherently connected with a desire, that is inherently linked with a certain belief. To have a belief one must have knowledge and judgement about a state of affairs, and a language to express it. A simple example to support his logic is the right to vote. A right to vote has only meaning if one has a judgement, a belief on voting and a desire to do so. If not then that right is a matter of indifference for that person. Babies and humans suffering from senile dementia lack knowledge about many states of affairs and consequently do not have corresponding beliefs and desires. Depriving them of these states of affairs is not a moral wrong, because they do not know what they are missing. One can only have a right if one has a want.

Apart from the apes, all animals lack the ability to have knowledge and judgement and hence moral rights cannot be given to them. Animals may be sentient but not sapient and therefore not equal to humans. Consequently, we do not have to treat them equally.

In Frey's theory one can clearly detect the legacy of Descartes (1997, V: 9-12). It is general accepted that Descartes is the philosopher who stated that 'animals are machines'. Unfortunately, this statement is often used outside its context. The theory of Descartes is far too sophisticated to put in a one-liner. When Descartes compares animals with machines, one must see it as part of his reductionistic view on nature. The body of humans and animals consists of different parts, such as bones, muscles and other tissues and organs that are connected with and interact upon each other. Therefore, the body resembles a machine. However, as Descartes states, the difference between a machine and a body is that the body is made by God and therefore incomparably better arranged, more adequate to move and more admirable than any machine of human invention. The distinct difference between humans and animals emerges from his famous ontology. Mind and body are separate substances: thinking (res cogitans) and extension (res extensa). It is only humans that are made of both substances. Although animals are special machines which might execute many things with equal or perhaps with greater perfection than any human being, they do not act from knowledge, but solely from the disposition of their organs. Animals do not have a mind, because they do not have knowledge and a language by which they can express their feelings and thoughts in words. Lack of knowledge and language are Cartesian criteria that regained their influence in Frey's animal ethics.

Peter Carruthers (1992) is another, and perhaps one of the most overt, followers of Descartes' 'animals are machines' theory. Animals do not have a *higher* consciousness that provides them with the awareness of their moral status and rights. They do not have the ability to interfere with the meaning and scope of moral rules. For a contractarian, as Carruthers is, this is an essential condition. Morality does not emerge from reason or feeling but from a set of rules created by humans to regulate their interactions. This set of rules determines our freedom and limitations. Moral rights and duties stem therefore from a social contract that is concluded by rational beings. Those who are not able to participate in the making of the social contract, cannot directly benefit from it. This applies not only to animals but also to the marginal cases such as (human) babies and the senile. However, Carrurthers claims that these marginal cases do have moral rights, and animals not, because some distinctions between humans and animals cannot be ignored. Firstly, there is no distinct line between a-rational, partly-rational and full-rational human beings, but there is a clear line between human and non-human animals. Secondly, not giving rights to the marginal cases would cause great social instability, because we are psychologically not able to bind ourselves to such a contract. A society without animal rights on the other hand would cause social protest but not social instability.

Despite the lack of animal rights there are some limitations on the human use of animals. These limitations are not directly based on the moral status of animals but indirectly out of respect for the feelings of animal lovers.

The essential relation between rationality and moral right is also the key issue in Roger Scruton's philosophy of animal ethics. Scruton (2000) claims that rights are linked with duties. In order to have a right one must be duty-bounded to respect the rights of others. Only a human meets this qualification and therefore the dilemmas between human interest and animal interest must in general be resolved in favour of the former.

Animals cannot have rights, since if animals have rights, we must regard them as criminals, because they do not fulfil the duty to respect the other animals' rights to live. That animals cannot have moral duties is not astonishing, because duties have only meaning for humans. One can only have a duty if one is capable of making rational choices. Hence, one must be a self-conscious and language-using being, who can give reason for his beliefs and actions and can enter into a reasoned dialogue with others. The relation between rights and duties is, according to Scruton, not a bi-implication: 'if rights, then duties', does not entail 'if duties, then rights'. This implies that, although animals do not have moral rights this does not mean that we have no duties towards animals. We do have duties towards some animals, in particular the domesticated, for their individual survival and well-being are dependent upon our care. Wildlife, however, cannot call for our duties, as their well-being and survival does not come under the laws of human responsibility but under the laws of nature.

The above anthropocentric philosophers have in common that they focus on the rational dispositions of humans and claim that only reason can determine what is morally right and wrong. Mary Midgley (1983) emphasises the human emotional dispositions of empathy and compassion. We are in general more emotionally attached to our fellow humans than to animals and therefore do not have to treat them equally.

Moral rights or wrongs are not a privileged product of the mind. Although our feelings can sometimes be diametrically opposed to our sound reasons, feelings can guide us perfectly to what we want or detest. Ethics is consequently not solely about having sound reasons, but also about having right emotions. A characteristic of our feelings towards moral care is that the more attached we are to beings, such as family, relatives and friends, the stronger our feelings are to protect them and to take care of them. Even if reason tells us that all humans are equal, we favour those to whom we are more attached and we give them, therefore, more privileges than others. This form of discrimination is prima facie morally accepted and justified.

Midgley expands her theory towards the moral status of animals. It might be true that humans and animals are equal, especially regarding the ability to experience pain and to suffer. However, we feel in general more empathy and compassion for our fellow humans than for animals. This justifies speciesism. But, according to Midgley, one cannot associate speciesism with racism as Singer does. Racism is always wrong, while speciesism is not. We do not need cultural or religious knowledge in order to give basic care to humans, but we do need knowledge of the species in order to provide animals with proper care. This shows that speciesism is not a form of racism.

For Midgley it is not wrong to use animals for human purposes, provided we acknowledge that (higher) animals are sentient beings and that we are aware of the morally substandard treatment of animals in certain sectors of society.

A characteristic of anthropocentric philosophers, radical or mild, is that they regard humans as superior to the animals and defend inequality. The anthropocentrist claims that the moral status of a being can only be based on dispositions which only humans have, such as reason, empathy and compassion. The mental abilities of animals and the level of their consciousness are either denied or classified as too limited to be of considerable moral importance. Anthropocentrists use inequality to justify or to promote the human use of animals; however, they call for restrictions when gross or unnecessary violations of the animal's welfare occurs.

4. Animal welfare and the moral status

One of the key concepts in the animal issue debate is welfare, because the moral status seems to be depended on the effect of having one. Although the welfare status of animals, i.e. vertebrates, is commonly accepted by many sciences and the public, within animal philosophy it is still in debate. Singer, Regan and Rollin have no shred of doubt that animals have the mental capacity to experience suffering, while Frey, Carruthers and Scruton challenge that perspective.

In the conceptual debate one is still looking for an adequate definition of animal welfare. Definitions of welfare range from objective denotations as 'living in harmony and coping with the environment' formulated by Lorz (1973) and 'an attempt to cope with its environment' given by Broom (1986) to a subjective one by Sumner (1996) as 'quality of life as it is experienced and valued by the animal itself'. The objective definitions of welfare are not too problematic because their parameters can be measured, such as pre-pathological states, stress, coping, fitness and adaptation. Subjective definitions, on the other hand, form a challenge to science, because welfare is interpreted in terms of mental states, such as emotions, wants, feelings and subjective suffering (cf. Bracke 2001). It is this kind of definition that challenges philosophy. In general there is no discussion as to whether (higher) vertebrates can perceive and express pain and pleasure, but dissent remains in the philosophical judgement as to whether animals can value an experience and whether that disposition is necessary to be conscious of pain and suffering.

Zoocentric philosophers find confirmation for the animals' ability to experience pain and suffering from the anatomical and physiological structure of the central nervous system. They reason from analogy, by claiming that if humans and higher animals share to a large extent the fabric of the body and elementary behaviour, then it is more than likely that they share some mental states as well. Vertebrates, in particularly mammals, have the same kind of brain and nerves as humans and we see in animals the same type of behaviour as ours in cases of desire, belief, pain and suffering. Sharing these behaviours in the same circumstances justifies the conclusion that higher animals are mental beings (cf. DeGrazia 1996).

Reasoning from analogy is not scientifically completely valid. Although a behaviour might indicate pain and suffering, such reasoning can generate fallacies too. Certain animal behaviour that looks like suffering does not correspond with the expected feeling or emotion, such as a terrifying screaming piglet that is lifted on his back or the mimicry of some wounded birds. On the other hand, a behaviour may look normal while other indications tell us that suffering is likely to occur, e.g. a buffalo that is attacked by lions but does not show any stress behaviour at all (cf. Harrison, 1991).

Besides the possible fallacies generated by the analogy argument, the fact that (higher) vertebrates have beliefs, desires and sensations is not at issue. Anthropocentric philosophers, such as Carruthers and Scruton, acknowledge this fact. But, Carruthers (1998, 2000) worked out a theory that challenges the mental scope of animals and even questions an animal's disposition to experience pain and suffering. The key in his theory is that consciousness is not related to language, as Frey holds, but to a faculty of thinking.

A subjective mental state implies consciousness, of which there are two kinds: a perceptual and a reflective one. A *perceptual* consciousness is the awareness of the (selective) stimuli one perceives from one's environment and to which one responds. This is a first-order consciousness we see in humans and animals. Vertebrates, especially mammals, are aware of the things they see, hear, smell and touch and they have some awareness about their responsive behaviour. The other form of consciousness is of a higher order and is known as *reflective* consciousness or self-consciousness. This type refers to the capacity for higher-order representation of the organism's own mental states.

Due to the higher-order character (thought about thought), the capacity for self-consciousness is closely related to the question about whether non-human animals have a concept of their own consciousness. So, one can be aware of something and one can be aware of the fact that one is aware of something. Having a belief or desire is, as such, not sufficient to have a welfare. One must be aware of the fact that one has a belief or desire. Only then can one be positively or negatively conscious of one's mental experience. Furthermore, such higher-order consciousness is not possible unless a creature has concepts that are necessary for thinking about his mental states. Reflective consciousness therefore requires the capacity to think about, and to conceptualise, one's own thoughts.

Carruthers maintains that there is little basis to think that any non-human animal can think about its mental state, with the possible exception of chimpanzees. Animals do have a feeling in the sense of a sensation, but that is a feeling without *a Feel*: without being conscious of that inward impression, state of mind, or physical condition. Conscious feeling or suffering cannot exist without concepts as part of a marked sentiment or opinion. Carruthers' reaction to the welfare definition of Sumner would be: animals can certainly sense some qualities of life, maybe they can experience them as well, but animals do not have the ability *to value* a quality of life. Any animal behaviour that appears to show the contrary might be nothing more than the result of an anthropomorphic interpretation.

It is almost incomprehensible that higher animals, zoologically compared with humans, do not have a welfare. We see definite welfare-related behaviour, but methodologically we know that behaviour alone is not sufficient validly to infer one's awareness of one's own mental state. Neither zoology nor philosophy can provide certainty that animals are 'conscious' of what they feel. This inability, however, does not imply that the moral cocoon that protects animals from human use is fragile. If the sciences are not able to present conclusive evidence on animals' experience of welfare, then room is permitted for different beliefs on this subject (cf. Vorstenbosch, 1997). In that case, the moral status of the animals does not depend on what it objectively is, but on what we subjectively believe. For example, we can simply believe that animals experience pain and suffering, because this will strengthen our relation with animals.

Next to such animal-related concepts as welfare and health, one finds moral concepts that are directly linked to the human-animal relationship. After all, one can be opposed to a certain instance of animal use because the animal as such has a negative welfare experience, or one can be opposed because we humans find it appalling, ill-mannered, depraved or sinful. When animal issues occur one can always ask oneself "who carries the burden?"; the animal directly because of the pain or suffering it experiences or humans indirectly because of the distress the animal practice causes them. The latter is getting more significant, because we humans have a natural tendency to project our feelings onto animals.

There are several vocabularies to assess the moral acceptability of the use of animals (cf. Schroten 1997). One can use the term *integrity* to emphasise the respect for the wholeness or unaffectedness, e.g. regarding cutting a dog's tail or ears or changing the animal's genetic makeup; *instrumentalisation* (reification) to denote the improper treatment of animals as if they were instrumental objects, for instance. in laboratory animal science; *naturalness* to show the dominance of technological influence in the domestication which furthers the distance of the animal with its natural origin such as in special breeds of cattle and pets; and *playing God* to express a concern, especially in genetic modification, regarding our moulding of life without having insight in the consequences and implications of this practice. These vocabularies may be morally convincing even in situations where animal suffering as such is not at stake.

These human-oriented concepts enrich the animal debate. They give us alternatives to replace or to supplement the multi-operational welfare concept. The moral status of animals does not, therefore, depend entirely on having a welfare. Regan is right to maintain that the moral status of animals is not entirely dependent on the their welfare but also on human attitudes towards their use. This variety of animal- and human-related concepts is another example of the diversity within the animal issue.

5. Moral deliberation

The determination of the moral status of animals is just a step towards moral decision making. Eventually we would like to know whether it is morally justifiable to use animals for a given human goal. Throughout the ages philosophy developed several strategies to determine the rightness of a decision, such as utilitarianism, deontology, virtue ethics and contractualism.

The most common and often used strategy is utilitarianism. Utilitarianism, nowadays also known as consequentialism, aims at a cost-benefit analysis in order to create the most favourable situation for as many subjects as possible. Pleasure, happiness or even preferences, on the one hand, and pain, suffering and dislikes, on the other hand, are quantified or qualified and put into a utilitarian calculus. Because we have a moral duty towards others to promote happiness and reduce suffering, the result must show whether our action serves the utilitarian objective of creating a state of affairs in which the overall good exceeds the overall bad.

Among the philosophers discussed, Singer, Frey and Scruton are the most inspired utilitarians in the field. Singer argues that in the production of the greatest ratio of pleasure to pain, one must take the suffering of every animal into account. By doing so, farm factories, hunting and some animal experiments will be immoral because the sum of the pain and suffering of the animals outweighs the sum of the pleasure humans gain from these animal products. Frey interprets the significance of animal suffering differently than Singer. What ought to be counted is not pain but desires. In that case the calculation will favour human interests, because animals lack the ability to desire.

With these examples the inherent problem of utilitarianism becomes apparent. The aspects of life one puts on both scales of the balance are open for selection and the standard to weigh these aspects has several interpretations. What is the key aspect of happiness and how to measure it, are questions utilitarianism faces. Besides, the most favourable outcome does not have to be morally the best option, as one can see by Scruton's justification of fox hunting by stating that it creates a lot of happiness not only for the hunters but also for the hounds and horses that chase their prey (2000, 117ff). More specifically, utilitarianism would justify experiments on human beings, if that would satisfy more interests than alternatives.

One trouble with utilitarianism is that it can not properly quantify animal welfare and human (moral) desires, and therefore marginalises its own utility.

The inherent flaws within utilitarianism open the path to the deontological approach. Deontological justification of an action is not based on utility or consequences, but on the action coherence with moral principles. Regan, who advocates deontology with his animal rights view, describes the essence very clearly. We can promote the welfare of the animals, but that does not alter the wrongness of our actions. As already mentioned, what is wrong is not the suffering of the animals but our attitudes towards the them. What matters is not the interest itself, but the individual that has the interest. Individuals have rights that ought to be respected regardless of the consequences.

The main problem with the deontological approach is that it is too rigid. It does not give much room to manoeuvre. Animal experiments and factory farming are morally wrong; full stop. The animal rights view has great difficulties with demarcating situations in which a certain use of animals is morally justified. Due to their rigidity, they easily create a position of stalemate, especially when two significant ethical principles are in conflict with one another. One cannot always uphold the principle of nonmaleficence, by not inflicting harm, and at the same time follow the principle of beneficence, by preventing and removing harm and promoting the good. In several situation the latter can only be pursued at the cost of the former, e.g. searching for a medicine against AIDS by using a great number of laboratory animals. The deontological approach confines itself to a limited room in the moral assessment of animal utilisation.

An alternative approach which avoids both utilitarian and deontological problems, is the virtue approach. In this line of deliberation one does let an action be guided entirely by the consequences nor by attendant principles, but by the excellence of a character or particular quality. The question is not 'what ought I to do?' but 'who should I be?' This question does not refer only to persons, but also to companies and the whole of society. A good character or trait, e.g. fairness, benevolence, truthfulness and generosity, does not need principles to guide it and will by its nature produce some favourable consequences.

Within the animal issue debate Stephen Clark (1997) is a clear advocate of this approach. The most important question for him is: how may we best order the community of which we are part? If we acknowledge that animals, by history, are a part of our community just as children are part of a family, then we must define the qualities of good stewardship just as we define the qualities of good parental care. Good stewardship towards animals does not entail using them for the unrestrained human urge to increase control over the world and to satisfy hedonistic lusts.

Although virtue ethics may be a workable strategy for moral deliberation, it has some difficulties too. Virtue ethics has a tendency to be conservative. Old traditional values have already proved their merit, so let's stick to them. This may narrow the openness for modern problems and hinder the search for creative solutions. Besides, the meaning and scope of a particular virtue are not indisputably defined. Known good traits, such as justice, prudence, conscientiousness and benevolence, leave some room for interpretation and are highly influenced by personality and personal situations.

Another deliberation strategy is to emphasise the social making of moral rules. In this strategy, called contractualism, what is right and wrong as well as the rules that go along with it result from an imaginary contract between rational agents. The contract is a hypothetical agreement as to the rules that govern the behaviour of the participants. Contractualism denies the validity of utilitarianism, deontology and virtue ethics, because these can be either too objectivistic or too intuitionistic, due to the fact that they do not result from mutual understanding. Peter Carruthers has already been described as an advocate of this strategy in his justification of his claim that animals have no moral rights. Those who are not able to participate in the making of the contract cannot directly benefit from it.

However useful this approach may be to overcome the problems of the other three approaches, it has some weaknesses too. Although contractualism is not an instrument to legitimate self-interests, it has a tendency to be too liberal. Individuals who are making the contract automatically make judgements about their best interests and about the means to satisfy their desires. It is not easy to make a contract that limits one's own pursuit for values, such as health, safety, prosperity etc. Self-limitation seems to be harder in the case of the use of animals and natural resources, because already existing moral and legal contracts, that impose limitations, mostly do not generate the expected reduction. In general, the social utility of nonmaleficence and beneficence becomes more hypothetical when one's own interests are at stake.

All the deliberation strategies have one thing in common: they are product-oriented. They all seek for a particular outcome: the best ratio, the best principle, the best trait or contract. The problem is that we cannot distinguish good from bad and right from wrong by using a list of rules or principles, by picking a virtue or creating a hypothetical contract, because reality does not allow such a rigid approach. Moral deliberation in reality is more an interplay between utilitarianism, deontology and virtue ethics. In our moral behaviour we always look at consequences, but also at the act and the motive. Utilitarianism rightly shows the impact of our actions, while deontology presents us with the different ways we might promote the best and virtue ethics develops our best character.

The question concerning the animal issue is whether it is wise to be product-oriented and aim at an outcome when the diversity of thoughts and practices is significant. What is good or bad depends mainly on a set of moral values from which we interpret the situation. This set of values differs from person to person. What is obligatory or desirable for one person can be merely permissible or even a matter of indifference for another (cf. Klaver *et al.*). The key in contemporary moral decision making seems to be that it is more aimed at the process of deliberation than at the outcome.

6. Working with diversity

The common fact of diversity is that all theories have some truth but lack overall satisfactoriness. The lacuna one theory creates can be filled by another. So far a pan-theoretical approach of the animal issue has not been created, and so far philosophy lacks the ability to settle the subject. Due to this shortcoming, the question might be raised: does philosophy have the task of sorting out the diversity, to melt pluralism to a monolistic approach? Maybe it is wise to answer with a no. The animal issue is more a public matter than a philosophical one. A diverse moral vocabulary makes moral monism undesirable and moral pluralism inevitable. In a pluralistic world, the task of the philosopher or ethicist is not to produce justifications but to elaborate and guide the process of consensus finding. Today's role of philosophy is to be a very powerful instrument to clarify and to develop moral vocabularies that further public deliberation.

Maybe the best thing we can do with the diversity is to take a pragmatic approach in which we consider the different theories of the philosophers not as justifications of a situation, a status quo or of a change, but as discoveries and elaborations of new ideas. All these ideas do serve a purpose. It makes us think better, it deepens our understanding of the complexity and it stimulates our imagination and solution finding (cf. McGee 1999).

Within the diversity of thoughts regarding the animal issue there may be some principles that will hold up after all. One of them is the acknowledgment that we humans have the ability to purposely damage and to promote the good that all living beings have of their own, by which they regulate and preserve their own life (cf. Tayor 1986). The other one is to be prima facie nonmaleficent by applying the principle of 'the benefit of the doubt' on all sentient animals, regardless of whether they can consciously experience pain and suffering.

References

Bracke, M.B.M., Spruijt, B.M. and Metz, J.H.M., 1999. 'Overall animal welfare assessment reviewed. Part 1: Is it possible? Netherlands Journal of Agricultural Science, 47, p. 279-291.

Broom, D.M., 1986. Indicators of poor welfare. British Veterinary Journal 142, p. 524-526.

Carruthers, P., 1992. The Animals Issue: Moral theory in practice. Cambridge, Cambridge University Press.

Carruthers, P., 1998. 'Animal Subjectivity', Psyche 4(3).

Carruthers, P., 2000. 'Replies to Critics: Explaining Subjectivity', Psyche 6(3).

Clark, S.R.L. 1997. Animals and their moral standing. London, Routledge: ch. 9, 13.

DeGrazia, D., 1996. Taking Animals Seriously: Mental Life and Moral Status. Cambridge, Cambridge University Press.

Descartes, R. 1997/1637. Discours de la Méthode. Hamburg, Felix Meiner Verlag.

Frey, R.G., 1980. Interest and Rights: The Case Against Animals. Oxfrod, Clarendon Press..

Harrison, P., 1991. `Do Animals Feel Pain?', Philosophy 66, p. 25-40.

Klaver, I., Keulartz, J., Belt, H. van den and Gremmen, B., 2002. Born to be wild: a pluralistic ethics concerning the introduced large herbivores in the Netherlands. Environmental Ethics, spring.

Lorz, A., 1973. Tierschutzgesetz. Verlag Beck, München, 272 p.

McGee, G. 1999. Pragmatic Method and Bioethics. In G. McGee ed., Pragmatic Bioethics. Vanderbilt University Press, London.

Midgley, M., 1983. Animals and why they matter. Athens, University of Georgia Press.

Regan, T., 1983. Case for Animal Rights. London, Routledge and Kegan Paul.

Rollin. B., 1981. Animal Rights and Human Morality. Buffalo, Prometheus Books.

Rollin, B.E. 1995 The Frankenstein Syndrome. Ethical and social issues in the genetic engineering of animals. Cambridge University Press, Cambridge.

Schroten, E. 1997. 'From a Moral Point of View: Ethical Problems of Animal Transgenesis'. In Louis Marie Houdebine (ed.), Transgenic Animals: Generation and Use. Amsterdam (Harwood Academic Publishers), p.569-574.

Scruton, R., 2000a. Animal Rights and Wrongs. Metro, London.

Scruton, R., 2000b. 'Animal Rights', Urbanities., 10/3.

Singer, P., 1973. Animals Liberation. New York, New York Review Book.

Singer, P., 1993. Practical Ethics second edition. Cambridge University Press, Cambridge.

Sumner, L.W., 1996.Welfare, Happiness and Ethics. Clarendon Press, Oxford, 239 p.

Taylor, P.W., 1986. Respect for nature: A theory of environmental ethics. Princeton New Jersey, Princeton University Press.

Vorstenbosch, J. 1997. "Conscientiousness and consciousness: How to make up our minds about the aninmal mind?". In M. Dol (*et al.*), Animal Consciousness and Animal Ethics. Assen ,Van Gorcum. p. 33-47.

Animal integrity

Henk Verhoog
Louis Bolk Institute, Hoofdstraat 24, NL-3972 LA Driebergen, The Netherlands

Abstract

The concept of animal integrity plays no role in traditional utilitarian and deontological theories of animal ethics. Most of these theories focus on the presence of animal consciousness, feelings and animal welfare. It is in the discussion about ethics and transgenic (genetically modified) animals that the concept has been introduced, for instance in the Netherlands. It is said there that this moral concept embraces those moral aspects which go 'beyond health and welfare'. In the broader literature about agriculture and gene technology (including crops) animal integrity is usually mentioned under the category of intrinsic concerns, whereas extrinsic concerns are related to the consequences of the technology (risk, benefits, etc.).

1. Introduction

Some zoocentric utilitarian thinkers have said that a violation of the integrity (wholeness or intrinsic nature) of an organism is only a moral issue, when the animal suffers as a consequence; otherwise it is an aesthetic issue. Virtue ethics, related to basic human attitudes towards nature, and deontological approaches referring to the human-animal relationship, are excluded by definition. This is different in a biocentric theory, in which moral respect is shown for the animal's 'nature' as such. This is the animal as it appears in everyday perception. The link with aesthetics is clarified by comparing perception in everyday life with perception in reductionistic, experimental science. It is shown that at least for virtue ethics there is a direct relation between an ethical attitude towards animals and how animals are perceived. The concept of animal integrity links subjective experience with the perception of the animal's wholeness.

2. Integrity and zoocentric animal ethics

First a point of clarification. I distinguish several bioethical theories such as: anthropocentrism, zoocentrism (sometimes called pathocentrism), biocentrism and ecocentrism (see Pompe, this book; Verhoog, 2000). Theoretically each of these theories can in principle have consequentialist, deontological and virtue-ethical interpretations. Philosophers such as Singer and Regan, for instance, are both zoocentric thinkers, in the sense that the presence of some form of consciousness (sentiency) is a necessary and sufficient condition for being morally relevant. But Singer interpretes this in a utilitarian way, and Regan in a deontological way.

Speaking about zoocentric ethics in this chapter, I refer to the consequentialist interpretation of zoocentrism. I give two examples. The first example is Bernard Rollin (1995). According to Rollin animals with some form of consciousness have basic interests, which are morally relevant and have to be protected. He introduced the concept of 'telos'. In an article from 1986 Rollin defines this telos as: "the set of needs and interests, physical and psychological, genetically encoded and environmentally expressed, which make up the animal's nature...It is the pigness of the pig, the dogness of the dog". In 1995 he calls it the species-specific nature of animals: "animals like humans have natures, and respect for the basic interests that flow from those natures should be encoded in our social morality". Zoocentric theory obliges us morally to take into account the interests that are believed to be essential and constitutive of the animal's nature. This is not a matter of being kind to animals, Rollin says; it is our moral duty.

It is important to realize that it is not the telos itself which is respectful, but the interests determined by it. Genetically engineering animals, for instance, is not wrong in itself, according to Rollin. Crossing species barriers, the creation of chimeras or the induction of leglessness in animals is not a morally relevant intervention because species are not morally relevant. Only individual animals who can suffer as a result of genetic engineering are morally relevant. Species cannot suffer. The animal's telos is not sacred. "I never argued that the telos itself could not be changed", Rollin says (1995). To change the telos of chickens through genetic engineering, so that they no longer have a nesting urge, means to remove a source of suffering for animals held in battery cages. They are better off than before. Rollin agrees that it may be better to change the housing conditions, but as long as this is not expected to occur in our present societies, it is better to decrease the suffering, even when this has to be achieved by means of genetic engineering.

For Rollin, the presence of consciousness (sentiency) is a necessary and sufficient condition for moral relevance. Sentient animals are said to have an intrinsic value. They cannot be looked upon as if they were mere instruments for human goals. Also for the animal, feelings of happiness are of intrinsic value to them (see Verhoog, 1992 for the different meanings of 'intrinsic value').

If the presence of consciousness is a necessary condition it implies that an entity without consciousness will not be morally relevant; moral agents will have no direct responsibilities towards them. To be a sufficient condition means that being conscious implies moral relevance, but being conscious is not necessarily the only criterion for inferring moral relevance.

Representatives of a zoocentric approach are not always clear as to their position in this connection. Take Garner (1993) for instance, who says:
"Put simply, if animals are regarded as mere unconscious machines we would seem to be justified in treating them in any way we choose. If, on the other hand, we accord them sentiency - the capacity to experience pain and pleasure - then it is clear that there ought to be at least some constraints on our behaviour towards them".

What about invertebrate animals? They are generally not regarded as sentient, but that does not mean that they are 'unconscious machines'. Even plants are not machines. According to Rodd (1990) "questions about mind and consciousness in other animals are of central significance for a theory of their moral status". Here again it is important to specify what is meant with 'of central significance'.

A second example of a zoocentric approach can be found in a critique by Sandøe, Holtug and Simonsen (1996) on an article by Vorstenbosch (1993) about the concept of 'genetic integrity'. When we genetically modify animals or plants we introduce genes from quite different, unrelated organisms (therefore we call the result a 'transgenic' animal). Vorstenbosch in this connection uses the concept of 'genetic integrity', which he defines as follows:
"In biotechnology the genetic integrity of the individual animal and of the species is central. We can define the genetic integrity of the animal as the genome being left intact. This seems to be a meaningful notion in view of the fact that we can clearly point out some factors or actions by which the genome would not be left intact".

In a reaction upon this article Sandøe *et al.* say that the demand to respect genetic integrity would also tell against selective breeding, and thus against domestication in general. According to them the idea of genetic integrity suffers from three main problems:

- genetic structures have changed continuously in the process of evolution; it is arbitrary to say that the present genetic make-up is in any way special;
- if the idea of genetic integrity is valid, it would not be a good thing to breed for increased health, to secure the health of domestic stock;
- there is no guarantee that the genetic structures which exist right now are more conducive to animal welfare than others.

Because of these reasons the authors think that the concept of integrity is not a useful concept. If there are any ethical limitations to breeding or domestication, they think it should be because of animal welfare reasons. The way of reasoning is similar to Rollin's zoocentric view. The second and third problem mentioned already presuppose that health and animal welfare are the main criteria. But the question was whether there are ethical issues which go beyond health and welfare. In organic farming, for instance, no problem is seen in breeding animals to increase health (resistance) and welfare, but in spite of this, gene technology is rejected, even if it could induce health or welfare (Verhoog *et al.*, 2003). As good utilitarians Sandøe *et al.* only look at the result (the genome is changed by traditional breeding and by genetic manipulation), not at the differences in the process. Ethics only comes in afterwards, when the transgenic animals are made already. One of the reasons for rejecting gene technology in organic farming is their choice for a biocentric (and ecocentric) approach. In this approach the human attitude and the way of interference are also important criteria, in addition to the consequences for the animal.

Against the first argument used by Sandøe *et al.* it can be said that it looks very much like a naturalistic fallacy. What we find in nature (continuous change of the genome) is used as an argument to justify an intentional human interference in the genome, which would never occur in nature. And, unless one thinks that animals have no moral status at all, man must have good reasons for interfering, irrespective of the speed of evolutionary changes. But more important perhaps, the fact that human beings also change during their life-time, both physically and psychically, does not prevent us from speaking about violations of their physical integrity at any specific moment. In biomedical ethics the integrity of the body is a very important morally relevant criterion. Something similar can be said about the genetic integrity of populations or species in the evolutionary process. Evolution is a slow process, contrary to the changes brought about by genetic engineering.

From the foregoing it will be clear that the concept of integrity does not fit into a zoocentric theory, as long as sentience is seen as a necessary and sufficient condition for moral relevance or moral status. The main reason is that it is not the 'nature' or telos of the animal itself which is morally important, but only the interests which follow from it, and their (dis)satisfaction. This is different in a biocentric theory, to which we now turn.

3. Integrity and biocentric ethical theories

That it is important to be specific about the question whether sentiency is a necessary or a sufficient condition for moral relevancy becomes clear when we look at the biocentric theory of Paul Taylor (1984, 1986). According to Taylor all living organisms have 'inherent worth', which he defines as follows: "The value something has simply in virtue of the fact that it has a good of its own. To say that an entity has inherent worth is to say that its good (welfare, well-being) is deserving of the concern and consideration of all moral agents and that the realization of its good is something to be promoted or protected as an end in itself and for the sake of the being whose good it is" (Taylor, 1984).

The domain of morally relevant natural entities is widened to all animals and also plants, indeed all living beings ('teleological centres of life'). Plants do not *have* interests such as sentient animals, but we can say that something which contributes to their good is *of interest* to them. Having a 'good life of its own' emphasizes the 'own-ness', the identity, or species-specific nature (telos) of the living entity. When being sentient is part of the species-specific nature, this will be taken into account when dealing with such animals. In this biocentric theory sentiency is clearly a sufficient, but not a necessary condition for moral relevance.

The concept of animal integrity fits into this biocentric ethical theory, and it is closely connected to the meaning of 'naturalness' in this theory. Rutgers and Heeger (1999) give the following definition of animal integrity:
"the wholeness and completeness of the animal and the species-specific balance of the creature, as well as the animal's capacity to maintain itself independently in an environment suitable to the species".

It is interesting to see that several of the criteria mentioned in this definition can be related to different levels of the 'nature' of an animal:
- The wholeness and completeness of the animal refers to the level of the individual animal. Integrity presupposes the existence of an 'organism', a living

whole with interconnected parts. It is the interconnectedness, the balanced harmony of the parts of the whole, which is somehow linked to the concept of integrity. Taking away the horns of cows, even if it is done painlessly, is not a morally irrelevant thing in a biocentric theory, because it violates the characteristic nature of cows. It somehow disturbs the organismic 'wholeness'. The moral relevance of 'individuality' (autonomy) is highest (reaches its highest stage) in human beings.

- The species-specific balance refers to the species-specific nature of the animal, the natural characteristics at the level of the species. When we say that animals should be able to perform their natural behaviours, we refer to this level. A species always fits into an environment which can be more or less specific, dependent on the species. The ability to adapt to a particular environment is part of the species-specific nature of an animal.
- The animal's capacity to maintain itself independently can be related to the third level of naturalness, related to the question: what does it mean to be an animal and not a plant, for instance? When, in discussions about housing conditions, we say that animals should be able to explore their environment, this refers to almost all animals. Therefore it goes beyond the species level.
- The last level of the 'nature' of an animal is that it is alive, just as plants are alive. It has characteristics of life, which it shares with all living beings (capacity to grow, to reproduce, self-regulation, etc.). In her dissertation about plant breeding and the intrinsic value and integrity of plants, Lammerts van Bueren (2002) distinguishes, in line with the levels mentioned here, phenotypic integrity, genotypic integrity, planttypic integrity and integrity of life.

We must now look at the implications of this biocentric view with respect to the genetic modification of animals. We have seen that Rollin accepted the concept of the animal's nature or telos, but its moral relevance is limited to sentient animals, and the interests determined by this telos. Moral relevancy of genetic engineering only arises when the individual sentient animal suffers as a consequence of it. In contrast to this, in a biocentric theory the animal's characteristic nature is itself morally relevant. I will give two examples in which first steps have been taken to implement the concept of animal integrity in public policy with respect to transgenic animals.

The first example refers to the 'no, unless' policy with respect to transgenic animals in the Netherlands (see Brom and Schroten, 1993; Verhoog, 2001). The making of such animals (also in the field of biomedical research) falls under the jurisdiction of a special law, the Animal Health and Welfare Act of 1992 (Ministry of Agriculture). That it does not fall under the Experiments on Animals Act of 1977 (Ministry of Health) is due to the fact that genetic modification of animals was seen as a violation of the integrity of the animal. And the latter criterion goes beyond the two criteria which are used in connection with animal experiments (health and welfare). The 'no', in the 'no, unless' policy refers to a deontological element in the ethical screening process: the making of transgenic animals is forbidden, unless it is done for a 'substantial goal' and the goal cannot be reached by alternative means (without the use of gene technology). To increase agricultural production by using transgenic or cloned animals would not be seen as a substantial goal in the Netherlands, in contrast to almost all research using animals (mice mainly) for biomedical research. With respect to this biomedical research, violation of the integrity of the organism is introduced as a third criterion besides health and welfare, but it is implemented in a utilitarian way (De Cock Buning, 1999). It was made operational by looking at the consequences of genetic modification:

- actual changes in the genome;
- actual changes in outward appearance (phenotypic);
- functional change in species specific behaviour;
- functional impairment of the ability to live autonomously.

In addition, the weighing up of the consequences for the animal against the goal of the experiment was done in a utilitarian way; the proportionality principle was applied. To fit into the utilitarian approach, violation of the integrity should be made measurable.

The second example is a report published in 1999 by the Danish Ministry of Trade and Industry. In this report it is said that the natural can be understood as synonymous with natural mechanisms. However, it can also be understood in a wider sense, i.e. as the overall coherency of which all organisms are part, and which has both a physical and a historical dimension. In this wider sense living organisms should be respected as parts of a spatial and temporal entirety. Nature and the overall coherence (mutual dependence) of life are perceived as vulnerable. This is why demands are made to respect the integrity of living organisms. Everything living has an integrity that can be encroached upon and destroyed. This entails a distinction between destructive and creative intervention, between considerate and reckless changes. The right to integrity concerns respect for dependence and encompasses ecosystems, plants, animals and human beings. Besides the utilitarian ethical view and discourse ethics (dealing with the social process of ethical reflection) the authors identify an ethics of integrity as being of importance when dealing with genetic engineering.

These examples show how the concept of integrity in a biocentric theory can be implemented in different ways:

- In a consequentialist approach we look at the consequences of human action, in our case the consequences for animals. In utilitarianism we make a kind of cost-benefit analysis, comparing the consequences for the animals wih the benefits for humans. Integrity can be implemented in this context when, as in Dutch policy, the harm done to the animals is extended to include violation of the integrity of an animal. It must be a kind of harm which can be measured, and it must allow for some degree of assessment.

- In the Danish report the normative principle of 'respect for the integrity of life' is emphasized. It comes close to respect for the 'Würde der Kreatur' (dignity of non-human organisms) in Swiss law (see also the discussion about this concept in the Journal of Agricultural and Enviromental Ethics 13/1-2). Vorstenbosch (1993) pointed out that human respect and animal integrity are closely linked in arguments. To a greater extent than with the concept of animal welfare, which can be affected by all kinds of natural circumstances, the concept of integrity directly refers us back to the moral responsibility of human beings for the state of the animals, whether they suffer from human actions or not. The concept of integrity, he says, invites us to choose a deontological ethical approach. In deontological approaches we evaluate the action itself, independent of the consequences, by asking whether it is in accordance with certain normative principles. If we accept the normative principle that we should respect animal integrity, than it is wrong not to do so. It is more of the nature of an either-or criterion. Such principles usually follow from a certain non-anthropocentric attitude towards nature: Nature as a 'partner' with humanity as a 'participant' in Nature (Zweers, 2000).

- Finally, we can try to connect the concept of animal integrity to virtue-ethics, where we look at the motives of the moral agent, and the virtues involved. Cooper (1998) follows this line of thought. He says that those who violate nature, violate themselves. He connects integrity with the human virtue of humility, which he defines as selfless respect for reality, for the animal 'fitting into its own being'. When humility is absent, it can lead to alienation, to a sense of being cut off from nature. Cooper's critique of utilitarianism is that it excludes human feelings. It further blinds us for the common experience that morality primarily involves a judgment of the persons acting, rather than of the consequences of the action.

In utilitarianism the attitude of humans towards animals, and the human-animal relationship, are nor relevant for the weighing of the consequences. A good illustration is the discussion with a biotechnologist I once had. This biotechnologist agreed with me that creating blind hens, so that they don't peck eachother, is a violation of their integrity, although their welfare is enhanced by it. But, he added, when the animal does not notice it, or does not suffer from it, then it is not important for the animal, it is just 'between the ears of the observer'. If a person feels that making a mouse with a human ear on its back is the wrong thing to do, because it violates the integrity of the mouse, while the mouse's well-being is not disturbed by it, than the person has a problem, not the mouse. This reminds us of Rollin who said of critique of gene technology, that this refers to an aesthetic, not a moral, issue. It is a problem of the subject, not of the object; and therefore it is 'subjective', it is in the eye of the beholder. This does not fit into the utilitarian approach. Utilitarians want criteria which are independent of human virtues.

4. Perception of animals in daily life and in science

At the beginning of this chapter I referred to Rollin who said that it is our moral duty to take animal interests into account, and that it is not a matter 'of being kind to animals'. It must be admitted that this is a strong manifestation of the zoocentric-utilitarian approach. When we can show, with scientific arguments, that animals suffer, than this is morally wrong, independent of the feelings of people. I have no problems with this, only with the wider exclusive claim that moral wrongness is restricted to these cases only. In the beginning of the development of the Dutch 'no, unless' policy, it was mentioned that gene technology is a qualitatively new step in the instrumentalisation ('makeability') of animals, in human scientific control over nature. This does not refer to the consequences for the animal, but is related to human attitudes towards nature, as if animals have no value-of-their-own (intrinsic value). Here I want to look at this aspect from another perspective.

I limit myself to those aspects of the violation of the integrity (and / or intrinsic value) of animals, which cannot be measured by looking at visible changes in the appearance of the animal. Here physical changes in the genome are also considered to be visible changes in the appearance of the animal. We have seen that such changes can be fitted into the utilitarian framework by extending the concept of harm. I focus on the human perception of animals and the relation of this to the human-animal relationship.

To start with, I want to mention the Swiss zoologist Portmann (Grene, 1968). Portmann was very much aware of the growing gulf between the biotechnical approach in biology and what he called the comparative holistic approach. In the latter approach one looks to 'authentic' phenomena, directly visible by the human senses, and compares these phenomena in order to find out the meaning within a larger whole. The other, reductionistic approach leads to causal analysis, to discover the mechanisms behind the scene, as it were. Knowing these mechanisms allows us to control nature. Portmann makes a distinction between a 'theoretical' and an 'aesthetical' function of the human mind. In the theoretical function the mind tries to transform qualitative phenomena to quantitative data by means of rational thinking. In the aesthetic approach the scientist does not try to explain what is directly given in our perception by means of mechanisms which are postulated inside an organism. Such an aesthetic and more intuitive perception of organisms can, according to Portmann, teach us something about what is 'essential'. The concept of intrinsic value is directly related to what is essential for the life of the organism involved. It sensitizes us to what is real about animals, independent of the use we may have of them.

In the reductionist-experimentalist approach, based on the Cartesian dualism between human subject and nature as material object, scientific knowledge of the primary qualities is considered to be knowledge of the real world, and the secondary qualities of our sensuous perception are subjective. Portmann takes the opposite view. What we experience in our world of everyday life comes first, is the primary world, and what the scientist finds in the laboratory is secondary, is derived. Portmann was concerned about the fact that in biology the emphasis shifted more and more to the causal-analytic approach. He was convinced that further alienation from nature, which is the result of this latter approach, can only be prevented when people have a direct experience of the richness, of the diversity of biological form, especially in scientific education.

The issue here is the growing discrepancy between the immediate perception of the animal in our daily life and the scientific, reductionistic perception of animals. I will illustrate this with a number of examples (taken from Verhoog, 1999 and 2002):

- Rollin (1990) describes in detail how, in connection with the study of animal consciousness, science has become increasingly remote from common sense and ordinary experience. Science rests on its own philosophical and ideological presuppositions, which are rarely examined. These presuppositions determine what counts as real, as fact, as legitimate data and explanations. According to these presuppositions, data about conscious experience or animal mentation are not considered as legitimate data; the mental phenomena have to be reduced to neurophysiological or chemical data. The tension between human experience in the world of everyday life and laboratory data is present everywhere in Rollin's book. The direct and often anecdotal evidence of animal consciousness and animal pain in our everyday life is denied in the very artificial setting of the laboratory, where the first aim of research is control. In addition, the moral relevance of experiential data about animal consciousness is largely absent in the laboratory setting; the objectification of science leads to the separation of science and ethics.

- Hearne (1987) gives another example of the discrepancy between our experience in the world of everyday life and the scientific objectification of animals. Hearne (1987) is a professional trainer of dogs and horses. Entering academia with an interest in philosophy, she was surprised to find that professors specifically denigrated students' language describing animals in subjective terms, that is, anthropomorphically. In Hearne's opinion the anthropomorphic language of everyday life, the language which is also used by trainers, is true to the nature of the animals as we experience them, and is, in that sense, perfectly objective. She refers here to the definition of objectivity as 'being true to the nature of the object studied'. In laboratory science objectivity means that the results are repeatable and controllable.

- Wieder (1980) describes how researchers studying animal behaviour in the 'behavioristic' tradition deal with chimpanzees outside of the experimental context. The chimpanzees are treated as if they are experiencing subjects, as embodied consciousness, and not as material objects (mere bodies). In this community-experience, subjectivity is apprehended directly in face-to-face encounters. Once these animals become objects of research, however, all subjective references are truncated. A new order of events is created, the order of pure objectivity that stands over and against the order of everyday life.

- Very illustrative is the work by Lynch (1988) and Arluke (1988, 1992). Lynch distinguishes between the 'naturalistic' animal of the common-sense perspective and the 'analytic' animal of the laboratory scientist. The naturalistic animal is the subject of anthropomorphic identifications. In the process of research this animal is transformed into the analytic animal, into data. In the scientific system of knowledge, the analytic animal is seen as the real animal. According to Lynch (1988) the laboratory procedures as such "assure the removal of the characteristics that make up the naturalistic animal". How this is done, and the ambiguities involved in this process, is also described by Arluke (1988, 1992). He (1988) writes about 'counter-anthropomorphism' when inanimate qualities are attributed to living things. Analytic animals are de-individualised and treated as anonymous beings. Social norms in the laboratory prevent scientists and animal technicians from treating laboratory animals as pets; they are instead treated as models, as supplies in grant proposals, etc. Arluke (1992) believes that this objectification or detachment is necessary for self-protection. Objectification breaks the interconnectedness between subject and object, thus moral constraints are nullified. Arluke (1992) suggests that this process of objectification rarely succeeds completely. Many animal experimenters have emotional difficulties with invasive animal experiments. However, in the laboratory setting there rarely is the possibility of openly discussing these problems. In the laboratory, in general, feelings about animal use remain private and extraneous to the 'real work' of the laboratory.

- In the history of cattle breeding, the animals own roles in the process of reproduction are completely taken away from them and brought under human control. The stages in this process are clear: artificial selection, artificial insemination, embryo transplantation, genetic engineering and finally cloning. Cloning is similar to vegetative reproduction. From a biocentric perspective this can be called a violation of the integrity of the animal at the level of animality. The animal is made plantlike as far as reproduction is concerned. This development also affects the farmer. Before science began to interfere in the process, animal breeding was a purely intuitive matter, based on the farmer's own perception of and experience with the cows. Since the rise of modern genetics, bit by bit the phenotypic properties of the animals are reduced to chromosomes, and finally to pieces of DNA (genes). Together with this, breeding comes to lie in the hands of specialised scientists and institutes. The experiential knowledge of the farmer becomes obsolete, and the farmer loses control over animal reproduction.
- My final example is from animal welfare research. Tannenbaum (1991) was one of the first authors who rejected what he called the 'pure science model' used by most researchers in the field of animal welfare. According to this model the scientist can do without value-judgements in animal welfare research. Animal welfare is believed to be a certain state of the animal which can be described objectively by scientists. Regarding ethical issues these researchers believe that everyone is entitled to an opinion, whereas making definitive statements about animal welfare is seen as the province of scientists. Sandøe and Simonsen (1992) say in contrast to this that scientific knowledge about animal welfare "does not by itself provide relevant, rational and reliable answers to the questions concerning animal welfare typically raised by the informed public". Strictly speaking, natural science can not say anything about the subjective experiences of animals, because the value-free method used excludes it. But it is exactly these subjective experiences which are at the central core of the lay public's definition of animal welfare.

Experiences cannot be measured directly; all that can be measured are objectively assessable parameters (pathological, physiological, behavioural). It is especially with the inference from measured parameters to the experiences of the animal that choices are made. The step back from measurement to judgment about the welfare state of the animal involves an interpretation which is not value-free. Sandøe and Simonsen are of the opinion that the concept of welfare itself lies beyond the general theoretical famework used by scientists.

The way of reasoning of Sandøe and Simonsen is taken a few steps further by Stafleu, Grommers and Vorstenbosch (1996). They distinguish several kinds of definitions of animal welfare (lexical, explanatory and operational definitions). Here I am especially interested in the difference between lexical and operational definitions. Typical lexical definitions come from our common sense perspectives of animal welfare, as they function in our world of everyday life. Examples are: welfare is a state in which an animal feels good, or absence of pain and suffering, or the famous definition by Lorz which says that animal welfare is a state of physical and psychological harmony between the organism and its surroundings. Lexical definitions also define the political and social frame of reference for scientific research, the social relevance of scientific approaches to animal welfare. Operational definitions define a concept in terms of specific experimental procedures, such as measurement of the corticosteroid level. A diversity of parameters may be developed at this level, parameters which again have to be interpreted to be made relevant for policy decisions about animal welfare. In this process from lexical to operational definitions not only the moral aspects but also the subjective feelings are lost in a diversity of objectively measurable parameters. These 'gaps' have to be bridged again if the scientific concepts of animal welfare are to be morally and socially relevant. But this bridging cannot be done objectively; it always involves an interpretation.

It is clear from these examples that the human-animal relationship changes in the process of objectivation which takes place in scientific methodology. Once the animal has become an object of scientific research, methodologically it only has an instrumental value. It is perceived quite differently from how it is perceived in common-sense experience. The latter experience comes much closer to what Portmann calls aesthetic and intuitive perception. It is characteristic of the aesthetic mind that it wants to leave the object as it is, to appreciate it for its own sake. In connection with living beings: it respects their intrinsic value. This suggests that there is a direct relation between the ethical attitude towards nature and how it is perceived.

5. Aesthetics and ethics

Hauskeller (1999) has argued that moral consciousness is triggered when we really see the otherness of living nature (or an animal), with all its specific details, in an aesthetic mode of perception. Through the aesthetic experience of beauty we can learn that there are things, which are worthy of preservation for their own sake. In the experience of the beautiful, nature appears to us in its immediate reality as an image ('Bild'). In such an image the inside (our consciousness) and the outside (the living organism which appears in consciousness) are no longer separated. To have this experience we must abstract from nature's instrumental value, its usefulness for us. Then one can experience nature's dignity. We therefore need an aesthetics which takes seriously immediate sensuous perceptual experience.

What is described here by Hauskeller comes close to what Cooper called the virtue of humility: selfless respect for reality; letting-be; an exercise to 'unself', to resign and to look away from one's own concerns, and take the ends of other creatures serious. Without humility we experience a sense of being cut-off from nature, a sense of alienation. Now one can also understand Cooper's statement that those who violate the integrity of nature also violate themselves. Inside and outside are no longer separated. That is why the concept of integrity directly appeals to human responsibility, as Vorstenbosch mentioned before.

Such a more aesthetic way of perceiving organisms can thus become a source of moral inspiration, in the sense that through it, we can become aware where we transcend certain moral boundaries. It will be clear that this moral inspiration relates to intrinsic concerns about genetic engineering: playing God, the unnaturalness, the violation of the animal's integrity or its intrinsic value. We have seen that these concerns have to do with our immediate sensuous experience of nature. And this experience is open to any person in the world of common sense. To be able to experience nature's intrinsic value, or better its inherent worth (dignity), one needs another mode of perception than the reductionistic one in experimental science.

Insofar as utilitarianism abstracts from the process, from the human-animal relationship, and only focuses on the result, on the consequences, we can say that there is an alliance between science and utilitarianism. The starting point for the argument in the last section was Rollin's zoocentric view that the problems that people have with gene technology as such (the intrinsic concerns) have to do with aesthetics and not with ethics. I suggest that what Rollin means with this is that it is not a problem for the animal, but for the human who perceives it. I further suggest that he means that perception is subjective, whereas the suffering of the animal is an objective fact.

In this way the whole process of human perception and action is excluded from ethical reflection; only the result counts. In the earlier example of cattle breeding we saw that the control over the breeding process was not only taken away from the animals, but also from the farmers. In classic cattle breeding the selection depended on each farmer's own immediate perceptions of the animal. With the increasing influence of science (population genetics, molecular genetics) the farmer's 'common-sense' perception is replaced by scientific perception. For utilitarianism what happens in this process, the transformation of human common-sense perception into scientific perception, is not important. This implies that for utilitarianism there is no relation between human perception and morality. I have tried to demonstrate that this is a very one-sided view of animal ethics, and ethics in general. As a result important moral concepts such as 'animal integrity' are not taken into consideration, not even in gene technology.

References

Arluke, A.B., 1988. Sacrificial symbolism in animal experimentation: object or pet? Anthrozoös II/2, p. 89-117.

Arluke, A.B., 1992. Trapped in a guilt cage. New Scientist 4 april, p. 33-35.

Brom, F.W.A. and Schroten, E., 1993. Ethical questions around animal biotechnology. The Dutch approach. Livestock Production Science 36, p. 99-107.

Cock Buning, Tj. De, 1999. The real role of 'intrinsic value' in ethical review committees. In M. Dol *et al.* (eds.), Recognizing the intrinsic value of animals, 133-139. Van Gorcum, Assen.

Cooper, D.E., 1998. Intervention, humility and animal integrity. In A. Holland and A. Johnson (eds.), Animal biotechnology and ethics, p. 145-155. Chapman and Hall, London.

Danish Ministry of Trade and Industry, 1999. An ethical foundation for genetic engineering choices. Copenhagen.

Garner, R., 1993. Animals, politics and morality. Manchester University Press, Manchester.

Grene, M., 1968. Approaches to a philosophical biology. Basic Books, New York/London.

Hauskeller, M., 1999. Auf der Suche nach dem Guten. Die Graue Edition, Kusterdingen.

Hearne, V. (1987). Adam's task. Calling animals by name. Heinemann, London.

Publications, London/Beverly Hills.

Lammerts van Bueren, E., 2002. Organic plant breeding and propagation: concepts and strategies. Doctoral Dissertation, Wageningen University and Research Centre.

Lynch, M.E., 1988. Sacrifice and the transformation of the animal body into a scientific object: laboratory culture and ritual practice in the neurosciences. Social Studies of Science 18, p. 265-289.

Rodd, R., 1990. Biology, ethics and animals. Clarendon Press, Oxford.

Rollin, B.E., 1986. "The Frankenstein Thing": the moral impact of genetic engineering of agricultural animals. In J.W. Evans and A. Hollaender (eds.), Genetic engineering of animals: an agricultural perspective, p. 285-297. Plenum Press, New York.

Rollin, B.E., 1990. The unheeded cry. Animal consciousness, animal pain and science. Oxford University Press, Oxford /New York.

Rollin, B.E., 1995. The Frankenstein Syndrome.Cambridge University Press, Cambridge.

Rutgers, B. and Heeger, R., 1999. Inherent worth and respect for animal integrity. In M. Dol *et al.* (eds.), Recognizing the intrinsic value of animals. Beyond animal welfare, 41-51. Van Gorcum, Assen.

Sandøe, P. and Simonsen, H.B., 1992. Assessing animal welfare. Where does science end and philosophy begin? Animal Welfare 1, p. 257-267.

Sandøe, P., Holtug, N. and Simonsen, H.B., 1996. Ethical limits to domestication. Journal of Agricultural and Environmental Ethics 9/2, p. 114-122.

Stafleu, F.R., Grommers, J. and Vorstenbosch, J., 1996. Animal welfare. Evolution and erosion of a moral concept. Animal Welfare 5, p. 225-234.

Tannenbaum, J., 1991. Ethics and animal welfare: "The inextricable connection". Journal of the American Veterinary Medical Association 198/8, p. 1360-1376.

Taylor, P.W., 1984. Are humans superior to animals and plants? Environmental Ethics 6, p. 149-160.

Taylor, P.W., 1986. Respect for nature. A theory of environmental ethics. Princeton University Press, Princeton.

Verhoog, H., 1992. The concept of intrinsic value and transgenic animals. Journal of Agricultural and Environmental Ethics 5/2, p. 147-160.

Verhoog, H. (1999). Animals in scientific education and a reverence for life. In F.L. Dolins (ed.), Attitudes to animals: views in animal welfare, p. 218-228. Cambridge University Press, Cambridge.

Verhoog, H., 2000. Defining positive welfare and animal integrity. In M. Hovi and R. García Trujillo (eds.), Diversity of livestock systems and definition of animal welfare, 108-118. The University of Reading, United Kingdom.

Verhoog, H. (2001). The intrinsic value of animals: its implementation in governmental regulation in the Netherlands and its implication for plants. In D. Heaf and J. Wirz (eds.), Intrinsic value and integrity of plants in the context of genetic engineering, p. 15-18. Proceedings Ifgene Workshop 9-11 May, Dornach.

Verhoog, H., 2002/2003. Biotechnologie und die Integrität des Lebens. Scheidewege 32, p. 119-141.

Verhoog, H., Matze, M., Lammerts van Bueren, E. and Baars, T., 2003. The role of the concept of the natural (naturalness) in organic farming. Journal of Agricultural and Environmental Ethics 16, p. 29-49.

Vorstenbosch, J., 1993. The concept of integrity. Its significance for the ethical discussion on biotechnology and animals. Livestock Production Science 36, p. 109-112.

Wieder, D.L., 1980. Behavioristic operationalism and the life-world: chimpanzees and chimpanzee researchers in face-to-face interaction. Sociological Inquiry 50/3-4, p. 75-103.

Zweers, W., 2000. Participating with nature. Outline for an ecologization of our world view. International Books, Utrecht.

The stockperson as a social partner to the animal? A stake for animal welfare

X. Boivin and P. Le Neindre
URH-ACS, INRA de Clermont-Ferrand/Theix, 63122 St-Genés Champanelle, France

Abstract

European husbandry is dramatically changing in respect to the human-animal interactions. Stockpersons have more and more animals and less and less time to take care of them. Then, farm animals need to be less dependent of direct contact with humans and become less familiar to the human presence and handling. In this paper, we will describe the consequences on such changes on human work and animal production and welfare. From empirical and scientific reports, we will also describe ways to improve the animal behaviour towards humans, using mainly examples developed on farm ungulates. The quality of the handling facilities, the importance of the psychological characteristics of the stockperson and the usefulness of a genetic selection process based also on animal responses to humans will be discussed. Furthermore, we would emphasize the importance of considering the farmers' interactions with their animals as a succession of events that lead to the human-animal relationship. From this perspective, the perceptions of the stockpersons by their farm animals and the mechanism that build these perceptions need to be fully understood. In the different farm species, the existence of imprinting and socialisation processes based on attachment mechanisms is empirically and scientifically discussed since many years. The idea of rapid, durable and efficient effects on latter human-animal interactions is practically very important for stockpersons. However if these phenomenon have been well described in birds and canidae, the amount of valid scientific data is still low in many farm species. Future areas of research will be suggested.

1. Introduction

In 2000 BC in Egypt, a calf was tied to the front leg of its dam as it sniffed it. Behind the cow, a stockman watched his young boy directly sucking milk from the teat. Four thousand years later, a dairy stockman, seated on a chair, milks his cows by hands. The cow also has her calf tied to her front leg. These two pictures look very similar.

The first one was engraved at the base of a pyramid; the second one was taken, only 20 years ago in a mountain pasture of the French Massif Central. This last picture is a touching scene of traditional family life, old-fashioned and mostly used as a display for tourists. Nevertheless this traditional way of milking a cow is still used in many less developed countries.

Over the last thirty years, the traditional French husbandry system of suckling cows has changed dramatically: During the seventies, a farmer cared for about ten cows. Most of these cows were tethered and the calves led twice a day for suckling under human control. This system is still commonly used and exposes calves to human contact at an early age in association with suckling. Nowadays, more and more calves are kept permanently with their dams in free-stable or in range conditions. Herd sizes have increased to about 70 cows per farm. Stockpeople have less and less time to spend taking care of their animals due to their increased working load and activities. By consequences, husbandry systems are developing where farm animals are less dependent of direct contact with humans and then become less familiar to the human presence and handling. Such conditions become closer to practices observed in countries such USA, Australia or Argentina where thousands of animals are reared together on large pastures with a minimal handling. However it is difficult to compare them to the European farmers practices which are different not only because of their mean herd sizes but also of their historical developments. All practices are rooted in the history of the relationships between farmers and farm animals from the early days. The advice given in Europe to the handlers for training their oxen in the beginning of XVIIIth century could still be in use: "be fair and patient with you animal". The advice is probably still valid but need to be revisited under the lights of our present scientific conceptions of the human-farm animal relationships.

In this paper, we will describe the consequences, of negative reactions of animals to humans and to handling. We will also describe ways to improve quickly, durably and efficiently the behaviour of animals towards humans. In particular, we emphasize the importance of considering the farmers' interactions with their animals as a succession of events that lead to the human-animal relationship. From this perspective, future areas of research will be suggested.

2. Fear of people, ease of handling, animal welfare, production and product quality

In this section, we will describe empirical reports and scientific evidence that demonstrate the importance of the stockperson-farm relationship on farmers' work, animal welfare and productivity for the last 30 years.

From the stockpersons' perspectives, accidents are unfortunately common during handling. They can lead to injuries or death of the person or animals involved. However, as far as we know, statistics on this topic are scarce. In New-Zealand, a survey reported that farmers ranked handling as the most dangerous working activity (Houghton and Wilson, 1994). In our experience, cattle stockpeople often enjoyed reporting anecdotal events, describing when they were pushed, threatened or charged by their animals. It is much easier to joke about an animal behaviour than to recognise the risk to have been injured or killed. Farmers consider dealing with an animals' reaction is a normal part of their working activities. However, working time and comfort for the handler can be quickly jeopardised if animals are insufficiently tamed. Many cattle or horses are sold or culled because of their inappropriate behaviour during handling or because of their reduced production induced by their fear of humans. Few studies have precisely evaluated these losses. However, the success of training programs to improve animal handling testifies the importance of the problem. In France, training sessions are performed with a high success by l'Institut de l'Elevage. More than 80000 handlers were trained since 1980 in such programs (Chupin and Sarignac, 1998). They are sponsored by an agricultural insurance company (la Mutualité Sociale Agricole) in order to reduce the number of accidents during handling.

From the animal perspective, research in laboratories or commercial farms showed that fear of humans generates acute or chronic stress during husbandry, transport and slaughtering. Many studies in poultry, pigs or cattle demonstrated that this stress affects animal production, animal health, meat quality and welfare (for review, Rushen *et al.*, 1999). The most important scientific contribution on the relationship between stockpersons' behaviour and animal welfare and production have been done by Paul Hemsworth and his group in intensive Australian productions. Experimentally, since the early eighties, they investigated the impact of different qualities of human contact on animal behavioural and physiological responses towards humans and on their growing rate and reproduction performances (For review, Hemsworth and Coleman, 1998). Clearly for example in pigs, periods of few minutes of contact with human can strongly affect all the parameters measured with dramatic reductions on animal welfare and production for the animals that were negatively handled (shouting, hits, electric shots). It is likely that pigs were able to generalise their experience with humans during the experimental treatment to any husbandry situations involving the stockpeople. Such generalisation then induced a chronic stress that affected all biological and productions parameters. The negative contact between humans and animals does not need to be systematic. When such contact was randomly provided to pigs at a low rate (20%) among positive interactions (e.g. petting), their effect on animal behaviour and welfare appears as similar as those observed in a consistent negative treatment (Hemsworth *et al.*, 1987). The unpredictability of the negative interactions is probably one of the major reasons of these experimental results. Implications for husbandry practices are obvious. Furthermore, from these experimental studies, they were interested by day to day interactions between stockpersons and farm animals in the context of intensive farms or abattoirs. In many farm species (poultry, pigs or dairy cattle), their research revealed a significant link between the stockpersons' behaviour towards the animals, animals' responses to humans and animal production or reproduction (Hemsworth *et al.*, 2003). A huge variability exists among stockpeople when they interact with their animals. For example, shouting, kicking or hitting are frequent when moving the animals. Such common interactions seem not necessarily perceived by the stockpersons as very negative for the animals. However, for example, Breuer *et al.* (2003) observed in dairy cows that moderate hits during repeated moving could induce higher fear of humans in general compared to handling performed more patiently and gently. In the same way, in very intensive husbandry systems such for veal calf production the human contact could appear unworthy. In the past, the young animals were often reared in individual pens and the human contact and handling reduced. However empirically, technicians working in veal calf production already reported that farmers could be classified in two categories. Some were true farmers who really took care of their animals, talking and petting them. Others were described as" petrol pump attendants" when providing the milk twice a day and trying to work as quickly as possible. These differences in the caretakers' behaviours were confirmed in experimental surveys (Lensink *et al.*, 2000a, 2000b, 2001). The behaviour of the farmers have consequences on animal's fear of human, stress during handling, transport and slaughtering and at the end, on meat quality in their calves.

In the following sections, we will focus our paper more on farm herbivores to illustrate more in details examples of research works that were done in this area of the human-animal relationships. This will also allow to discuss more precisely the ways to improve the human-farm animals relationship.

3. An example of research work: a study of beef cattle docility

A large part of our work on this field of research was done on the ease of handling of beef cattle. Adult cattle are large, heavy and strong and then potential sources of danger to handlers. Cattle are prey animals. They have strong fear reactions and efficient anti-predator strategies such as strongly gregarious behaviour and active defence (attack) against predators. These adaptive responses are important in the wild, but are much less desirable in husbandry, in particular when shown towards handlers. There is a need to understand and minimize these behaviours. As a first step, we developed a tool to measure an animals' reaction to humans and handling. For this purpose, we developed, with a French technical institute (l'Institut de l'Elevage), a test procedure so-called "the docility test" (Boivin *et al.*, 1992b, Le Neindre *et al.*, 1995.). This test was inspired from practical situations where there is a potential for both, fearful animals and their handlers to be injured. The animal is sorted from a group of peers and led to a 5x5m adjacent pen. The human attempts to lead and restrain the tested animal for 30 consecutive seconds in a corner of the pen, at the opposite side of the pen from where its peers are kept. The tested animal has the choice of accepting the human constraint more or less rapidly, not accept it and sometimes react very strongly to the handling by trying to escape the test pen or attacking the handler. Both these behaviours jeopardize the safety of the animal and its handler.

Experiments comparing different husbandry systems of beef cattle were performed in the same herd (Boivin *et al.*, 1994, Le Neindre *et al.*, 1996). During their first three months, some of the animals were reared "traditionally", involving the cows and their calves being led twice a day for suckling under human control. Other animals were reared in free stables or in range condition. In these last two groups, calves were permanently reared with their dams. After the early rearing, animal were reared together on pasture and later half of them were tied during winter time or reared outdoors. Their reaction to humans and handling since five months of age to 20 months of age were measured.

Our experiment highlighted a possible constraint that occurs when shifting from the traditional system to the free-stable or range husbandry conditions: animals may become more fearful, difficult to handle and dangerous to themselves or towards the handler. In our experiment, most of the traditionally reared animals accepted the constraint of the docility test and were never aggressive towards the handler. By contrast, animals from free-stable, and even more, from range conditions showed a high level of fear during the docility test and could sometimes threaten or attack the handler. Husbandry conditions have a strong impact on animals' behaviour towards humans with strong consequences both for animal welfare and farmers' work. Even in husbandry systems where human contact is strongly reduced, contact is still an important component of high standard animal production. In the next section, we will discuss how to improve response of farm animals to humans.

4. Good tools for good workers

Many husbandry or veterinary practices require that an animal be caught or restrained. Such practices are usually perceived as fearful or painful by the animals. Their association with human presence and behaviour can give a strong negative image of the handler for the animals (for review, Rushen *et al.*, 1999). Such operations are not necessarily seen by the stockpeople as very stressful to the animals. For example, feed-lot steers seem to perceive sham-branding to be as aversive as hot-branding or freeze-branding suggesting that handling per se is stressful to the animals. Negative husbandry practices can also influence the animals' further reactions. For example in Australia, farmers mules some of their lambs (surgical removal of the skin from the breech area of the lamb to protect them against blowfly strike). This operation may not appear to be very painful as sheep express few behavioural reactions during the procedure, however, physiological data show that animals react strongly to mulesing (Fell and Shut, 1989). Additionally, more than three months later, sheep continued to avoid the person who held the sheep during the operation, more than another stockperson who was not involved.

When handling facilities are not well designed or in good condition, handling can take longer. Under these circumstances, animals attempt to escape more, requiring greater efforts from the handler. The risk of injury to both animals and humans increases and, as a consequence, negative human behaviours towards the animals (shouting, hitting, kicking...) also increase. It is important to understand that when handling procedures are performed quickly, softly and easily, they have positive consequences on the subsequent human-animal relationship. Our knowledge on the design of handling facilities has been greatly improved by the work of Temple Grandin (Colorado State University) and here, in France, by the work of Jean-Marie Chupin (Institut de l'élevage). For example, in cattle handling facilities, they emphasized the importance of building curvilinear laneways where the crush is not visible at the end of this corridor, and that the lighting and slope of the corridor have an important role on the ease of driving cattle. Many parameters, based on our knowledge of animal's behaviour have to be considered to build good working handling facilities. For more details, we advise you to read the text from Grandin (1970) or Chupin *et al.* (2001).

5. Good workers for good animals: the importance of the psychological characteristics of the stockperson

Since Seabrook (1972), many scientific reports have showed the importance of the stockperson's psychology and working activity in her/his relationship with farm animals. Over the past thirty years, the personality of the stockpersons (for example introvert, extravert, self-confidence...), their job satisfaction, their self esteem, their working load and organisation have been shown to have strong impact on animals' behaviour and welfare but also on their productivity and quality of the product (Seabrook, 1972, 2001, English *et al.*, 1992, Hemsworth and Coleman, 1998, Lensink *et al.*, 2001, Porcher, 2001...). As a spectacular example related to the safety of people, Renger (1975) observed the handling of dairy bulls by different stockpersons. It is often said that stockpeople collecting sperm of bulls in station stay for less than 6 months or more than 20 years (Chupin, personal communication). This difficulty of handling such large and dangerous animals was found to be related to the personality of the handler. Renger (1975) observed that a calm and self-confident handler scarcely faced defensive reactions from the bulls. An unsure, inexperienced or nervous person had much more difficulties. If a stockperson was excitable, unbalanced or had a tendency to be violent, he/she had a greater chance of inducing very dangerous reactions from the bulls. The different human cues that induce strong reactions from the bulls towards the stockpersons are also important (for review, Boivin *et al.*, 2003). Aspects of communication, predictability and controllability of human behaviour from the animal's point of view are certainly involved but have until now not been so widely studied.

For the past 15 years, the importance of the attitude of the stockpersons towards animals and their work have initiated many studies on several species (pigs, poultry, dairy cows and veal calves), either at the farm or abattoirs (Hemsworth, 2003, Lensink *et al.*, 2001, Waiblinger *et al.*, 2002). In particular, negative attitudes of farmers towards animals or toward work with animals are related to their negative behaviour towards the animals If stockpeople interactions with animals are more often negative than positive and if they occur regularly during the production activity or the slaughtering process they can induce acute or chronic stress. This affects animal welfare, production and meat quality. The advantages of working on attitudes are that they are learned and maybe more easily susceptible to change compared to some other psychological human characteristics (Hemsworth and Coleman, 1998).

6. Using the genetic variability for selecting the animals

It is necessary to take into account the genetic variability and possibility of selecting animals that are the most adapted to husbandry constraints, in particular to human handling. Many studies have demonstrated a genetic basis of the animal's response to humans (for review, Boissy *et al.*, 2002). We will illustrate this aspect with our work on beef cattle docility. We have tested thousands of limousine cattle using the docility test in collaboration with "France Limousin Sélection". The first work was published in Le Neindre *et al.* (1995). A heritability between 0.2 and 0.3 was estimated. Despite thousands years of domestication and elimination of the most reactive animals, it is surprising that an important variability still exists on a criteria that is so basic and important for the human-animal relationship. The interaction between the genetic background and husbandry conditions is strongly suspected to explain such variability. In traditional husbandry systems, the human-animal interactions are frequent. It is likely in this conditions that individual experiences have hidden the genetic variability of the response to humans. This variability could appear much more strongly in rearing conditions where human contact is less numerous. Temple Grandin observed that our European "continental" breeds can more abruptly react to handling (e.g. rushing or jumping to the walls) than Anglo-Saxon breeds (personal communication). Differences in the history of the breeds reared spending more or less time in range conditions may explain this variability. As described in the introduction of this paper, the present tendency of husbandry systems is to increase the number of animals per farm, to decrease the number of stockpeople and to leave the calves together with the dam. This should probably contribute to the expression of the genetic variability if an active genetic process is not performed on this criterion. The genetic selection to improve docility or temperament of beef cattle is nowadays a worldwide practice. Numerous tests have been developed in different laboratories around the world. In addition to the docility test we developed, the crush test, measuring the agitation of the animal in the crush, has been largely used in research (for review, Grandin *et al.*, 1997, Burrow *et al.*, 1997). An individual test, measuring the flight distance has been developed in New-Zealand and Australia (Fisher *et al.*, 2001). A very simple test measuring the speed to exit a crush has also been used in Australia (Burrow and Corbet, 2000). Researchers used also a qualitative scale measuring the reaction of an animal to an approaching human (Kuehn *et al.*, 1998). All these situations of measurement have revealed genetic variability but their scientific validity need to be improved, in particular by comparisons on the same animals.

A last perspective to improve the stock persons work with the farm animals is a better understanding of the stockpeople-farm animals relationship from the animal's point of view and what are the factors that affect it.

7. The importance of considering the human-animal interactions as a part of an inter-individual relationship

Farmers, economists or psychologists all have their own conceptions of the human-farm animals' relationships. As ethologists, we try essentially to describe this relationship from the animals perspective.

As a first step, it is important to remember the role of the stockperson and how an animal perceives their activities. The first objective of stockpeople is to produce and earn their living from their work. To achieve this, they feed the animals, directly or through different mechanised systems where humans are not associated with the food reward. Stockpeople also move or handle the animals, directly or with the aid of handling facilities. Furthermore, stockpeople manage the animals' environment and perform sanitary or management practices such as castration, dehorning...). Finally and probably the more important activity is watching the animals. Stockpeople have a huge liberty to organise their time and their interactions with the animals. They can talk to them and traditionally give them names. They can watch them at a distance, without entering the herd or the group or pass through them. They can also individually approach them, reward them with salt and concentrate, touch them and even pet them.

Most of these interactions occur repeatedly, throughout the animals' life. Such interactions between individuals that know each other build what the ethologists call an inter-individual relationship, often described for intra-specific social relationships (Hinde, 1976, Estep and Hetts, 1992). Such relationships are studied by describing the representation of each partner by the others. Behind this concept of relationship, we need in particular to consider the emotional state induced by the interactions between the partners (Aureli, 2001) and the mechanisms that build the animal's perception of the stockpeople. This is also particularly important for our understanding of the relationship between the stockpeople-relationship and animal welfare (Veissier *et al.*, 1999).

It is important to consider the environment in which these interactions between stockpeople and animals occur. Most of the domestic species are gregarious and social. From birth and under natural conditions, the young calf, for example builds a social network, first with its dam and later with the other member of the herd. In husbandry conditions, humans regularly affect the social environment of the animal, often from birth. They also often interact later during special periods of the animals' life (weaning, parturition...). All these factors have to be taken into account when researchers try to understand the stockpeople-farm animal relationship and the animal's perception of humans.

Estep and Hetts (1992) after Hediger (1965) categorised the different relationships between humans and animals. The stockperson can be perceived by the farm animal as a predator, a neutral stimulus, as a source of food and water and in some cases as a social partner or even more a conspecific. We will consider this last category on the following parts of this paper.

8. The stockperson as a social partner for the farm animals?

Empirically, humans frequently talk about their animals as if they were perceived them as social partners, even as conspecifics. For dog trainers, the owner should behave as the leader/dominant of the pack. For horse trainers, empirically developing "ethological" methods, the horse should be "imprinted" to the humans or as the dogs, the human should behave as the dominant animals of the herd. In cattle husbandry, the stockperson should be the leader of the herd. He should be perceived at the same time as "dominant" and "friend" of the animals. In France, these principles are for example tought by l'Institut de l'Elevage during training sessions for cattle handlers. These principles are in accordance with the way African Fulani nomads build their relationship with their cattle (Lott and Hart, 1979). They lived permanently with their animals, spending a lot of time stroking the animals at a young age and reacting strongly (shouting and charging) to any threatening signals from the part of the animals. For Lott and Hart (1979), such human behaviours look very similar to beef cattle intra-specific social behaviours. These authors suggested we inspire our occidental husbandry techniques from these traditional methods. However what scientific knowledge on animals' perception of the human supports these empirical considerations?

9. Some scientific concepts and facts

The concept of imprinting developed by Lorenz (1935) is well known. Imprinting can today be considered as an attachment phenomenon during sensitive periods in birds or mammals that is expressed later on in the development of filial or sexual preferences (Bretherson, 1992). The concepts of attachment and sensitive period are essential here. Lorenz considered that this period was not sensitive but critical and the imprinting process was viewed as irreversible. However for the scientists working in this area, imprinting has no more its critical and rigid aspect and subsequent "imprinting" can also occur. It is also relevant in this chapter to mention the important work from Scott and al. (for review, 1970, 1992) working with dogs on the concept of socialisation to conspecifics or to humans. Their scientific work is still valid, in particular for dogs' training. Socialisation can be defined as the building of a social relationship (including the human partner), mainly based on a process of attachment and adjustments to the other members of the social group, in particular during sensitive periods of the animals' life. Once again, attachment processes and the existence of sensitive periods in particular to human contact are mentioned. So, the concepts of imprinting and socialisation are closely related. Nevertheless, one important difference is probably in the adjustment that should occur between individuals during the socialisation process. For example in the dog, the dam inhibited the biting behaviour of her cubs which will be less likely to bite the human owner in later life.

Imprinting has been studied in detail in birds but much less in mammals (Janzen *et al.*, 1999). Kendricks *et al.* (1998) studied the filial and sexual imprinting in goats and sheep. It seems to exist in males of the two species but much less in females. Sambraus and Sambraus (1975) described sexual imprinting towards humans in boars, rams or bucks reared in complete isolation from their own species. Such imprinting is not useful, except maybe in a few cases such as artificial insemination where it could be interesting that the human was really considered as a conspecific. To increase the probability of success, inseminators sometimes used the clothes of the familiar stockperson, possibly using familiar olfactory cues. Imprinting to olfactory cues could be proposed to explain this strategy. However, more simply, the familiar odours could reduce the novelty of the inseminator, decreasing animals' fear of unfamiliar humans and increasing reproductive performance.

The process of socialisation to humans has been mainly studied in canidae. However, the number of studies on other species, especially farm species, appears extremely limited. Mostly, they investigated the possible existence of sensitive periods to human contact, especially in early age, in farm ungulates such as sheep, goats, cattle or horses. We will describe some scientific evidence of such periods of contact is the next chapter. By contrast to empirical assertions, the scientific evidence of such periods in early age in horses are lacking (Lansade *et al.*, 2004, Hausberger *et al.*, 2004). Could the stockpersons really be perceived as social partners (leader, dominant, and friends) by the farm animals? This question is still unanswered, even with dogs, as few conclusive studies today exist (Rooney *et al.*, 2001, 2002).

10. Existence of sensitive periods to human contact in the farm ungulates' life

We mentioned in the previous chapter, the importance of sensitive periods of human contact in the animal's life. We will illustrate this concept with two experiments taken among others suggesting their existence in the farm ruminants life.

We already mentioned in the introduction how the husbandry system of the calves during the first three months can influence their reaction to humans and handling up to 20 months of age (Boivin et al., 1994, Le Neindre et al., 1996). Outdoor animals and traditionally reared calves had for example a similar heart rate when alone in a crush cage. However and despite seven-teen months spent together after their 3 months of age, calves kept outdoor showed a significantly higher heart rate when the handler was standing in front of the animals. These results are similar with those obtained by Lyons et al. (1988, 1989). They observed that dam-reared goats were much more fearful towards humans, even as adults than artificially-reared ones that were in addition petted during their first ten days by their caretaker.

In another experiment performed in Danmark, dairy calves were fed from a bucket and petted at different periods during early age (Krohn et al., 2001). Control animals did not receive any additional human contact. We measured the animals' reaction to humans when they were 40 days old. When the human contacts were provided from day one to four, calves approached the humans much more quickly than the control ones. By contrast, later human contact (between day 6 to 9 or 11 to 14) did not appear as efficient as early contact.

All together, these results suggest that the days following birth could effectively represent a more sensitive period to human contact in the animals' life. The social environment and the type and duration of contact are probably important factors that could influence the latter human-animal interactions.

11. The importance of the social network in the development of the human-animal relationship

The social environment of farm animals is often managed by the stockpeople, especially for young animals. The current and strong tendency is to leave the young animals with their dam. This reduces the amount of work involved in the care of young animals and allows the development of the young dam relationship to improve animal welfare. This is true in meat production systems but also for dairy calves that could benefit from the human proximity (Krohn *et al.*, 1999). However what influence does the presence of the dam play on the development of the human-animal relationship?

In an experiment performed on artificially reared lambs, we compared animals that received human contact (petting and feeding) during the first six days of their life with unhandled control animals (Boivin *et al.*, 2002). Half of the animals were reared in the presence of their dam and one twin without having the possibility to suck her (separation with a open-barred fence). The other half was reared in the presence of their twin only. During the human contacts, animals showed an increased motivation to interact with their familiar stockperson irrespective their maternal environment. However, 6 weeks later, only the animals reared in the absence of their dam showed a stronger affinity towards their stockperson compared to control ones. Such results seem to indicate that the presence of the dam could be a limiting factor of the development of a positive relationship between animals and their caretakers. Similar results were also observed with dairy calves (Krohn *et al.*, 2003).

11.1 Research perspectives

Results presented above suggest the existence of a sensitive period for human contact in the farm animals' life. Adequate human contact during these periods could decrease animals' fear of human, efficiently, durably and with a minimum of time, improving both animal welfare and human work. However experimental or field studies are for the moment limited in their quantity but also in their real application on farms. In particular, long term effects are rarely tested or demonstrated (for review, Burrow, 1997). Often scientists test the animals' motivation to interact with humans (approach/avoidance test) but more rarely in practical situations for the farmers. Moreover, negative effects of human contact such as sexual imprinting should also be investigated.

The previous studies also demonstrate the strong impact of the social environment in early age on the development or durability of the human-animal relationship. The results allow us to question the origins of systems that separated the young from their dam and, as for example in suckling cattle, move the calf for suckling under human control. Housing conditions involving animal tethering and farmers' needs to watch calves' suckling are commonly proposed to explain such practices. Nevertheless, it is important to question the role of the dam-young separation in the latter quiet reactions of the calves to stockpeople and handling. Such separation could also help the stockperson to integrate into the social world of the young animal. The current husbandry systems leave the calves for longer periods with their dams with possible strong consequences on later human animal relationships. Works by Hausberger *et al.* (2004) suggest that it could be useful to increase the human contact with the dam in the presence of her young. Dam's reactions to humans could act as a model for later reaction of the young animal to their stockperson. It may also be important to consider other particular periods in the animals' life without the influence of the dam. We can mention the period following parturition or the period of weaning (artificial and definitive separation between the dams and their young) which have been suggested as sensitive to stockpeople contact (Hemsworth *et al.*, 1987, 1989, Boivin *et al.*, 1992a). These works should be confirmed and further investigated. Finally, it is important to notice the consistency in the type of human contact used in experiments that positively affect the human-animal relationship. Imposed human presence, stroking and association with feeding were almost always involved. What are their respective importances? By which mechanism are they affecting the human-animal relationships (simple habituation or learned helplessness, simple reward conditioning or attachment processes...)?

11.2 Attachment as a possible mechanism of development of the stockperson-farm animals relationship?

One of the commonly used tools to evaluate an animals' reaction to humans is without doubt the standard human approach test developed by Hemsworth and coll. in pigs studies (Hemsworth and Coleman, 1998). The animal is put alone in an arena test for a short "familiarisation period" in this unknown environment. Then, the experimenter enters the arena and stands motionless. The approach/avoidance responses towards the human are recorded. Animals that were negatively handled during a prior experimental treatment (hits, electric shocks, shouting...) approached and interacted with the experimenter during the test less than controls or positively treated animals (petting). The increase in cortisol was also higher during this test for the negatively treated animals compared to the animals from the other treatment, suggesting that animals were more fearful of the human. However what was really measured for the animals that were approaching the experimenter? Is it simply an indication of the absence of fear or could such behaviour reflect other motivation such as reward or reassurance provided by the human presence in a stressful situation? Indeed, the initial "familiarisation period" to the arena test could also be perceived as very stressful for highly social animals. In the same way, what is it happening when the experimenter leaves the test arena? Could such a situation induce a separation distress as was the case in experiments testing animal imprinting or attachment (Kraemer, 1992)? We recently tested these questions with artificially reared lambs with which we controlled human contact and social environments (for review, Boivin *et al.*, 2003). We compared animals that received human contact in association with feeding in early age (physical contact, hand and bucket feeding three times a day for five days) to control animals that did not received such contact. These animals approached the experimenter more often and durably in the test arena by contrast to control animals that were not motivated to contact the human. Moreover when the experimenter left the pen, they vocalised significantly more than control ones. Such results suggest a possible attachment for the familiar stockperson or that humans can be perceived as a social substitute that may reduce isolation stress.

12. Conclusion

The applied objective of our current research is to obtain quickly, efficiently and durably, animal's responses that facilitate work by stockpeople and improve the welfare of farm animals and their productivity. The empirical idea that the stockperson could integrate into the social world of the farm animals, as a dominant, friend, and leader has not been demonstrated. However, the concept of inter-individual relationship between human and animals gives an interesting and large theoretical framework to explain animals' behaviour towards stockpeople. This theoretical framework highlights the need to take into account all the interactions that occur between the two partners of the relationship from their first encounter. It insists on the necessity to further investigate the perceptions of each partner during their development, and then the animal's perception of ourselves. It also highlights the need to consider further emotional and cognitive abilities of farm animals. The concept of socialisation integrates all the mechanisms that allow the animals to retain information and that we should understand. Such mechanisms can be "simple" such as habituation or Pavlovian conditioning and have been extensively studied in the past. However the definition of socialisation includes more "complex" learning processes associated with a sensitive period and attachment phenomenon which are much less known in our farm species.

Changing the human-animal relationship requires also a change in the behaviour of the stockpeople. We mentioned the work demonstrating the importance of daily interactions with the animals and their key human factors. Research in particular on stockpeople's attitudes towards the animals and their work with the animals have demonstrated their impact on animal welfare and production. They create a good basis to build training program for farmers/handlers (Hemsworth and Coleman, 1998, Hemsworth *et al.*, 2002). For the moment, they are limited to few species and types of production and need to be generalised. In the general conceptual framework presented in this paper, we feel that it is important to consider daily interactions in the general context of the husbandry practices (practices for food distribution and animal watching, practices for tethering, dehorning, use of dogs...). It is also important to consider the way farmers' work with the animals from an early age. It seems that a large variability exists in husbandry practices, used by farmers to improve animal docility. A systematic description of this diversity and its effect on the human-animal relationship is required, in enlightening the existing scientific knowledge on farm animals' behaviour. Farmer training should assist them to think further about their daily interactions with their animals, and the efficiency and organisation of their practices that could transform the human-animal relationship. This research may increase the comfort and job satisfaction of the stockpeople and consequently improve farm animal welfare.

References

Aureli, F. and Schaffner, C.M., 2002. Relationship assessment through emotional mediation. Behaviour, 139, p. 393-420.

Boissy, A., Fisher, A., Bouix, J., Boivin, X. and Le Neindre, P., 2002. Genetics of fear and fearfulness in domestic herbivores. 7th World Congress on Genetics Applied to Livestock Production. 19-23 August, Montpellier, France, p. 32

Boivin, X., Boissy, A., Nowak, R., Henry, C., Tournadre, H. and Le Neindre, P., 2002. Maternal presence limits the effects of early bottle feeding and petting on lamb's socialisation to the stockperson. Appl. Anim. Behav. Sci., 77, p. 311-328.

Boivin, X., Le Neindre, P. and Chupin, J.M., 1992b. Establishment of cattle-human relationships. Appl. Anim. Behav. Sci., 32, p. 325-335.

Boivin, X., Le Neindre, P., Chupin, J.M., Garel, J.P. and Trillat, G., 1992a. Influence of breed and early management on ease of handling and open-field behaviour of cattle. Appl. Anim. Behav. Sci., 32, p. 313-323.

Boivin, X., Le Neindre, P., Garel, J.P. and Chupin, J.M., 1994. Influence of breed and rearing management on cattle reactions during human handling. Appl. Anim. Behav. Sci., 39, p. 115-122.

Boivin, X., Tournadre, H. and Le Neindre, P., 2000. Hand-feeding and gentling influence early-weaned lambs'attachment responses to their stockperson. J. Anim. Sci. 78, p. 879-884.

Boivin, X., Le Neindre, P., Boissy, A., Lensink, J., Trillat, G. and Veissier, G., 2003. Eleveur et grands herbivores: une relation à entretenir. INRA Prod. Anim., 16, p. 101-115.

Boivin, X. Lensink, J. Tallet, C. and Veissier, I. Stockmanship and farm animal welfare. Animal Welfare, 12, p. 479-492.

Bretherton, I., 1992. The Origins of Attachment Theory: John Bowlby and Mary Ainsworth. Developmental Psychology, 28, 5, p. 759-775.

Breuer, K., Hemsworth, P.H. and Coleman, G.J., 2003. The effect of positive or negative handling on the behavioural and physiological responses of nonlactating heifers. Appl. Anim. Behav. Sci., 84, p. 3-22.

Burrow, H.M., 1997. Measurements of temperament and their relationships with performance traits of beef cattle. Animal Breeding Abstracts 65, p. 477-495.

Burrow, H.M., Seifert, G.W. and Corbet, N.J., 1988. A new technique for measuring temperament in cattle. In Proc. Aust. Soc. Anim. Prod., p. 154-157.

Chupin, J.M. and Sarignac, C., 1998. How to train cattle breeders to handling of bovine? In Veissier I., Boissy A. (Eds.), Proceedings of the 32th International Congress of the international Society for Applied Ethology, Clermont-Ferrand, France, p. 117.

Chupin J.M., Houdoy D., Carotte G. and Perrin M., 2001. Elaboration d'un recueil de prescriptions techniques pour la conception et l'équipement des bouveries d'abattoirs. Renc. Rech. Ruminants, 8, p. 129-132.

English, P., Burgess, G., Segundo, R. and Dunne, J., 1992. Stockmanship: Improving the care of the pig and other livestock. United Kingdom. Farming Press. 190 p.

Estep D.Q. and Hetts S., 1992. Interactions, relationships and bonds: the conceptual basis for scientist-animal relations. In Davis H. and Balfour D. (Eds), The Inevitable Bond: Examining Scientist-Animal Interactions. Cambridge University Press, Cambridge, United Kingdom, p. 6-26.

Fell, L.R. and Shutt, D.A., 1989. Behavioural and hormonal responses to acute surgical stress in sheep. Appl. Anim. Behav. Sci. 22, p. 283-294.

Fisher, A.D., Morris, C.A. and Matthews, L.R., 2000. Cattle behaviour: comparison of measures of temperament in beef cattle. Proceedings of the New Zealand Society of animal production 60, p. 214-217.

Grandin, T., 1997. Assesment of stress during handling and transport. Journal of Animal Science 75, 249-257.

Grandin, T., 1997. The design and construction of facilities for handling cattle. Livestock Production Science, 49, 2, p.103-119.

Hausberger, M., Henry, S. and Richard, M.A., 2004. Expériences précoces et développement du comportement chez le poulain. 30ème journée de la recherche équine., 3 mars 2004, Paris, p. 155-164

Hediger H., 1965. Wild animals in captivity. London: Zool. Soc., 56, p. 154-183.

Hemsworth, P. H., Barnett, J. L., Tilbrook, A. J. and Hansen, C., 1989. The effects of handling by humans at calving and during milking on the behaviour and milk cortisol concentrations of primiparous dairy cows. Appl. Anim. Behav. Sci., 22, p. 313-326.

Hemsworth, P.H., 2003 Human-animal interactions in livestock production. Appl. Anim behave. Sci., 81, p. 185-198.

Hemsworth, P.H., Barnett, J.L. and Hansen, C., 1987a. The influence of inconsistent handling by humans on the behaviour, growth and corticosteroids of young pigs. Appl. Anim. Behav. Sci., 17, p. 245-252.

Hemsworth, P.H., Coleman, G.J., 1998. Human-Livestock interactions: the stockperson and the productivity and welfare of intensively farmed animals. CAB International, New York, NY, USA, 158 p.

Hemsworth, P.H., Coleman, G.J., Barnett, J.L., Borg, S. and Dowling, S., 2002. The effects of cognitive behavioural intervention on the attitude and behaviour of stockpersons and the behaviour and productivity of commercial dairy cows, J. Anim. Sci., 80, p. 68-78.

Hemsworth, P.H., Hansen, C., Barnett, J.L., 1987b. The effects of human presence at the time of calving of primiparous cows on their subsequent behavioural response to milking. Appl. Anim. Behav. Sci., 18, p. 247-255.

Hinde R.A., 1976. Interactions, relationships and social structure. Man, 11, p. 11-17.

Houghton, R.M. and Wilson A.G., 1994. The prevention of injury among farmers, farm workers, and their families: a programme for development of interventions for rural communities: farm survey findings (N)3 in a series of four reports). University of Otago consulting group, Dunedin, p. 50-55.

Janzen, M. I. D., Timmermans, P. J. A., Kruijt, J. P. and Vossen, J. M. H., 1999. Do young guinea pigs (Cavia porcellus) develop an attachment to inanimate objects? Behav. Proc., 47, 1, p. 45-52.

Kendrick, K.M., Hinton, M.R., Atkins, K., Haupt, M., Skinner, J.D. (1998). Mothers determine sexual preferences, Nature 395, 229-230.

Kraemer G.W., 1992. A Psychobiological theory of attachment. Behav. Brain Sci. 15, 493-541.

Krohn, C.C., Boivin, X. and Jago, J.G., 2003. The effect of presence of the dams during handling of young calves. Appl. Anim. Behav. Sci., 80, 263-275.

Krohn, C.C., Foldager, J. and Mogensen, L., 1999. Long-term effect of colostrum feeding methods on behaviour in female dairy calves. Acta Agr. Scand. Sect.. A, Anim. Sci., 49, 57-64.

Krohn, C.C., Jago, J.G. and Boivin, X., 2001. The effect of early handling on the socialisation of young calves to humans. Appl. Anim. Behav. Sci., 74, 121-133.

Kuehn, L.A., Golden, B.L., Comstock, C. R., Hyde, L. R. and Andersen, K. J., 1998. Docility EPD for Limousin cattle. In: Beef Program Rep. Colorado State Univ., Fort Collins. P. 73-77.

Lansade, L. Bertrand, M, Boivin, X. and Bouissou, M.F., 2004. Effects of handling at weaning on manageability and reactivity of foals, Applied Animal Behaviour Science, 87, 1-2, p. 131-149.

Le Neindre, P., Trillat, G., Sapa, J., Ménissier, F., Bonnet, J.N. and Chupin, J.M., 1995. Individual differences in docility in Limousine cattle. J. Anim. Sci., 73, p. 2249-2253.

Le Neindre, P., Boivin, X. and Boissy, A., 1996. Handling of extensively kept animals. Appl. Anim. Behav. Sci, 49, p. 73-81.

Lensink, B.J., Boivin, X., Pradel, P., Le Neindre, P. and Veissier, I., 2000a. Reducing veal calves' reactivity to people by providing additional human contact. J. Anim. Sci., 78, p. 1213-1218.

Lensink, B.J., Fernandez, X., Boivin, X., Pradel, P., Le Neindre, P. andVeissier, I., 2000. The impact of gentle contacts on ease of handling, welfare and growth of calves and on quality of veal meat. Journal of Animal Science 78, p. 1219-1226.

Lensink, B.J., Veissier, I. and Florand, L., 2001. The farmer's influence on calves' behaviour, health and production of a veal unit. Anim. Sci., 72, p. 105-116.

Lorenz, K., 1935. Der Kumpan in der Umwelt des Vogels. Zeitschrift Ornit., 83, p. 137-213, 289-413.

Lott, D.F. and Hart, B.L., 1979. Applied ethology in a nomadic cattle culture. Appl. Anim. Ethol., 5, p. 309-319.

Lyons, D.M., 1989. Individual differences in temperament of dairy goats and the inhibition of milk ejection. Appl Anim Behav Sci 22, p. 269-282.

Lyons, D.M., Price, E.O. and Moberg, G.P., 1988. Individual differences in temperament of domestic dairy goats: constancy and change. Animal Behaviour, 3, p. 1323-1333.

Porcher, J., 2001. Le travail dans l'élevage industriel des porcs. Souffrances des animaux, souffrance des hommes. In Burgat F. and Dantzer R. (Eds). Les animaux ont-ils droit au bien-être. INRA Editions, Versailles, France, p. 25-59.

Renger, H., 1975. Agressive Verhalten von Bullen dem Menschen gegenüber. Diss. Med.Vet., München, Germany.

Rooney, N.J., Bradshaw, J.W.S. and Robinson, I.H., 2000. A comparison of dog-dog and dog-human play behaviour. Applied Animal Behaviour Science 66, p. 235-248.

Rooney, N.J. and Bradshaw, J.W.S., 2002. An experimental study of the effects of play upon the dog-human relationship. Applied Animal Behaviour Science 75, p. 161-176.

Rushen, J., Taylor, A.A. and de Passillé, A.M., 1999. Domestic animals' fear of humans and its effect on their welfare. Applied Animal Behaviour Science, 65, p. 285-303.

Sambraus, H.H. and Sambraus D., 1975. Prägung von Nutztieren auf Menschen. Zeitschrift Fur Tierpsychology 38, p. 1-17.

Schwartzkopf-Genswein, K.S., Stookey, J.M. and Welford, R., 1997. Behavior of cattle during hot-iron and freeze branding and the effects on subsequent handling ease. J. Anim. Sci. 75, p. 2064-2072.

Scott, J.P., 1970. Critical periods for the development of social behaviour in dogs. In Kazada, S., Dennenberg, V.H. (Eds.), The Postnatal Development of Phenotype. Academia, Louvain la Neuve, Belgium, p. 21-31.

Scott, J.P., 1992. The phenomenon of attachment in human-non human relationships. In Davis, H. and Balfour, D. (Eds.), The inevitable bond: examining scientist-animal interactions, Cambridge University Press, Cambridge, United Kingdom, p.72-92.

Seabrook, M.F., 1972. A study to determine the influence if the herdsman personality on milk yield. J. agric. Labour Sci., 1, p. 45-59.

Seabrook, M.F., 2001. The effect of the operational environment and operating protocols on the attitudes and behaviour of employed stockpersons. In Human-animal relationship: stockmanship and housing in organic livestock systems. Hovi, M., Bouilhol, M. (Eds). Proceedings of the 3rd NAHWOA Workshop, University of Reading, Clermont-Ferrand, France, October 21-24, 2000, p. 7-15.

Veissier I., Sarignac C. and Capdeville J., 1999. Les méthodes d'appréciation du bien-être des animaux d'élevage. INRA Prod. Anim., 12, 2, p. 113-121.

Waiblinger S., Menke, C. and Coleman, G., 2002. The relationship between attitudes, personal characteristics and behaviour of stockpeople and subsequent behaviour and production of dairy cows. Appl. Anim. Behav. Sci. 79/3, p. 195-219.

Associations are changing public opinion and breeding methods to improve animal well being

Anne Vonesch
Chambre de Consommation d'Alsace, Alsace Nature, SociétéProtectrice des Animaux de Strasbourg, France

Abstract

This chapter is based on inside experience of associations, mainly in Alsace, but also in neighbouring countries. It relates what associations are doing to improve farm animal welfare. The distress of farm animals is a challenge for communication.

Consumer Associations in Alsace have built up partnership with breeders. The result was a successful alternative line of pork production, based on food security, animal well-being, and an affordable price. Animal welfare became an important and durable issue because Alsatian consumers have taken their time to clarify the link between food safety, animal well-being, and sustainability. Ethics appears as a principle of precaution. Environmental protection associations are more and more interested and involved in farm animal welfare issues. Animal protection associations work on a European level. They have evolved from pity to competence and efficiency, concerning advancement and implementation of law, farming methods, and welfare labels. They need guidance from animal friendly experts.

Teaching humane ethics means to educate a new generation of farmers who communicate with their animals and who know how to manage farming systems where animals can "be themselves".

1. Introduction

This chapter does not aim to be a sociological study. It relates years of experience shared with others, within associations that gave me responsibilities[58]. It deals with facts, mainly from Alsace, and also beyond France in neighbouring countries. The issue is that of activity destined to improve farm animal welfare. The present paper considers what associations achieve, in an evolutionary way, as a result of their communication with people and with animals[59].

Before relating these facts, I want to begin with some very general ideas about these associations as far as their relation to animals is concerned, followed by some considerations on the dilemma of communicating on animals' distress.

Consumers are at first interested in what they have on their plate, and marketing sells them a reassuring 'image' of farm animals. But they may also meet with living animals, and beyond food safety they are motivated by ethics.

Environmental protectors communicate with nature, and they have strong feelings for wild animals that should possibly be left undisturbed by human activity. They accept death as part of the natural food chain and population control. But they may create hospitals to save injured wild birds.

Animal protection organisations, drawing concrete conclusions from their relationship with animals, are having to face the 'rules of the game' forced upon them by law, science, media, market forces, and politics. Under strong emotional pressure to achieve results, the militants work hard, getting more and more creative while it appears that the dice are loaded. Competence has evolved from compassion. Four aspects will be more particularly emphasized: the associations' impact on European legislation and its implementation, examples of further advancement, and experimentation of economic viability of welfare.

---

[58] As a member of the Food Committee and the working group 'Collective Restauration' at the Chambre de Consommation d'Alsace, and specialized in breeding conditions, I also represent consumers in the standing group animal welfare at the European Commission. I have been given responsibility for piloting of the agriculture network of Alsace Nature, and I am member of the directing committee of the agriculture network of France Nature Environnement. I am supported by the Society for the Protection of Animals in Strasbourg in the campaign "Breeding with Dignity".

[59] As J. Porcher says: "...It is affectivity, empathy, sympathy, friendship that allow us access to the world of the other, human or not...." (Eleveurs et animaux, réinventer le lien. p.202) This is far away from the one-way communication of marketing .

Farm animal experts have a crucial influence on technical practice, mentalities and innovation. What do associations expect from experts ?

2. Farm animals' distress: a challenge for communication

Militants fighting for the animal cause experience intense communication with animals. Often they are not taken seriously. They are charged with being 'emotional'. I am convinced that the heart of the matter is not excess of emotion. The problem may be lack of knowledge and experience, lack of diplomacy, lack of all sorts of capacities and means, but there is no incompatibility between high scientific competence on the one hand, and emotions such as love, empathy and compassion on the other. There is no valuable reason to discredit such feelings. Communicating emotions with animals supports the moral contract which keeps us going. We cannot turn back, having looked into the eyes of the young sow who trembles, imprisoned in a cage. Her place in creation would be to explore forest and clearing in the dew of morning. "I cannot do anything for you, but tomorrow I will do everything I can to get your sisters out of there". A cow at the market stands in front of her calf to protect it. We know about fear and pain, separation and death. We must follow our path, despair in our heart, as for a suffering child when we must at all costs find a doctor who will comfort and heal. However, if we want to be efficient and not rejected by those to whom we address our message, we must keep smiling.

But how can we reach the public on the subject of the daily distress of farm animals in our proud, wealthy villages with their clean friendly farmers, our neighbours, and sometimes our politicians? Passers-by will be likely to stop, if information stands of associations display striking photos such as cows being loaded on to a ship and suspended by one leg. Extreme television documentaries shake public opinion - once, twice, and then people switch off. Why put up with something when we have no influence in the matter? An explosion of emotion can prick our consciences, and create a movement of opinion. The sale of meat can be reduced for a few days. But all that does not yet allow for construction after the shock.

Is it possible over a long period to be involved in animal suffering without risking one's psychological balance? Maybe not - unless there is some hope, and some consoling result. Facing extreme suffering, there has to be an opportunity to save animals. For instance the association 'Animals Angels' (www.animals-angels.de) has the guiding principle "we are with the animals". Its teams are too often helpless witnesses. Cruel scenes can reappear in their night-mares. But from time to time they will buy an animal for rescue and let it experience happiness, confidence in life, and friendship with people. There must be somewhere in the world nice farmers who always are friendly towards their animals and who feel happy when their animals have fun. There must be agricultural college teachers who abandon the teaching of factory farming, and who innovate by educating professionals to breed healthy animals that are allowed to "be themselves".

3. Consumer Associations in Alsace: from food safety to ethics

We are now observing the birth of a solid consensus and constructive action for the well- being of farm animals. First we look at the different actions that took place, and than we try to analyse what made them possible.

It happened in Alsace, thanks to the Chambre de Consommation d'Alsace (CCA, Chamber of Consumers, www.cca.asso.fr). The CCA is a regrouping of associations that deal with consumption at regional level, for common services and actions. The different preoccupations and priorities of the member associations are expressed through a lively associative democracy. The CCA has 18 salaried workers. The associations handle 53 000 requests each year, 40 % being for the CCA, and they publish a magazine "Le Consommateur d'Alsace" printed in 12 500 copies. The CCA's Food Committee, built up and directed by Rési Bruyère, was particularly dynamic.

The interest for farm animal welfare first came up, when the CCA joined a petition started by local animal protection associations (SPA). It called for husbandry systems to preserve animal welfare, environment, and health. That was in 1996, at the time of the first BSE crisis. The CCA added to the text of the petition that consumers should get information on breeding methods. 36 521 signatures, 3/4 from Alsace, proved it to be representative for public opinion.

After a press conference, the CCA was contacted by a young pig breeder, Thierry Schweitzer, who said: "I would like to breed pigs according to the demand of modern society - what exactly do you want me to do?" The first meetings with him led to successful cooperation. A triple objective was defined: food security, animal wellbeing, and an affordable price, which means cheaper than organic pork. Guidelines were defined together, with support from the local SPAs, to meet this objective: healthy feeding, without animal meal, without systematic antibiotics, without GMOs; bedding and freedom of movement for all animals at all ages with a ban on cages for sows, for gestation as well as for farrowing, and surfaces per animal about twice the minimum required by the law; and the abandonment of tail-docking and toothclipping.

As far as castration is concerned, the use of anaesthetics is introduced. The cost overrun for the consumer, in large-scale distribution, should not be more than about 0,76-1,52 Euros per kilo, which is largely respected. The producer should obtain a reasonable price, that is to say 0,23 Euros per kilo of carcass more, as compared to the standard. The result was an alternative line of pork production, the trade mark called 'Thierry Schweitzer' (www.thierry-schweitzer.com). The partnership between breeders and associations serves as a sales argument, not as a support for one person or trade mark, but as a support to a satisfactory list of specifications which they put together themselves.

The success of this action has resulted in a follow-up project called "Sustainable breeding with solidarity". In this framework, the Institutional Catering Working Group of the CCA has put together an exhibition on the theme of "Eating Responsibly", a pedagogical support for pilot menus in institutional catering. The triple objective is therefore food security, animal well-being, and solidarity. The desired content of a farm assurance scheme is explained in the "Eating Responsibly" Exhibition. Traceability is an indispensable, but insufficient, basis. The guarantees must deal with animal feed in order to ensure both the health of the animals and the absence of contamination or risky residues. The guarantees must also deal with the manner of breeding, that is to say the well-being and the health of the animals. They should ensure respect of the environment, and lastly allow for a reasonable remuneration for the breeder, and an affordable price for the consumer. An independent control system is essential. There is a very large consensus today on the fact that in the food area it is no longer the cheapest price which has the highest demand, but a certain quality, for everyone - and not food on two levels, one healthy and secure for the rich, the other cheap and risky for low budget households. This exhibition has been shown in various collective restaurants, and mainly in schools and school restaurants. Dependant on the motivation of the teaching staff, classes had a commented introduction to these issues. Often people expected information on healthy diet or on taste and gourmet culture. The issue of consumer responsibility and ethics seemed totally new, and the young people as well as their teachers discover striking facts about how animal are treated, though the exhibition shows only standard production without any sensationalism. Whenever possible, the exhibition was co-ordinated with pilot meals prepared with food from sustainable agriculture. Organic meat was too expensive for school restaurants. The CCA proposed a viable compromise between credible sustainability criteria and affordable price.

Consumers have said what they do not want. They wrote to the Chief Administrative Officer of the Region[60] and they said in a press conference[61] that they no longer want industrial pig-breeding. They voted against a major farm assurance scheme because the list of specifications was not satisfactory concerning feeding and welfare. Official signs of quality are not necessarily an efficient tool for animal well-being. Consumers are represented in the Regional and the National Label and Certification Committee whereas environment and animal protection are not. In this framework, the representative of the consumers demanded and obtained that "Label Rouge" pork charcuterie, ham, dried sausage etc. should obligatorily be manufactured not any more from industrially bred pigs but obligatorily be from pigs bred according to the List of Specifications "Label Rouge" for pork meat, as from 2005. Then the production line succeeded in getting authorization for production of "Label Rouge" pork to take place on slatted floors, without bedding, like factory pigs.

The CCA informs consumers. It's magazine has published comparison of different varieties of pork, based on comparison of the farm assurance schemes. The aim was a very simplified evaluation which takes into account aspects of food safety, welfare, environment and price. This is meant to be a first pragmatic attempt to get familiar with credible sustainability criteria.

Consumer information on eggs and on breeding conditions for laying hens is another constant field of activity. A comparison of different labels has been published in July 2001, and brought up to date by an article in December 2003. Consumer information has been experimented within a supermarket at the eggs' shelf. A project on consumer information concerning beef is still running, another one on organic agriculture is published in March 2004.

Such an explicit commitment to animal well-being is new in France. National consumer associations have diverse attitudes. Thus, the President of the Union Fédérale des Consommateurs Que Choisir (Federal Consumers' Union What to Choose, www.quechoisir.org) supports the idea of "reasoned agriculture" (which is correct standard practice), promoted also by the major Farmer's Union and the

---

[60] Letter signed by the President of the CCA, from January 7 2000, demanding " the discontinuation of extension and creation of industrial pig farms in Alsace,... producing a sort of pork which consumers reject more and more... concerning welfare... on slatted floor without litter... fixed, sow tethered or confined in a farrowing cage, insufficient surface... stress... all elements which, when they become known by consumers, entail perplexity and rejection...".

[61] Press conference of Consumer organisations together with Alsace Nature and the SPA of Strasbourg, December 2000.

pesticide industry. The grounds[62] for this are nevertheless judged to be insufficient by several other large consumer associations, who insist that agriculture should become sustainable. They give a much more exacting sense to 'sustainable'. Professional pork and poultry production magazines prefer to publish consumer opinions denouncing ecological and animal protection associations[63] - regardless of their being unrepresentative.

What are the reasons that made it possible that animal welfare became an important and durable issue for consumers in Alsace?

Alsatian consumers have taken their time to go to the root of the matter. Let us accompany them. Buying habits are often in contradiction with opinion. But after there have been information campaigns on the subject, opinions could change. Buying habits evolve according to experiences and new opportunities. Life experience is worth more than paper or words. Time is needed.

Information on the quality/price ratio is privileged by classic consumerism. As for meat, opacity reigns over breeding conditions. In Alsace, consumer associations have denounced this. Just before the mad cow crisis, they were told that traceability of meat would be impossible. The predominant worry was, of course, the health of the consumer. Thanks to major pressure by public opinion, intensive work on food safety is being carried out today on a scale never seen before, notably in hygiene, with generalised setting up of the HACCP (Hazard analysis and critical control point) and security controls destined to protect the consumer against BSE (Bovine spongiform encephalopathy) and other serious infections. But what exactly is the link between food scandals and ill-treatment scandals, between food safety and animal well being? It's a long and winding road towards joining the two together. In the beginning of this work on animal welfare, some consumers said: "we don't care how the food is produced, as long as it is safe". But most consumers consider that there are moral limits on the way we treat animals. The moment one refuses simplistic or demagogic slogans, it is necessary to adapt to complex reality. In actual fact, a lack of scruples is by its very nature global, and public opinion is right to link the two.

---

[62] The grounds of 'agriculture raisonnée' (reasoned agriculture) for the qualification of farms has been approved by decree, followed by several orders, in Spring 2002. As far as animal welfare is concerned, the constraints are no more than respect of the law and correct sanitary practice.

[63] Filières avicoles, février 2001, p.122 "Eric Avril (FO consommateurs): le bien-être? Un artifice pour rendre l'Europe moins compétitive". In this article, Eric Avril, general secretary of his consumer organisation, identifies himself with the point of view of the French poultry industry: "Animal welfare mostly interests only pressure groups, which generally speaking are minorities, and act under Anglo-Saxon influence". But it is not revealed that, in Alsace, a consumer representative belonging to the same association, very concerned about water quality and price, and not indifferent towards ethics, regularly votes against industrial breeding farms in the Committee concerned by authorisation procedures.

There was for instance a boycott campaign on calf meat fattened on hormones - which was efficient as far the suppression of hormone treatment was concerned. But this missed the point, that liquid feed and deliberate iron deficiency imposed on calves, is against nature - as is all animal meal given to cows - and that their detention in narrow individual boxes on slatted floors is inhuman. It was necessary to wait for the campaigns run by animal protection associations calling for a new European directive for the protection of calves, to hear of their distress. But the public already associated veal with the idea of scandal.

Another example is laying hens in battery cages. The first reaction is to link food scandal and ill-treatment and to consider these eggs as revolting, infected with salmonella and full of drugs. Then we have the pat replies of the profession, who explain in detail the sanitary measures taken in industrial breeding. There have been two attempts at dialogue with battery-hen producers in Alsace, concerning the "Bureland" ("Farmers country") trademark. Consumers came to the conclusion that these producers did not want them to visit their egg production plants, nor to listen to them ; they only wanted to "communicate" on their products. A trainee was assigned to work on the eggs. He benefited from greater transparency from the "Matines" company.

Alsatian consumers have more easily access to information from bordering countries. A Swiss agriculture education video series on the behaviour of pigs in various systems[64] has opened the eyes for several persons who then contributed to developing an alternative pork production line. When a visit to an intensive pig farm was organized for consumer representatives in Alsace, they were not convinced by cage comfort and slatted floor hygiene. Information from the Swiss egg production, combining good sanitary quality and respect of the behavioural needs of hens, was most useful. Consumers in Alsace demand free-range egg products for collective catering, but the 'rules of the game' which apply in the E.U. make it hard to have an influence on egg production lines.

Do consumers, who have become town-dwellers, ignore the realities of farming? It is fashionable to denounce the fact that children of today do not know where milk comes from. It is true that the economic and technical logic ruling over breeding overtakes by far any normal imagination. As far as farm animals are concerned, our citizens have not, however, lost their memory. Many have experienced bonds with farm animals and have made personal observation on what animals choose to do. EV has told us about her chicken that she taught to peck. AH remembers her little

---

[64] Comportement du porc domestique (Behaviour of domestic pig) by Thomas Sommer, sold by Centre des moyens d'enseignements agricole, CH 3052 Zollikofen

calf that ran to her for comfort when he got a fright in the meadow, and her hens that ran to greet her when she came back at the weekend. MH dreams of going back to the land and being close to the animals; he bears a grudge against those who were responsible for his father giving up his farm. RB, who loves eating meat and charcuterie, saw blood running out of a truck transporting live cattle.

A field visit where one can see with one's own eyes what is going on, is most important. Thus, a visit to a free-range laying hens farm showed that there were hens with badly trimmed beaks. Consumers' reaction was very strong, and henceforth, for them, this aspect must be part of all quality research in egg production.

We have been lucky, in Alsace, to work with Rési Bruyère at the CCA. At the farm, no label on any bag of food escapes her attention. The breeder, who does not know what raw materials are contained in the food just delivered, had better watch out! She puts a lot into trying to understand the manner of breeding. She was extremely disturbed by the detention of sows in individual cages. "Did the sow tell you she is unhappy?" was the mocking reaction from professionals. The consumers did not give way[65], and the report of the Scientific Veterinary Committee at the EU vindicated them[66].

The credibility of people and democratic team work, building up confidence, will be decisive. Most people do not want animals to suffer. The main problem with animal protection associations is confidence. Confidence depends on honesty and very much on competence: years of learning and listening.

Progressively, the link between food security and animal well-being was clarified. Consumers find out that animal production has even more impact on the health and the welfare of humans and animals, by environmental issues, breeding conditions, employment and working conditions. Indeed, tolerating exemptions from morality for the sake of profit generates risks. Consumers expressed it through the following propositions[67]: - replacing the dictatorship of profit where only feed conversion counts, by a more moderate and healthy productivity; reducing the amount of drugs used in breeding; ensuring the physiological feeding of animals; avoiding stress and providing conditions favourable to good health which includes

---

[65] The representatives of consumers in the Certification Committee of Alsace voted in 1997 against the Farm Assurance Scheme of 'Burehof' pork produced on fully slatted floors, with sows kept in individual crates.

[66] The welfare of intensively kept pigs. Report of the Scientific Veterinary Committee. Adopted 30 September 1997

[67] "Manger responsable" Exhibition, inauguration at the Robert Schuman High School in Haguenau in November 2001

movement and access to outdoor climate, and, in particular, good immunity; this is in everyone's interest, consumers, breeders, animals, the environment.

Ethics in consumption is a strong new orientation, rooted in various commitments. "We are all responsible" says the exhibition produced by Alsatian consumers. Militants usually want to work for collective benefit. Many are to be found in their neighbourhoods, some are specialized in helping families. Help for people with excessive debt is a permanent concern. The promotion of ecological consumption and ecological mobility are regular activities. Involvement for ethical consumption strongly supports fair trading between North and South. A project on ethical investment is in operation. Promoting "sustainable breeding with solidarity" fits perfectly into these activities. Ethics appears as a principle of precaution.

4. Environmental protection associations: from biodiversity to well-being

In this area, we should observe the impact which environmental protection associations can have on the evolution of farm animal production systems. First we look at different actions, mainly in Alsace, and than we try to understand what makes them adopt the cause of farm animals.

"Alsace Nature" (http://alsace.nature.free.fr) is the regional federation of nature and environmental protection associations in Alsace. In 1996, Alsace Nature co-ordinated a detailed and comprehensive publication on agriculture and nature. The issue of farm animal welfare was included, and I was put in charge of it.

Alsace Nature is opposed to factory farming, but experience shows that opposition only has a chance of succeeding if the local officials run it with determination. The question of animal well-being has no place in the authorization procedure for industrial breeding installations. For the investigating commissioner - often a retired military officer - the company head is someone who inspires confidence.[68] Scientific arguments, even if taken from European reports, which are at the basis of all decisions of the European Commission, do not weigh heavily in the face of professional speeches. The investigating commissioner relies on personal

---

[68] The report approving a project for 214 000 young hens in cages in 67160 Riedseltz, 19.6.1999, says: "The reputation of seriousness and know-how of the farmer is even better established, since without these criteria the company would have closed its doors a long time ago".

judgement, intuition.[69] The only option is the application of texts. This is what the Chief Administrative Officer of the region does. The ethical point of view is systematically out of bounds.

Alsace Nature strongly commits itself to the development of organic agriculture. The image of organic agriculture is sometimes more idyllic than the reality. But the advantage of the message is that it is simple, being the best possible for welfare and environment. In spite of its distrust for intermediary solutions, Alsace Nature is ready to support concrete progress at a middle point between organic and conventional agriculture if the arguments are good.

Alsace Nature cooperates with the Alsatian consumer organization CCA, and they had a common press conference, to oppose industrial pig breeding and to support the Thierry Schweitzer pig production.

In November 2003, Alsace Nature organized a tri-national meeting on "animal welfare and sustainable agriculture", together with its German and Swiss partner associations. The CCA participated. The variety of contributions, political, technical, environmental, and economic, yielded a major dialogue for better comprehension of difficulties and possible solutions.

How did Alsace Nature come to adopt this new cause? Respect for life, in the tradition of Albert Schweitzer, is part of its culture. It has always, provided support for the fight to obtain respect for animals[70]. Its ethical base is to preserve nature, not only because it is indispensable to man, but also for nature itself. The militants of Alsace Nature are tortured souls, having seen so many meadows being ploughed up, orchards torn up, natural paths disappear under urbanisation. They have a very precise knowledge of pollution, in particular pollution of subterranean and surface water, which in Alsace is much more due to maize and to wine than to manure.

---

[69] The report, 7.1.1997, approving a project for 100,000 hens in cages, says: "Having never penetrated a battery cage building before, I was rather positively surprised, compared to what I imagined and what one hears. The animals do not give any impression of unhappiness....they eat and drink ... they are not beaten nor ill-treated... ... in fact they can not walk around nor use their wings nor see daylight. But they do not know that these possibilities exist and will never know it..... I have the strong conviction that Mr. R... with his 30 years experience of the hen and the egg is a breeder in whom we must have confidence, to run a farm of more than 100 000 laying hens respecting the elementary rights of animals, ruled by the law in force...."

[70] In 1996, Alsace Nature coordinated a detailed and comprehensive publication on agriculture and nature. On my demand, the issue of farm animal welfare was included, and I was put in charge of it.

Every time the "bio" market increases, animal well-being increases. However, love for the meadows that have to be saved has obscured the rest. Given that the farmer keeps 70 hectares of meadow, and as much as possible in extensive management. Who cares if the cattle leave the cowshed or not? As Alsace Nature is most concerned by the conservation of exceptional biotopes with rare birds and flowers, there has been a general tendency, in the plain of Alsace, to associate nature protection with prohibition of pasture. Interest in "normal" nature and landscape, such as intensively used pasture around dairy farms, is new, and goes along with new opportunities of the European common agriculture politics.

Members of Alsace Nature often find themselves in the role of opponents, recognised for serious competence. They act independently, according to their conscience, because the association endeavours to be independent in spite of getting subsidies. It is known to defend public interest, as for instance preservation of ground water[71]. This tradition enables the association to attack subjects and defend points of view in advance of its time, and interfering with economic interests - which is indeed the case for animal welfare.

At the national level, the struggle of associations against pollution by factory farming is virulent, in particular in Brittany. The "Coordination Nationale contre les Elevages Industriels" (CNCEI) (The National Coordination against Industrial Breeding, http://perso.wanadoo.fr/coordination.nationale) was launched to combat smell and pollution. It has taken up the cause of animal well-being, and keeps its member associations well-informed. "France Nature Environnement" (FNE, www.fne.asso.fr), which regroups regional environmental protection federations, has also taken up the cause and includes animal well-being in its demands to the Ministries concerned with agriculture. It appears with evidence, that the officials and militants of these movements work with profound humanity for a planet where life can evolve in all its diversity. It is not that they do not have respect for farm animals, but rather that they lack of time to deal with all the urgent dossiers.

Sustainable agriculture has its first List of Specifications, produced by the farmers and militants of the Réseau Agriculture Durable (RAD) (Sustainable Farming Network, www.civam.org), without waiting for the inertia of Ministers to be converted into action. Animal well-being is not put first as in Alsace, but the conditions are favourable for it. One of the leaders of this movement is André

---

[71] The ground water used by the tourist town of Obernai is at the very limits of potability because of nitrate and pesticide pollution. Under the 548 hectares of prairie, for whose protection Alsace Nature had battled with the greatest difficulties, nitrates are only 10mg/l, a little above natural level. On 29 January 2002, three mayors came to ask the opinion of Alsace Nature about the consumption of drinking water in that protected area.

Pochon, farmer and author, militant for meadows and for cows, which have a grass cutter in front and a fertiliser spreader behind. Listen to him: "History will judge; it will be hard on a generation who exhausted the planet's resources, and polluted water, air, earth, sea. But will the instrumentalization of domesticated animals for the ends of industrial production not be more important in the law-suit that will take place? Whatever happens, agricultural managers who stick to industrial breeding methods do no favours to their principals. It is high time to open people's eyes, not only for ethical reasons, but also for economic ones. Thanks to the awakening of citizen-consumers, animal well-being becomes an indispensable sales argument."[72] Tongues are loosened.

5. Animal protection associations: from pity to competence

Local associations can intervene in individual cases of cruelty. But here we look at actions made by animal protection associations for legal, technical, and economic progress. Major campaigns take place on a European level ; French associations participate. Are they efficient ? Neighbouring countries have experience of successful national campaigns. What made it possible ? Case studies on sows and laying hens and examples of welfare labels give us a hint.

5.1. European law

The first idea is that cruel breeding conditions must be banned. Major campaigns take place on a European level (www.eurogroupanimalwelfare.org). A lot of energy has been devoted to this option. Since the protection of farm animals is harmonized at European Union level, the agent of these bans should be the European Commission, with the opinion of the European Parliament, and the final approval of the Council of Agricultural Ministers. The three authorities are targeted by campaigns of associations, in Brussels, Strasbourg and member states. The rate of progress of the Commission is roughly one directive on protection of farm animals every two years[73], on condition that the Presidency is a country that is motivated and ready to meet all the inherent delays to procedure, as was Sweden in the case of sows.

Is this associative action for better European law efficient or inefficient? As far as inefficiency is concerned, it has to be considered that the faults and compromises

---

[72] This text was published in the daily paper "Ouest France" (Britanny) on 3 July 2002.

[73] Directive on transport 1995, Protection of calves 1997, Protection of laying hens 1999, Protection of sows 2001.

of a Directive, over and above the announcement, leave in place breeding systems that are recognised as bad, such as slatted floors without litter for veal calves and pigs, anaemic veal calves, battery cages for hens, farrowing crates for sows.

As far as efficiency of the Directives is concerned, it has to be considered that intensive installations are losing some of their superior cost-effectiveness, and that alternative systems are gaining increased interest from breeders, technicians, and researchers. Many farmers choose a well functioning alternative system and a good image rather than the minimal respect of law. At the very least, condemnation of the actual systems through public opinion has now materialized.

5.2. Enforcement of the law

Intensive breeding installations are closed to the public, and associations don't get in. Concerning for instance veal calves in France, the first control campaign produced no sanctions[74]; deep in the country people are not even informed that there is a ban on muzzling calves for white veal, and veterinary surgeons are notoriously implicated in white veal production[75]. It has been announced that the control of egg production plants will generate no sanctions either, the application of the directive will be delayed, and the egg producer are campaigning to abolish this Directive.

Another aspect where associations are active is the respect of specifications, particularly for labelled quality products. Associations in Germany have visited alternative egg farms and denounced a certain practice of so-called free range egg production, cheating the consumer.[76] However, this sort of cheating can be legal, as veterinary services in France specified that law only imposes doors to go outside for so called free range hens; it does not impose that hens really walk out through these doors[77].

With regard to transports, militants from associations drive behind the trucks transporting live animals. They can prove that legal transport time, obligatory route planning, resting periods, watering, and fitness for transport are not respected[78].

---

[74] Information from the Bureau de la Protection Animale, Ministère de l'Agriculture.

[75] La Semaine Vétérinaire n°1057 25 mai 2002 relates as 'further education', on an entire page, a study concerning parasitism on muzzled (as usual) Limousin veal. What about the law prohibiting muzzling (Journal Officiel de la République Française 27.1.1994 and 14.12.1997), and the welfare needs of calves?

[76] Kritischer Agrarbericht 2001, www.bauernstimme.de

[77] Letter from the Veterinary Service of the Bas-Rhin, 20 Septembre 2000.

[78] Report from the Commission to the Council and the European Parliament on the experience acquired by Member States ....concerning the protection of animals during transport. COM(2000) 809 final Transport of live animals, FVE Position paper, www.fve.org.

Some forewarding agents regularly overload their trucks.[79] Thus associations are an essential source of information for the European Commission.

As far as abattoirs and markets are concerned, far too often law is not respected. Lack of time causes brutality. Cows remain waiting in the abattoir during the weekend, without care. It happens that lambs, calves, pigs, are hung up for bleeding after inefficient stunning. This is in France the conclusion of the delegates of the "Oeuvre d'Assistance des Bêtes d'Abattoir" (OABA, www.oaba.asso.fr), which has a contract allowing its delegates to enter, in exchange for discretion on results. The OABA delegate, a regular witness to ill treatment, was beaten up in the cattle market of Chateaubriand, which led to a court case and a discharge of the accused, all done very discreetly. The PMAF (www.pmaf.org) has filmed ill treatment of animals in French markets. It needed only two weeks to produce shocking testimonies that were shown on TV. Greek associations work hard to obtain change in Greek abattoirs and organise education of the employees with the help of Eurogroup.

Having demonstrated "the low degree of priority"[80] given by member states to the protection of animals, the associations were once again launching a campaign against long distance transport of animals and restitution of exports of living animals. They have efficiently put together overwhelming proof in order to demonstrate to the Commission the need to act. However, in spite of more than 1,5 million of signatures submitted to them already in 1991, against the advice of the European Parliament, and in spite of periodic initiatives and announcement, there is no significant restriction on life export, mainly to Mediterranean countries. However, some local situations have improved, thanks to militants being present on checkpoints and on markets, thanks to competence, dialogue, co-operation, and training, thanks to slightly improved official control.

5.3. Further advancement of certain countries

In certain countries constructive and well thought out actions prepare and anticipate European evolution.

A first example is individual crates for dry sows that the EU has timidly decided to phase out by 2012. The United Kingdom, Denmark, the Netherlands, Sweden, Finland and Switzerland had already made these illegal before this directive, or started phasing them out. Moreover Sweden and Switzerland are also working

---

[79] For instance the Ligue nationale luxembourgeoise pour la protection des animaux has proved severe and repeated overload of Dutch trucks transporting pigs for slaughter.
[80] Report of the Commission COM (2000) 809.

towards liberty of movement for farrowing sows. In Switzerland farrowing crates will be phased out in July 2007, and newly built accommodation must allow the sow to turn around. Such advances would not be possible without a strong mobilization of public opinion, thanks to long term hard work on the part of the associations, the commitment from some scientists and veterinary services, and also the experience of farm assurance schemes which put animal well-being first. There is also scientific testing of new breeding systems, including their ability to meet behavioural needs, before they get authorized. Associations demand such testing, anticipated by Switzerland, in the EU.

Another example is the German advance, banning battery cages for laying hens as from 2007[81], whilst Europe is content to announce a law to replace classic cages by so-called "modified" cages that are in fact a way out for cage breeding. Ever since the sixties, when battery cages first invaded egg production, they were denounced, but most people did not know or did not want to believe it. The description "KZHühner", concentration camp hens was attacked in a court case, but former internees of concentration camps said that the term was suitable, and it was authorized. Farm animal experts, mainly ethologists, worked out scientific argumentation. The VgtM (Verein gegen tierquälerische Massentierhaltung, - Association against ill treatment and mass breeding of animals, www.vgtm.de) created in 1972, was the first association specializing in farm animals to mobilize public opinion. The President of the Bundesverband Tierschutz (www.bvtierschutz.de) is himself an alternative egg producer. The major German organizations federated in the Bündnis Tierschutz (www.tierschutzbund.de) discussed with the production line the Lists of Specifications for free range breeding, controlled and labelled by the "KAT" (www.kat-cert.de) system of control of German and imported free-range eggs, set up by the profession. An essential stage had been the judgement of the Constitutional Court in July 1999, that declared null and void the order that authorized cages for battery hens, ruling that cages inflicted significant and durable lack of welfare - which means suffering - on laying hens, and that their behavioural needs were repressed in an excessive manner. Lawyers and ethologists furnished the arguments. All animal protection associations were mobilized. Demonstrations, symbolic actions, and sit-ins, accompanied the legal battle. Opinion polls condemned cages. The consumption of free-range eggs increased, and hard discount supermarkets introduced them into their range of products. The decisive vote to abandon cages took place in the Bundesrat in October 2001. However, the battle continues, as another vote in November 2003, initiated

---

[81] www.freiheit-schmeckt-besser.de "liberty tastes sweet" is the official site of the Ministery concerning the new order about hens.

by the Land which has the most industrial animal production, has modified the first one to give another chance to cages now called "appartments".

Firstly these examples perfectly illustrate that associations intervene at all levels; opinion campaigns, making public expert knowledge, legal measures, good alternative breeding techniques, farm assurance schemes, labelling and control. Secondly, in order to succeed, major personal investment is needed from high level specialists, motivated by ethics, and political good will is needed to implement the ethical demands of citizens.

5.4. Economic viability

Economic viability is a major concern for associations. On the international level, it has been the subject of several reports by Eurogroup and its member organisation RSPCA[82]. Associations have played an important part in obtaining the agro-environmental subsidies. Some anti-ecologist reactions have turned into acceptance. They are lobbying for subsidies for animal welfare, which will make welfare constraints more tolerable.

In European countries, associations support farm assurance schemes that give priority to animal welfare. Once technical viability of alternative breeding methods is satisfactory, the products have to be sold. Some welfare label are very exacting, with a small number of farms under contract, others are less constraining and reach a significant part of the market. Thus labelling prepares the way from an experimental pilot system to generalization. It is a chance for professional information and public debate. It anticipates better legislation.

The Swiss "KAG-Freiland" trade mark is a very strict pilot-project, concerning 260 farms, with mostly direct sale. Animals are outside every day.

The Swiss "Coop-Naturaplan" comprises 1400 pork farms and about 90% of the sale of pork by Coop, the second largest retailer of Switzerland. An accredited organism that belongs to the Swiss Protection of Animals (www.protection-animaux.com) operates control.

---

[82] "Conflict or concord? Animal welfare and the World Trade Organization" illustrates problems and possible solutions concerning domestic and wild animals.
"WTO Food for thought. Farm animal welfare and the WTO" proposes different solutions.
"Hard boiled reality", 2002, analyses the economic consequences of abandoning battery cages.
"Into the fold: bringing animal welfare into the CAP", 2002, analyses opportunities for welfare payments.

The German "Neuland" is a trade mark created by five major environmental, animal, consumers, alternative farming, and fair trade associations. It has very interesting specifications, but has not developed at the same rate as the Swiss labels[83].

The "Freedom Food" (www.freedomfood.co.uk) in the United Kingdom belongs to the RSPCA has attained a considerable part of the market, mainly eggs and pork.

The "Thierry Schweitzer" pork trade mark, based on partnership with associations, started in 1998. With only four farmers producing the pork it enjoys a notable commercial success, attaining more than 15% of the pork sold in a big hyper-market.

"KT-Freiland" in Austria is an organic trademark, run by an association, which insists on animal welfare.

"Tierschutzgeprüft" (Controlled by animal protection) in Austria guarantees credible labelling of alternative eggs and supports its own more exacting specifications.

Swiss "Natura Beef" (www.svamh.ch) is a producers label, supported by animal protection. It is young beef about 10 months old, coming directly from the dams' herd, "red as beef, tender as veal", and has become a great commercial success.

In many cases, animal welfare is a major sale argument for organic products. But welfare products are not necessarily organic; if not, they are cheaper. These examples show that the animal welfare sales argument can work, even very well. Particularly in Switzerland and in the United Kingdom long term and professional information and debate have been made available to the public.

Some retailer chains include ethics towards animals in their programme. This supposes an excellent preparation of public opinion. Coop and Migros, the biggest retailers in Switzerland, have sold no cage shell eggs since 1996. Coop has an ethical programme on 4 "labels of confidence" for man, animals, and environment. It began in 1993 and attained more than 1000 Million CHF of turnover in 2001.

---

[83] Why? a higher-level prestige image in butcher shops? strict limitation of production size on the farm? management? or simply because in Germany alternative farmers can practice direct sale and do not need a special label?

Some British, German, and Austrian retailers have also eliminated cage eggs. The British associations have even succeeded in eliminating the sale of fat liver by supermarkets[84].

The Swiss Coop co-finances research to find a solution for castration of piglets, as it has decided to ban castration without anaesthesia as soon as possible, which is a result of successful campaigning by Swiss animal Protection.

There is a major need for "good companies", for good directors, for good sale persons, who in their profession - transformation, distribution, technical advice, etc. -, use the small amount of room for negotiation that they possess, in the middle of constraint and economic pressures, to give decency a chance. This small chance they provide can have a snowball effect.

Launching animal friendly products is most difficult, when ethical production is rare and logistics are expensive because of small volumes, and people do not yet know what it is about. This makes the success of the "Thierry Schweitzer" brand so remarkable. There had been an extraordinary coincidence: The Director of an Auchan hypermarket took personal pleasure in promoting the "Thierry Schweitzer" pork. He let associations enter his hypermarket, and even a live sow with her piglets in a large littered pen: not only the children were delighted, but also adult women and men were standing there, observing, maybe dreaming, meeting and communicating with this live animal that is behind the ham package.

6. Farm animal welfare experts: from commercialising life to changing the system

Associations demand more funding for research in organic agriculture and alternative animal friendly breeding systems. Experts with high personal commitment for animals are exceptional. Some contribute with some good arguments or good new techniques to improve animal welfare, some raise fundamental questions, as did R. Dantzer and P. Mormède (1979). Dr. J. Spranger from the FIBL (Research Institute for Organic Agriculture, www.fibl.ch) considers that pathology of domestic animals is mostly caused by human failure, also in the psychic domain. His ethical principles would not allow him to submit an animal to a poor social environment for research. Prof. D. Fölsch has demonstrated that even

---

[84] British associations had found foie gras from Alsace, "made without force feeding", sold by Waitrose. As members of the French OABA, we investigated in Alsace and found no such production. In May 2002 Waitrose wrote to 'Advocates for Animals' to confirm that Waitrose does not sell any foie gras.

severely injured or ill hens continue to lay eggs, showing that productivity does not indicate welfare (Fölsch, 1977). It appears that even a single insignificant animal like a hen or a rabbit deserves a gentle and healing treatment.

The image of experts is suffering[85]. It would be so easy for veterinary surgeons or agronomic professors to explain to politicians that there is need for change. Associations can be even more distressed to find out that 'research' is carried out to defend foie gras production, battery cages, individual crates, and minimal progress, instead of adapting new systems to the animals' physiological and behavioural needs. There has never been any consumer demand for genetic manipulation of farm animals, one-sided selection criteria and certain reproductive techniques, but there is tremendous demand for 'natural' products.

Among animal production scientists it is common belief that a scientist should not express ethical commitment towards animals, but let others take ethical decisions; thus, if he has a moral commitment for animals, he is no longer an objective and credible scientist. This would mean that a medical researcher who has a strong ethical commitment towards his patients is not a credible researcher, which clearly makes no sense. Maybe there is some confusion about what the ideal of 'independence' could mean. Being partial for the poor and the suffering, for the dumb and the helpless, is valued quite differently by the public as opposed to being partial for the rich and the mighty. The domination of the scientific community by the economic community is a major problem in animal production.

Associations need information and advice by experts. Usually experts are friendly and ready to help, answer questions and give a copy of a publication[86]. The welfare pork Thierry Schweitzer in Alsace would not exist if years ago a Swiss researcher[87] had not accepted to give advice. Others kindly take the trouble to explain that the best thing to happen to hens is to be selected in order to adapt to cages; and since there are so many children suffering in the world, why care as much about calves?

When Prof. Fölsch was in Zurich, developing alternatives to battery cages for laying hens, he already supported the "KAG" (Consumers' working group,

---

[85] Armand Farrachi Les poules préfèrent les cages. Quand la science et l'industrie nous font croire n'importe quoi. Ed. Albin Michel. "Hens prefer cages. Science and industry would have us believe anything."

[86] Robert Dantzer, Magali Hay and Armelle Prunier have been particularly helpful to find a way to experiment anaesthesia for castration of piglets on the Thierry Schweitzer farm, and Isabelle Veissier for facts on calves, J-M Chupin on abattoirs and farm buildings, J Capdeville on cows' welfare, and many others deserve gratitude for sharing their knowledge.

[87] Dr. Wechsler is now directing the Centre for proper housing of ruminants and pigs, FAT, CH 8356 Tänikon I want to thank him for this first lesson in farm animal ethology.

www.kagfreiland.ch), the first association to build up a farm assurance scheme for free range eggs, which today still owns the most exacting trade mark for animal welfare. He returned to Germany to teach farm animal ethology. It so happened that he used his precious time to teach associations, because their part in society is too important to allow them to do poor work through incompetence. His Department organized a meeting on pedagogic access to farm animals for children (Simantke and Fölsch, 2000), where amongst other organisations, associations from different countries presented their education programme.

Associations can become professional and engage persons who have done scientific work. Motivated agronomic engineers and veterinary surgeons can be at the spearhead of progress. In Switzerland, this has led to co-financing of research projects[88]. In the United Kingdom, animal protection regularly finances research.

When experts organize themselves to associations for animal welfare, there is hope for change. For instance in Austria, it was the group "Kritische Tiermedizin" (Critical Veterinary Medecine, www.agkt.de) that has launched the most exacting organic farm assurance scheme for welfare in Austria (www.freiland.or.at). It has permanent support from scientists, and organizes a colloquium once a year on science and the practice of farm animal welfare.

Created in 1978, the IGN (International Society of Livestock Husbandry, www.ignnutztierhaltung.ch) aims to promote breeding, on a scientific basis, with respect of animals. It awards a prize every year for the best scientific work for better welfare, and publishes a booklet every 3 months with abstracts of scientific articles that are particularly decisive on welfare. It is a precious source of information for associations.

Northern Europe also has this kind of scientific tradition, better known and influential, such as "Universities Federation for Animal Welfare" (www.ufaw.org.uk), founded in 1926, based on the belief that animal problems "must be tackled on a scientific basis, with a maximum of sympathy and a minimum of sentimentality", emphasizing its independency. Animal friendly farm advisers can organize themselves to develop high welfare systems for breeding in farm practice. The German BAT (Beratung artgerechte Tierhaltung, www.bat-witzenhausen.de) for instance is a dynamic group of consulting engineers.

---

[88] The Swiss Protection of Animals and the KAG regularly finance research on welfare, by themselves, or together with public or private funding, sometimes with the FIBL: for instance on aviaries, perching space for broilers, optimal free range management for pigs, broiler or turkeys, endoparasites of poultry, welfare of rabbits, free range fattening of male laying hybrides, loose housing for cows with horns, anaesthesia for castration of piglets, fattening of young boars...

7. Education for ethics: from zootechnics to the practice of wellbeing

Jocelyne Porcher has concluded that the level of agronomic education determines a differentiation in the affection farmers have for their animals: the higher their education, the less affection they show. Society finds it certainly difficult to accept such a perversion of education.

Teaching ethics towards animals should include communication with animals - not only anthropocentric anthropological discourse about them, but observation and "listening" to how this animal experiences and feels its world, natural or man made, other con-generic animals, and man.

Why not let "case studies" become "case actions"? Action and initiative are essential to applied ethics. It is interesting to hear what different people think about ethical conflicts, but even more interesting to find out what one can do to avoid pain and mutilation, to allow normal behaviour, to improve transport and culling, to convince consumers, etc..

Teaching humane ethics and standard factory farming is a contradiction. So the best "case action" would be to convince colleagues to replace conventional zootechnics by developing and teaching humane production with animals that can move freely, go outside of buildings, have comfort, appropriate feed and optimal health, and live in harmony with other animals of their own species and with mankind. Associations will be very happy to meet and to give support to this new generation of farmers.

References

Dantzer, R. and Mormède, P., 1979. Le stress en élevage intensif, Masson/INRA

Farrachi, A., 2000, Les poules préfèrent les cages. Quand la science et l'industrie nous font croire n'importe quoi. Ed. Albin Michel.

Fölsch, D., 1977. Die Legeleistung - kein zuverlässiger Indikator für den Gesundheitszustand bei Hennen mit äusseren Verletzungen, Tierärztliche Praxis 5, p. 69-73.

Hoop, R.K., 2002. Tiergesundheit in der schweizerischen Geflügelwirtschaft - Rückblick, aktuelle Schwierigkeiten und Ausblick Arch. Geflügelk. 66 (3), 114-118

Pochon A., 2001, Les sillons de la colère. Ed Syros

Porcher, J. 2001, L'élevage, un partage de sens entre hommes et animaux: l'intersubjectivité des relations entre éleveurs et animaux dans le travail. Thèse. Institut National Agronomique de Paris-Grignon.

Porcher, J. 2002, Eleveurs et animaux, réinventer le lien. Editions Presses Universitaires de France

Prunier A., Hay M., Servière J., 2002. Evaluation et prévention de la douleur induite par les interventions de convenance chez le porcelet, Journées de la Recherche Porcine, 34, 257-268

Scientific Veterinary Committee, The welfare of intensively kept pigs. Report adopted on 30 September 1997

Simantke, C., Fölsch, D., 2000. Pädogogische Zugänge zum Mensch-Nutztier-Verhältnis, Tierhaltung Band 26 Ökologie.Ethologie.Gesundheit, Gesamthochschule Kassel

RSCPA, Eurogroup: Conflict or concord? Animal welfare and the World Trade Organization" illustrates problems and possible solutions concerning domestic and wild animals.

RSPCA, Eurogroup: WTO Food for thought. Farm animal welfare and the WTO

RSCPA, Eurogroup, 2002, Hard boiled reality

RSCPA, Eurogroup, 2002, Into the fold: bringing animal welfare into the CAP

Sommer, T., Comportement du porc domestique (Behaviour of domestic pig), video serie, sold by Centre des moyens d'enseignements agricole, CH 3052 Zollikofen

Spranger, J., 1998, Die Bedeutung der Mensch-Tier-Beziehung in der Alternativmedizin, Kurs 98.254, Landwirtschaftliche Beratungszentrale CH-8315 Lindau

Studer, H., Schweiz ohne Hühnerbatterie. Wie die Schweiz die Käfighaltung abschaffte. ISBN 3-905647-12-5

Ulich, K., 1998, Die Freilandhaltung von Legehennen, VgtM

Vonesch, A., 1996, Conditions de vie des animaux d'élevage: réalités et perspectives. LA NATURE pour la reconversion de l'espace rural, Bulletin de la Société Industrielle de Mulhouse n°835, p.49-61.

Trade regulations, market requirements and social pressures effects on introducing animal friendly livestock production systems

Tadeusz Kuczynski[1] and Stefan Mann[2]
[1]Department of Environmental Engineering, University of Zielona Gora, Poland, and
[2]Swiss Federal Research Station for Agricultural Economics and Engineering, Taenikon, Switzerland

Abstract

In the paper the most important forces which can affect introduction and development of high animal welfare standards in animal housing are presented. Introduction of appropriate regulations at the national and international level is only one and probably not the most important factor involved in their efficient introduction to animal production systems. Of much more importance appears to be consumer's demand for higher-priced food produced at high animal welfare standards. This fact is commonly acknowledged even by institutions involved in creating regulations and their efficient introduction. Another important factor seems to be the farmer itself including all the connections with his suppliers, like: genetic, feed, pharmaceutical and equipment companies and their economical interests. To more or less extent all of them must find their potential profits from changes in the production process.

There is also tremendous effect of trade globalisation, most spectacularly illustrated by WTO policies, which do not allow putting restrictions on import because of nature of production process.

At present the most appropriate actions which could be recommended for developing more animal friendly production systems seem to be: educating the consumer on ethical values of animal welfare, labelling food according to its method of production, subsidizing research into low-cost animal friendly production systems, negotiation with WTO (i) to introduce at least minimal requirements to production process, (ii) to get approval for domestic support to compensate farmers for additional cost of investments aimed to improve animal welfare (iii) to introduce obligatory food labelling regarding meeting requirements for animal welfare.

1. Introduction

In order to enhance animal welfare on farm, in transportation or at slaughter it is crucial to implement new production systems and replace systems with an insufficient capacity to live up to the needs of animals and farmers. Implementation of the new, animal friendly systems strongly depends however on their economical justification.

The following chapters will shed light on the questions that can provide high standards for animal welfare as a core element of production and how these aspects can be integrated in production systems.

Probably most important issue in understanding success or failure of introducing such programs in any production process, is to properly identify:
- the actors on the scene;
- their actual roles in production process;
- their interests and
- possible exogenous influences, which can affect their way of acting.

2. Animal welfare

2.1 Definition of animal welfare

There is no consensus between scientists in a precise animal welfare definition. Most scientists more or less accept the definition by FAWC (2002) based on five freedoms:
1. Freedom from hunger and thirst, by ready access to fresh water and a diet to maintain full health and vigour.
2. Freedom from discomfort by providing an appropriate environment including shelter and a comfortable resting area.
3. Freedom from pain, injury and disease by prevention, rapid diagnosis and treatment.
4. Freedom to express normal behaviour by providing sufficient space, proper facilities and company of the animals own kind.
5. Freedom from fear and distress by ensuring conditions and treatment, which avoid mental suffering.

The definition given above is very broad and does not provide enough bases for assessing animal welfare on site, particularly in relation to all animals on global basis (Stricklin *et al.*, 2000). There are many other, more detailed, definitions of animal welfare that usually require a complex and fully multidisciplinary approach (Gonyou, 1986; Broom, 1991).

A realistic multidisciplinary assessment of animal welfare should consider such factors as (Hurnik, 1988; Ahnberg and Rodriguez-Martinez, 2001; Tosi *et al.*, 2001; McGlone, 2001; Rushen, 2003):

- Conditions related to specific production system: (i) space allowance per animal, (ii) indoor environment, (iii) providing freedom of movement, (iiii) animal-stockperson contact.
- Level of productivity: (i) direct animal productivity, (ii) human labor requirement, (iii) cost of production, (iiii) longevity.
- Behavior: (i) eating/drinking/activity time, (ii) aggressive interactions (tail/ear biting), (iii) abnormal behaviors, including stereotyped behaviors, (iiii) cleanliness.
- Physiology: (i) changes in hypothalamic-pituitary-adrenal (HPA) axis functioning, (ii) blood pressure, heart rate, respiratory rate.
- Health and immunity: (i) overall incidence of disease, (ii) mortality rate, (iii) level of immune protection, (iiii) the need of preventive drug use, (iiiii) drug and antibiotics use in intensive animal husbandry.
- Anatomy: (i) bone strength and rate of injury, (ii) wounding, especially of skin (e.g., bites or abrasions).

The way animals look and behave is probably the most frequently used indicator in assessing animal welfare by the farmers. Something like standard patterns of animal reactions to human contact, typical for most of the animals in a herd, as well as standard patterns for individual animals is being formed. The farmer as a first sign of problems often recognizes reactions different than the "standard ones".

Another frequently used animal welfare indicators are various physiological measures of stress, especially changes in hypothalamic-pituitary-adrenal (HPA) axis functioning, based on assumption that changes in HPA axis functioning is a generalized response to stressors of all kinds (Rushen, 2003).

The assumption behind using general welfare indicators is *"that all challenges to animal welfare will ultimately affect the animal in the same way, and thus they will allow us to measure the sum of the effects of the different challenges to the animal"* (Rushen, 2003).

As a result, behavioural or physiological indicators of welfare have to be considered in the context of variability of animal genetics, housing, feeding and management conditions.

There have been a numerous attempts to use the duration of resting behaviour for welfare assessment of cattle in different housing systems (Chaplin, 2000). Their reliability may be questioned, however, since a substantial part of lying time can be considered not as "essential" but an "optional" behaviour, done when there is a lack of competing behaviours to perform (Rushen, 2003).

Similarly, stereotypic behaviour is often considered as a good indicator of animal welfare and their incidence has been used in attempts to compare levels of welfare in different types of housing systems, with conclusions that stereotypic behaviour in sows was most frequent when they were confined to small spaces (Broom *et al.*, 1995; EU Scientific Veterinary Committee, 1997).

Other research proved however that stereotypic behaviour was strongly related to feeding problems (Rushen *et al.*, 1993). Changes in diet appeared to have much greater effect on the incidence of stereotypies than changes in environment (Terlouw *et al.*, 1991). Robert *et al.* (1997) found that using of appropriate diets can virtually eliminate stereotypic behaviour, even in closely confined sows.

The problems with interpreting stereotypic behaviour as welfare indicators are well illustrated by the results of Dailey and McGlone (1997) who compared the behaviour of gilts housed intensively indoors to gilts housed in large outdoor paddocks. A higher incidence of oral/nasal behaviour (which are the most common components of stereotypies) was found in the outdoor gilts.

There are relatively few behaviours which can give direct information that an animal is not feeling well. Typically we have sets of behaviours, which we consider as normal: lying, staying, eating, rooting, playing, etc. All of them can be normally seen at one time in a group of animals. On the other hand we do not consider "lying" as normal animal behavior when the feed is loaded to the trough. One disadvantage of automatic feeding system is that feed is being given to all animals at the same time what makes impossible to observe the reactions of single animals to this event.

There has been much evidence in the U.S. of large pig operations with automatic sow feeding where the sows were neglected and not observed each day or for several days at a time. Some owners in the U.S. do not install automatic feeding for sows when they had hired help because they wanted to force their helper to walk down the row of sows in order to feed them and at the same time they would be able to observe each sow (Muehling, 2001).

The methods that can be easily used for assessment of animal welfare by the farmers in their everyday work appear to be not very useful when the assessment is carried out by people from outside to check if animal welfare at the farm meets suitable requirements, connected with official regulations or labelling rules.

For this reason there is recently growing interest in retailers as well as suitable authorities more in checking out the potentials of the farms and farmers for providing high animal welfare than in assessing animal welfare itself.

There are indirect systems for on farm evaluation of animal welfare based mainly on checking conditions connected with animal housing and management which affect freedom of animal movement and social contacts with other animals, condition of flooring for lying, standing and walking building environment, the intensity and the quality of a care given to animals by a farmer (Bartussek, 1999; von Borell, 2001; Ofner *et al.* 2001).

Some scientists claim that such approach, although taking into account a lot of factors, which can actually affect animal welfare, create a risk that some important factors will not be considered. The other weakness is not taking into account the possible interactions between various factors since animal welfare depends on all of them as a whole. To avoid these weaknesses, the necessity of introduction of some indices directly describing animal welfare has been suggested (Capdeville, 2001).

Taking into account the complexity of animal welfare assessment methodology and difficulties with its overall assessment, there are suggestions that specific welfare indicators should be developed to assess each type of specific challenge instead of attempting to discover general measures that can be used to measure the effect of any and all challenges to animal welfare (Rushen, 2000; Dawkins, 2002).

This approach seems to be particularly justified when assessment of animal welfare is carried out for checking to what extent the official regulation or labelling standards are being obeyed since the assessment time is in this case very limited. *"Comparing housing systems is not so much a process of "measuring" animal welfare but rather a process of making a judgment, by combining as many sources of information as possible, of the risk that specific welfare problems will occur in each type of housing system"* (Rushen, 2003).

When trying to work out the most appropriate method(s) for animal welfare assessment we should be conscious that it is not purely or even mainly scientific issue but also ethical one (Tannenbaum, 1991; Fraser, 1999) and consider to whom the assessment is addressed. The scientific knowledge about animal welfare *"does not by itself provide relevant, rational and reliable answers to the questions concerning animal welfare typically raised by the informed public"* (Sandøe and Simonsen, 1992). Educating the consumers on the scientific details of animal welfare as well as food labelling will surely lead to raising their awareness of the existing problems in animal production but most probably will not result in increased readiness to pay more for food produced in animal friendly production systems. To succeed in this field we should rather try to reach peoples' value systems, their feelings and emotions, and this seems to have more with ethics than with science.

Finally, animal welfare should be considered together with other issues connected with animal food production that may influence social acceptability of specific production systems, like: promoting rural development, protecting the environment, natural resources and local biodiversity, preventing global warming, providing for high food quality and safety, working conditions, etc. The variability of these issues and fact that they are often mutually interrelated create additional problems when we try to address animal welfare as one of the most important societal concern related to animal production process.

2.2 Animal welfare and WTO policy

"Globalisation, free trade and WTO trade rules are seen as impacting negatively on valid societal concerns such as the environment, public health, consumer protection (especially relating to food products) and animal welfare, to name but a few of the major issues." (Lamy, 1999). It would be rather difficult to expect anything else when food safety is the only aspect of sustainable approach to animal production officially recognized by WTO as Sanitary and Phytosanitary Agreement reached at its Uruguay round.

Following the fetish of free trade and global liberalisation, the WTO set up requirements that any restrictions on import of animal products must be supported by sound scientific evidence or internationally recognized standards proving that the products can create a risk for human health (Lindsay *et al.*, 1999; Byrne, 2000). While this prescription is meant to discourage members to come up with non-tariff trade barriers, in effect they are also discouraged to implement stricter quality criteria for their own producers, as this always distorts domestic competitiveness.

WTO rules leave not too much room for import restrictions connected with process or production method:

"If countries were given a green light to place sanctions on every product made in a way determined to be objectionable, trade barriers could proliferate. For example, manufactured imports could be banned because pollution emitted at the factory of origin did not meet the domestic standards of the importing country. Or agricultural goods could be barred because the government of the importing country objected to the methods of farming and raising livestock of the producing country." (Lindsay *et al.* 1999).

However, there is well known evidence from recent years when import restrictions connected with character of production process did not find WTO understanding. In fact, the restrictive position of the WTO currently makes any control or inspections checking the production process in the countries from which import takes place, useless.

In the so-called shrimp-turtle case, the WTO ruled against a U.S. ban on shrimp from countries not adequately, according to US environmental regulation, protecting sea turtles from being caught and killed in shrimp nets. The WTO did not question the reliability of scientific justification of the ban but that it concerned production method, not the product itself (Lindsay *et al.* 1999).

Any import restrictions connected with the production process can only be potentially introduced if requirements of international agreements are explicitly not met (Trading into the Future, 2001).

Some examples of existing provisions in the WTO agreements dealing with non-trade concerns have been listed below (Trading into the Future, 2001):
* GATT Article 20: policies affecting trade in goods for protecting human, animal or plant life or health are exempt from normal GATT disciplines under certain conditions.
* Technical Barriers to Trade (i.e. product and industrial standards), and Sanitary and Phytosanitary Measures (animal and plant health and hygiene): explicit recognition of environmental objectives.
* Agriculture: environmental programmes exempt from cuts in subsidies
* Subsidies and Countervail: allows subsidies, up to 20% of firms' costs, for adapting to new environmental laws.
* Intellectual property: governments can refuse to issue patents that threaten human, animal or plant life or health, or risk serious damage to the environment (TRIPS Article 27).

- GATT's Article 14: policies affecting trade in services for protecting human, animal or plant life or health are exempt from normal GATT's disciplines under certain conditions.

When the issue is not covered by an environmental agreement, WTO rules apply. The WTO agreements are then interpreted in a way that (Trading into the Future, 2001):
- Trade restrictions cannot be imposed on a product purely because of the way it has been produced.
- One country cannot reach out beyond its own territory to impose its standards on another country.

At Doha Ministerial Conference from the November 2001 member governments commit themselves to comprehensive negotiations aimed at (The Doha Mandate Explained, 2001):
- Market access: substantial reductions in various trade restrictions confronting imports,
- Exports subsidies: reductions of, with a view to phasing out, all forms of these,
- Domestic support: substantial reductions for supports that distort trade.

At the conference however, for the first time the general note was accepted *"of the non-trade concerns reflected in the negotiating proposals submitted by Members"* and confirmation given *"that non-trade concerns will be taken into account in the negotiations as provided for in the Agreement on Agriculture"* (The Doha Ministerial Declaration, 2001). The term "non-trade concerns" was not defined.

One of the three categories of proposals on possible changes in the provisions of Annex 2 of the Agreement on Agriculture ("Green Box") relates to adding new types of programs or payments under the Green Box, addressing the question: *"Whether other proposed amendments or additions to the provisions of Annex 2 of the Agreement on Agriculture should be included, such as compensatory payments for higher animal welfare or other production standards or payments to address other non-trade concerns?"* (Harbinson, 2002).

As it can be seen this is the first time when "animal welfare" has been placed officially in WTO working documents.

The most recent modifications of scope of paragraph 12 of Article 20 encompass environmental programs/animal welfare payments (WTO Negotiations on Agriculture, 2003), with the following assumptions:

- Eligibility for such payments shall be determined as part of a clearly-defined government environmental, conservation or *animal welfare* program and be dependent on the fulfillment of specific conditions under the government program, including conditions related to production methods or inputs.
- The amount of payment shall be limited to the extra costs or loss of income involved in complying with the government program.

The last proposition is at the early stage of negotiation, What probably would be of particular importance is strong determination of the European Commission not to deviate from their current standpoint that non-trade concerns must find consideration in the results of the Doha round (Fischler, 2003a).

Mandatory food labelling, which may appear to be crucial in introducing animal friendly production systems, is also presently under discussion at the Doha round. *"While some participants consider that their specific proposals regarding improved consumer information and criteria and guidelines for the implementation of mandatory labelling for food and agricultural products should be dealt with in the framework of the negotiations on agriculture, other participants insist that the TBT (Technical Barriers to Trade) Committee is the appropriate forum to address labelling issues"* (Harbinson, 2002).

There is understanding that *"Members shall develop a common understanding, interpretation or guidance, on the criteria and guidelines for the implementation of mandatory labelling requirements for food and agricultural products"* (Harbinson, 2002). The key point is that labelling requirements and practices should not discriminate - either between trading partners or between domestically produced goods or services and imports (Trading into the Future, 2001).

With existing provisions for Doha round aiming to give WTO more attention to non-trade concerns, there are still significant doubts among the scientists as well as the politicians whether it would really bring expected results.

According to Hobbs *et al.*, (2002), the WTO, as presently constituted, is not able to deal with consumers' demands for animal welfare protection and more generally with ethical issues of livestock production, primarily since it gives governments some flexibility to respond to producer demands for protection but denies any flexibility to deal with demands for protection concerned societal concerns.

The issues connected with societal concerns introduced into future WTO negotiations do not seem to satisfy EU officials, remaining *"unbalanced by concentrating on purely trade aspects while neglecting non-trade concerns. For us, any agreement on agriculture should enable trade liberalisation to take place in such a way as to permit policies to promote rural development, protect the environment and local bio-diversity, and grant support to meet animal welfare standards, as well as to pursue other non-trade objectives"* (Fischler, 2003b).

2.3 Animal welfare and consumers

Animal welfare is an excludable good. That means farmer A may well keep his pigs in an animal-friendly way while farmer B does not. In result, at least one of the two criteria for public goods (Samuelson, 1954) - non-excludability - is not fulfilled and animal welfare has to count as a private good, so that a market for animal welfare may well develop. Indeed, at least in most Western societies one can find two, more or less separate, markets for animal food, one with food originated from animals that are kept at minimum-standards according to the national regulation, the other with food originated from animals kept with particular care of their well-being.

This situation is depicted in Figure 1 by applying the traditional price-quantity approach of microeconomics to animal welfare issues. Providing animal welfare can significantly increase the cost of production and therefore has a supply function S_{aw} that lies above the supply function of "standard meat" S_s. On the other hand, there are two demand curves: One part of consumers (yet usually the majority) does not care too much for animal welfare issues and demands standard meat (demand function D_s) which leads to the equilibrium E_s with a comparatively low price p_s. The other part demands only meat from animals with particularly high husbandry-standards (D_{aw}), which in effect leads to a higher price p_{aw}. The share of this part of consumers and their demand function probably has a lot to do with attitudes to animal welfare in various countries or even regions that may depend on religion, culture but also proper education.

Social acceptance for higher prices, however, will probably to largest extent depend on the country's wealth, income of its average citizen and the amount of money spent by him on food in relation to other goods. Consumer's demand for meat from animal-friendly production and the connected acceptance for increasing prices of food produced with special regard to providing for animal welfare seems to be the most important force affecting the market and thus the farmer's readiness to produce in that way.

The leading role of the market in driving development of animal friendly methods of production is well understood by European officials, responsible for introduction of suitable regulations. Commissioner Byrne's address on October 7th, 2000 General

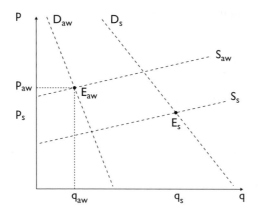

Figure 1. A bifurcated market for meat

Legend of Figure 1
ps - Price for standard animal product
paw - price for animal product with added animal welfare value
qs - quantity of standard animal product
qaw - quantity of animal product with added animal welfare value
Ss - Supply function for standard animal product
Saw - Supply function for animal product with added animal welfare value
Ds - Demand function for standard animal product
Daw - Demand function for animal product with added animal welfare value
Es - Market equilibrium for standard animal product
Eaw - Market equilibrium for animal product with added animal welfare value

Assembly to the Association of European Poultry Slaughterhouses mentioned animal welfare problems more in connection with its market consequences than in terms of official regulations.

"Animal welfare, as you may well be aware, is a subject of increasing concern amongst European consumers... Your customers are increasingly demanding higher standards. You would be very misguided to ignore these demands. The consumer, as you well know, has a habit of getting what he or she wants. If you cannot meet their demands, others will" (Byrne, 2000)

Less optimistic are results of 3 years comparative research recently conducted in five European countries: England, France, Germany, Ireland and Italy, which revealed that consumers were uninformed and did not understand the meaning of animal welfare. When asked, the consumer equated animal welfare with food safety and quality. Together these terms were perceived by consumers to reflect a "natural way" of food production (Ouédraogo, 2002).

Concern for animal welfare appeared to be very low, if not totally absent, when consumers were asked what food issues had been of concern to them. Contrary, when asked specifically on animal welfare, they usually expressed a high concern (Ouédraogo, 2002).

Such results suggest that animal welfare is at present ranked rather low, in relation to other societal concerns connected with food, but can also be interpreted that there is a lot of room for improvement, particularly by proper consumers' information and education, which makes them to realize the ethical aspects of animal production.

A prerequisite necessary to encourage consumer to pay for higher animal welfare standard is appropriate labelling of food with regard to animal production process.

2.4 The rationale of government interventions

Considering the results of consumer research given by Ouédraogo (2002) it is rather difficult to assume that consumers will be ready to pay for providing high animal welfare by themselves, unless they become more conscious of the well-being problems faced commonly in animal production.

An agricultural system is sustainable when its benefits outweigh costs, (Stricklin *et al.*, 2000), speaking in a holistic way that includes the value of non-market goods like animal-welfare or the environment. However, the attempt to maximize benefits and to minimize costs would mean probably something different for producers, consumers or decision makers.

The first would be looking for maximizing the direct profit in economical terms, while the second would be trying to optimize the subjective (which may include high animal welfare standards) value of the product, considering at the same time its price. For the third "maximizing benefits and minimizing costs", might be considered in long-term consequences.

This is apparently a level on which most initiatives connected with introduction of high animal welfare standards in animal production take place at present.

Probably the real challenge now is how to effectively bring people to long-term understanding of "maximizing benefits and minimizing costs".

There is growing concern from the farming sector that the costs of providing for high animal welfare standards are being ignored by authorities.

One way to provide for improving animal welfare in modernized or newly constructed farms is subsidies. That goes back to the concept of individualistic merit goods by Brennan and Lomasky (1983) and Erlei (1992). They suggest that a person has two competing preference orders that determine buying behaviour: market preferences and reflective preferences as their moral beliefs. By being in favour of subsidies for animal welfare, the consumer wants the state to change market conditions so that the market behaviour can be brought into accordance with reflective preferences.

Figure 2 provides a simplified example how such subsidies work. Starting with the situation as depicted in Figure 1, government intervention changes market equilibriums. By providing subsidies to animal-friendly husbandry systems, the respective supply curve shifts to the right towards S'_{aw}. This leads to a lower price p'_{aw} and, in turn, to a larger share of consumers who are willing to pay the shrunk price margin for meat from animal-friendly husbandry systems. An increased turnover of such meat q'_{aw} decreases then the remaining demand for standard meat and shifts the demand curve to the left D'_s. The resulting lower price for standard meat p'_s is certainly one of the unwanted side-effects of a subsidy for animal welfare measures, while the decreased quantity of standard meat (q'_s) is an intended effect.

There is strong determination in European Commissions for both Health and Consumer Protection and Agriculture, Rural Development and Fisheries to support all actions connected with improving quality of food understood as a result of specific process of production: *"High welfare standards raise the costs of production for producers, but this is exactly the reason why we introduced within the rural development concept the possibility to compensate these additional costs, and in the forthcoming agricultural negotiations within the WTO the European Union will defend the justification for such payments"* (Byrne and Fischler, 2001).

The second option of providing for high animal welfare standards is introduction of appropriate obligatory regulations. This goes back to Buchanan's (1991) concept of constitutional economics as re-interpreted by Tietzel and Müller (1998). While Buchanan interpreted the constitution (or, in a broader sense, the law) as a tool for citizens to prevent politicians and bureaucrats from doing unwanted things, Tietzel and Müller saw laws as a tool to instruct politicians what it is that society wants.

Hence, every regulation that forces farmers to keep certain animal-welfare standards may be seen as such linesetting. Introducing of such standards instead of splitting the market for animal welfare, leads to the situation as depicted in Figure 3. As every farmer has to provide added value in form of increased animal welfare restrictions, the market is "re-united" again. A lifted supply curve S'$_s$, caused by additional costs through animal welfare regulation, leads to equilibrium with less quantity and a higher price. Thus, animal welfare regulation is likely to reduce the quantity of meat consumed and to increase meat prices.

The most important issue of forcing regulations is that they should not result in significant increase of production cost. The attempt should be to minimize controls and restrictions to the areas where it cannot be avoided (Stricklin, 2002).

Another important actor on the scene may be non-governmental organizations. They may have an important effect on making way for the official regulations and subsidies. Social organizations and pressure groups, even if only representing particular minority interests, may put considerable pressure on the government.

2.5 Animal welfare and livestock industry

Official regulations, putting more and more constraints on production process, is probably not the best way to encourage the farmer to change the way he used to produce, particularly when he is not economically rewarded for his attempts. Most crucial issues here will be:
- Development the new - more animal friendly and comparable regarding the production cost - technologies,
- Pressure from the market (by higher price for better product),
- Pressure from other producers involved in food chain (equipment, genetic material and feed suppliers, food processors, etc.)

It is already possible to see the leading firms involved in animal market seriously consider various actions to make the production methods more animal friendly.

Cargill Inc, USA has been developing the programs for animal production management, which make it possible to account for various, animal welfare indicators (Kuczynski, 2001). DeLaval (Sweden) has adapted a new policy for animal welfare looking for the best animal husbandry practices for future use as well as identifying attitudes, policies, practices and legislation towards housing, management, feeding, breeding, transportation and slaughter (Ahnberg and Rodriguez-Martinez, 2001). MacDonald's, Austria uses eggs only from the laying hens farms, which meet the "TGI 35 L" Animal Needs Index (Ofner *et al.*, 2001).

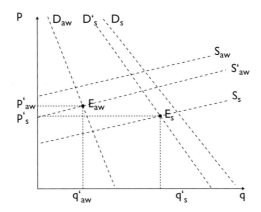

Figure 2. *Effect of a subsidy for animal-friendly practices.*

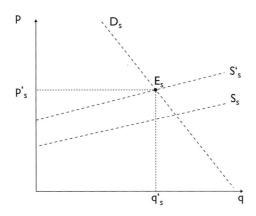

Figure 3. *Effect of animal welfare legislation.*

Legend of Figure 2-3
ps - Price for standard animal product
paw - price for animal product with added animal welfare value
qs - quantity of standard animal product
qaw - quantity of animal product with added animal welfare value
Ss - Supply function for standard animal product
Saw - Supply function for animal product with added animal welfare value
Ds - Demand function for standard animal product
Daw - Demand function for animal product with added animal welfare value
Es - Market equilibrium for standard animal product
Eaw - Market equilibrium for animal product with added animal welfare value

3. Conclusions

The action which can lead to efficient introduction and development of high animal welfare standards should be addressed mostly to consumers whose driving force is that they pay money for the products they want to buy and thus potentially affect everything connected with animal production which can be of subjective value to them.

We can affect consumers' preferences by:
- Introducing a reliable and efficient system of animal product labelling with regard to animal welfare in the production process. A voluntary system of product labelling gives the consumers information on the production method and freedom to decide whether to purchase the product or not. This form of action preceded by proper control and evaluation of production process in countries from which we want to import seems to be the only one which at least at present could be accepted by WTO (Lindsey *et al.* 1999).
- Further educating consumers on animal welfare and its ethical dimension. There is a lot of room for improvement in this area. Consumers should be more conscious of the real nature of animal production processes. Interventionist whole institution educational models based on training citizens in self-evaluative action research strategies can appear to be very helpful in improving understanding of the nature of the animal welfare issues in food production.
- Identifying and popularising effects of animal friendly housing methods on product quality; there are already some evidences. Environmental enrichment by peat and straw made pork tenderer than it was at barren environment (Mossberg, 2001). The bacon from pigs kept in straw courts was of a superior eating quality compared to other housing systems (Mossberg, 2001). Environmental enrichment was also found to positively affect the tenderness and flavour of calf meat (Andrighetto *et al.*, 1999).

The other party interested is a farmer who gets his reward from consumers as incentives paid for getting what they exactly want. Consumer's expectations are reaching farmers mainly through retailers and processors and can further influence his approach to sustainable animal housing systems and their management. One should understand, however, that to start animal friendly production, farmers have to be sure or at least believe that it makes a good alternative for traditional production and they will not end up bankrupt. We can affect farmer's attitude to keep high animal welfare standards by:
- Identifying and popularising beneficial effects of animal friendly housing methods on production economy. Many complaints on lower production

efficiency in these systems reflect difficulties within the transition period (from old to new technology). The transitional character of worsening the results has been confirmed experimentally by JSR Healthbred, one of the leading world pig genetic companies, when to obey introduction of new regulations on pig welfare they were forced to change their sow housing systems from individual stalls with no straw to group sow housing systems with straw, eventually developing a new system with economic performance equivalent to the old intensive system (Kuczynski, 2001). Investment cost for group housing system for sows, might be lower than for individual stall systems (Mossberg, 2001).

- Encouraging research on development of cheap and effective technologies, which can guarantee high animal welfare standards.
- Working out the system of animal products labelling with regard to animal welfare in order to affect the readiness of the consumer to pay higher prices for products ranked higher and subsequently encourage the producers to meet labelling requirements.
- Trying to get WTO approval for at least some minimum regulations for animal welfare standards at the farms, which could be obligatory in global scale, and would give importer right to stop an import from the farms, which do not follow these regulations.
- Attempting to get WTO approval for domestic support to compensate farmers for additional cost of investments aimed to improve animal welfare and introducing obligatory food labelling regarding meeting requirements for animal welfare.

To be able to effectively influence the attitude of consumers (by labelling), and producers (by introducing animal friendly housing systems, subsidies or just proper regulations), we must be capable to properly and relatively fast assess to what extent specific housing systems can provide for high animal welfare standards from one side and to accurately test the new production systems, which being not more expensive than the old ones, can produce at least equally efficiently, providing for high animal welfare standards.

References

Ahnberg, A. and Rodriguez-Martinez, H. (Eds.), 2001. Summary of group discussions and reports in Workshop on "Animal Welfare in a Global Perspective - with Focus on Diary Animals", Hamra Sweden, p. 57-59.

Andrighetto, I., Gottardo, F., Andreoli, D. and Cozzi, G., 1999. Effect of type of housing on veal calf growth performance, behaviour and meat quality. Livestock Production Science, 57, p. 137-145.

Bartussek, H., 1999. A review of the animal needs index (ANI) for the assessment of animals' well-being in the housing systems for Austrian proprietary products and legislation. Livestock Production Science 61, p. 179-192.

Boyazoglu, J., 1998. Livestock farming as a factor of environmental, social and economic stability with special reference to research. Position paper. Livestock Production Science 57, p. 1-14.

Brennan, G. and Lomasky L., 1983. Institutional Aspects of „Merit Goods" Analysis. Finanzarchiv 41, p. 183-206.

Broom, D.M., Mendl, M.T. and Zanella, A.J., 1995. A comparison of the welfare of sows in different housing conditions. Animal Science 61, p. 369-385.

Broom, D.M., 1991. Animal welfare: concepts and measurement. Journal of Animal Science 69, p. 4167-4175.

Buchanan, J., 1991. Constitutional Economics. New York.

Byrne, D., 2000. Safety, the most important ingredient of food: challenges for the poultry industry - General Assembly of the Association of European Poultry Slaughterhouses, Maastricht, 7 October.

Capdeville, J., 2001. Welfare of dairy cattle and relation with the housing conditions in a cubicle system. In Proceedings of International Symposium of the 2nd Technical Section of C.I.G.R. on Animal Welfare Considerations in Livestock Housing Systems, October 23-25, 2001, p. 87-97

Chaplin, S., 2000. Resting Behaviour of Dairy Cows: Applications to Farm Assurance and Welfare. Ph.D. Thesis, University of Glasgow, Glasgow, p. 144.

CoE, 1993. Protocol of Amendment to the European Convention for the Protection of Animals kept for Farming Purposes. Treaties and Reports. Council of Europe Press. Strasbourg.

Dailey, J.W. and McGlone J.J., 1997. Pregnant gilt behavior in outdoor and indoor pork production systems. Applied Animal Behaviour Science 52, p. 45-52.

Dawkins, M.S., 2002. How can we recognize and assess good welfare. In Broom, D.M., Coe, C.L., Dallman, M.F., Dantzer, R., Fraser, D., Gartner, K., Hellhammer, D.H. and Sachser, N. (Eds.), Coping with Challenge: Welfare in Animals Including Humans. Dahlem University Press, Berlin, p. 63-78.

Erlei, M., 1992. Meritorische Güter - Die theoretische Konzeption und ihre Anwendung auf Rauschgifte als demeritorische Güter. Münster: Lit

Farmer, C., 1997. Both energy content and bulk of food affect stereotypic behaviour, heart rate and feeding motivation of female pigs. Applied Animal Behaviour Science 54, p. 161-171.

FAWC, 2002. Farm Animal Welfare Council, 2002; Foot and Mouth Disease 2001 and Animal Welfare: Lessons for the Future.

Fischler, F. and Byrne, D. 2001. Food Quality in Europe - an open debate about the future of agriculture, food production and food safety in the Europe.

Fischler, F., 2003a. WTO and Agriculture: EU takes steps to move negotiations forward. Europe - Magazine of the European Union 04/03.

Fischler, F., 2003b. Address at Trade, reform and sustainability: meeting the challenges, Agra Europe Conference London, 31 March 2003.

Fraser, D., 1999, Animal ethics and animal welfare science: bridging the two cultures. Applied Animal Behaviour Science 65, p. 171-189.

Gonyou, H.W., 1986. Assessment of comfort and well-being in farm animals. J. Anim. Sci. 62, p. 1769-1775.

Harbinson, S., 2002, WTO Negotiations On Agriculture. Overview. Committee on Agriculture. Special Session. TN/AG/6, 18 December 2002.

Hobbs, A.L., Hobbs, J.E., Isaac, G.E. and Kerr, W.A., 2002. Ethics, domestic food policy and trade law: assessing the EU animal welfare proposal to the WTO. Food Policy 27, p. 437-454.

Hurnik, J.F., 1988. Welfare of farm animals. Applied Animal Behaviour Science 20, p. 105-117.

Kuczynski, T., 2001. Market requirements and social pressures influence on introduction and development of animal friendly livestock housing systems. In Proceedings of International Symposium of the 2nd Technical Section of C.I.G.R. on Animal Welfare Considerations in Livestock Housing Systems, October 23-25, p. 41-51.

Lamy, P., 1999. Intervention of. Trade Commissioner - Assembly of Consumer Associations in Europe, Conference, 18-19 November.

Lindsey, B, Griswold, D.T., Groombridge, M.A. and Lukas, A., 1999, Seattle and Beyond "A WTO Agenda for the New Millennium". Trade Policy Analysis No. 8 November 4.

Mossberg, I., 2001. Animal welfare regulations in Sweden and the EU - with special reference to pigs. In Proceedings of International Symposium of the 2nd Technical Section of C.I.G.R. on Animal Welfare Considerations In Livestock Housing Systems, October 23-25, p. 15-26.

Muehling, A.J., 2001. Farm animal welfare: the contemporary context (Regulations - reality or fiction) - roundtable discussion at International Symposium of the 2nd Technical Section of C.I.G.R. on Animal Welfare Considerations in Livestock Housing Systems, October 23-25.

Ofner, E., Amon, B., Amon, Th. and Boxberger, J. 2001. Assessment quality of the TGI 35 L Austrian Animal Needs Index. In Proceedings of International Symposium of the 2nd Technical Section of C.I.G.R. on Animal Welfare Considerations In Livestock Housing Systems, October 23-25, p. 79-86.

Ouédraogo, 2002. Consumers' Concern about Animal Welfare and the Impact on Food Choice: Social and Ethical Conflicts. In Workshop "Teaching Animal Bioethics in Agricultural and Veterinary Higher Education in Europe" INPL - Nancy.

Rushen, J., 2000. Some issues in the interpretation of behavioural responses to stress. In Mench, J., Moberg, J. (Eds.), The Biology of Stress. CAB International, Wallingford, United Kingdom, p. 23-42.

Rushen, J., 2003. Changing concepts of farm animal welfare: bridging the gap between applied and basic research. Applied Animal Behaviour Science 81, p. 199-214.

Rushen, J., Lawrence, A.B. and Terlouw, E.M.C., 1993. The motivational basis of stereotypies. In Lawrence, A.B., Rushen, J. (Eds.), Stereotypic Animal Behaviour: Fundamentals and Applications to Animal Welfare. CAB International, Wallingford, United Kingdom, p. 41-64.

Samuelson, P., 1954. The Pure Theory of Public Expenditure. Review of Economics and Statistics 36, p. 387-389.

Sandøe, P. and Simonsen, H.B., 1992. Assessing animal welfare. Where does science end and philosophy begin? Animal Welfare, 1, p. 257-267.

Scientific Veterinary Committee, 1997. The Welfare of Intensively Kept Pigs. Health and Consumer Protection Directorate-General, European Commission. Doc XXIV/B3/ScVC/005/1997.

Speroni, M., 2001. Animal welfare in an Italian context, in Workshop on "Animal Welfare in a Global Perspective - with Focus on Diary Animals", Hamra Sweden, p. 47-56.

Stricklin, W.R., Algers, B. and Vikinge, L., 2000. Bioethics and Sustainable Agriculture, EURSAFE Symposium, Copenhagen.

Tannenbaum, J., 1991. Ethics and animal welfare: the inextricable connection. Journal of the American Veterinary Medical Association, 198/8, p. 1360-1376.

Terlouw, E.M.C., Lawrence, A.B. and Illius, A.W., 1991. Influences of feeding level and physical restriction on development of stereotypies in sows. Animal Behaviour 42, p. 981-991.

The Doha Declaration Explained, 2001. The November 2001 declaration of the Fourth Ministerial Conference in Doha, An unofficial explanation of what the declaration mandates. Qatar.

The Doha Ministerial Declaration, 2001. The November 2001 declaration of the Fourth Ministerial Conference in Doha, Qatar Doha, 9-14 November 2001.

Thompson, P.B. and Nardone, A., 1999. Sustainable livestock production: methodological and ethical challenges. Livestock Production Science 61, p. 111-119.

Tietzel, M. and C. Müller, 1998, Noch mehr zur Meritorik. Zeitschrift für Wirtschafts- und Sozialwissenschaften 118, p. 87-127.

Tosi, M.V, Canali, E., Gregoretti, L., Ferrante, V., Rusconi, C., Verga, M. and Carenzi, C., 2001, A descriptive analysis of welfare indicators measured in Italian dairy farms: preliminary results. Acta Agriculturae Scandinavica, Sect.A, Animal Sci., Suppl. 30, p. 3-4.

Trading into the Future, 2001. The World Trade Organization 2nd edition revised, March 2001, WTO Publications, World Trade Organization, Centre William Rappard.

Verhoog, H., 2000. Defining positive welfare and animal integrity Diversity of livestock systems and definition of animal welfare. Proceedings of the Second NAHWOA Workshop, Cordoba, 8-11 January, M. Hovi and R. Garcia Trujillo Eds., p. 108-119.

Von Borell, E., 2001. An evaluation of "indexing" welfare in farm animals. In Ellendorff, F., Ladewig, J. (Eds.), Proceedings of Sustainable Animal Production Conference 5: Health and Welfare in Farm Animals. Institute of Animal Science and Animal Behaviour, Mariensee, Germany, 2000.

WTO Negotiations on Agriculture, 2003. Revised First Draft of Modalities for the Further Commitments; paper, circulated to member governments on 18 March 2003, ahead of the 25-31 March negotiations meetings.

The human-animal relationship in higher scientific education and its ethical implications

Caroline Vieuille and Arnaud Aubert
Département des Sciences du Comportement, Université de Tours, Parc de Grandmont, 37200 Tours, France

Abstract

Biology and animal production manuals have in common to tackle animals from an instrumental viewpoint and to exclude the ethical issues raised by the use of animals (for experimentation or production).

A study conducted in our university reveals that the students' representation of animals consists in assuming that the more an animal is perceived as "intelligent" (which is related to its symbolic representation), the more concern has to be granted to its suffering.

Most of scientific manuals refer to animals as valid models of humans' emotional processes (i.e. stress, anxiety, depression). However, little or nothing is said about animal suffering and ethical issues of animal experimentation. Moreover, some French animal production manuals deal with welfare or with animal's emotions, but as above, animal ethics remains considered under an instrumental standpoint.

The lack of training in ethics contributes to settle-down such pre-formed representations. Moreover, procedures involving animals are often transmitted as a set of "customs" and "habits" conditioning the integration of young professional. This takes to consider ethics as superfluous external constraints to which "experts" have to submit to meet the "naïve" demands of public.

In conclusion, we point out our own difficulties as scientists to construct another relationship with animals, and about the interest of including these difficulties into educational practices.

1. Introduction

The term of 'ethics' relates back to moral principles (Trésor de la langue française, 1980, VIII:245), a notion referring to the values that are usually practiced in one society. Moral and ethics grossly present the same direction. But ethics would have a prerogative for theoretical consideration: it would question about the sources, freedom, values, ends of actions, dignity, relations with others, and the concepts surrounding these difficult issues. Ethics describes and defines the principles, while moral prescribes the constraints on any human activity, that are essential for protection of life and respect of people (Quéré, 1991).

Applied to the human-animal relationship and to animal treatment, ethics asks the question of moral concerns about the treatment of animals. Ethics does not formulate predictive statements and cannot be tested in laboratory, unlike science. However, the ethical principles are bounded to the scientists, as to all human actors. The importance of a humane care and treatment of animals used for research and teaching is increasing and constitutes in western countries a responsibility we all share. Henceforth, each time research and teaching involves animal participation, it is essential that the welfare of those animals is properly considered and protected. It places upon the investigators a primary responsibility to observe the ethical, professional and legal principles involved. We propose to tackle animal's treatment through the place given to ethics into scientific textbooks, through animal's present representations and through the inherent conflicts about using laboratory animals or farm animals.

2. Place afforded to ethical questions

2.1 In manuals about animal production

Needless to remind that southern European countries tend to define animals "by default" (Amengaud, 2001). That is by the absence of soul, reason, language, self-determination (...), in the outcome of Christianity, a belief that places man at the top of the creation and gives him a power of domination on other beings. Such a view about the human-animal relationship probably favoured the development of productivity-oriented animal production sciences. It is particularly the case of France, where farm animals were likened to 'animal machines' (Sanson,1882), whose concern is still about controlling it's biological functions, in the service of economy. There is nothing surprising about the fact that a teaching discipline, which was created in order to prepare to breeding professions, did not include ethical questions. It clearly gives a predominant place to the financial dimension: for example, the French term of "zootechny" is defined as "the science of animal production. It gives as purpose the study of rules that enable to yield a profit from capital through animals" (Leroy,1882). Nowadays, its' goals are the transmission of knowledge about the functioning of animals and farms and training in farming practise, in decision making and in technical management (Guyomarc'h, 2002). In French zootechny manuals, Bernard Denis (1996) reported that the words "moral", "welfare", "comfort", remained extremely rare and when mentioned, they were used indirectly, in sentences about sanitary conditions of animals.

2.2 In biology manuals

Animal experimentation has been considerably exploited to clear up the general laws of biology whatever they concern physiology, metabolism, heredity, immunity and so on (Chapouthier, 1990; Bockaert, 1998).

But how ethical questions are raised by the experimental use of animals taken up by scientific handbooks? What does it say about human-animal relationship?

This question is generally not much dealt with (or even not at all) and seems to attest the uneasiness of scientists in front of their animals use's ethical implications. In most of general biology, physiology or ethology handbooks published in French language during the last ten years (therefore published later to the French ratification (i.e. 09/02/1987) of the European Convention for the protection of animals used in experiments), the questions of animal's experimental use and its ethical implications are generally missing. Terms referring to "ethics" are rarely present (or totally absent such as "suffering"), even in manuals detailing animal models of stress, depression and anxiety.

From a sample of thirteen of these biological manuals, we found that only two of them touch up ethical concerns about animal experimentation (cf. Table 1). Moreover, only in a single bibliographical case, legal and institutional references are provided. This general lack of reference to ethical concerns and guidelines is detrimental to the development of ethical questioning since it participates in giving the impression that ethics belongs to an ideological debate devoid of rationality and henceforth unrelated to scientific activities.

Table 1. Summary of the content of some selected issues from a sample of thirteen French-language manuals commonly used by French students and published between 1993 and 2002. The column "Ethics" indicates if the issue is treated, and in such a case, the length it was granted. Marks indicate that the issue is treated in the manual, whatever the length of the treatment.

Title	Ed.	Year	Topics			
			Ethics	Pain	Stress - Emotion	Suffering
Anatomie et Physiologie Humaines***	DeBoeck	1993		√	√	
Manuel de biologie pour les psychologues	Dunod	1993		√	√	
Biologie	DeBoeck	1995		√	√	
Neurosciences	Pradel	1997		√	√	
Biologie et physiologie animales	Dunod	1998				
Psychobiologie***	DeBoeck	1998		√	√	
Introduction biologique à la psychologie	Bréal	1999	2 pages*	√	√	
Manuel d'Anatomie et de Physiologie	Lamarre	1999		√		
Neurosciences***	DeBoeck	1999		√	√	
Biologie animale T2	Dunod	2000				
Neurosciences	Dunod	2000		√	√	
Le comportement animal: psychobiologie, éthologie et évolution ***	DeBoeck	2001			√	
Ethologie	DeBoeck	2002	<1 page**	(√)		

*Four bibliographical references are listed; legal and institutional references are mentioned.
** No bibliographical or legal references are given.
***These manuals are translated from english

3. Perception of animal

3.1 Public opinion

Present public image of animals contains a wide range of symbolic representations, pets being perceived as life companion, while other animals can be identified as belonging to other categories. Such mergers are related to the provided use of one animal, rather than to the species he belongs to (Burgat, 1999). Thus, animals-for-farming, animals-for-discovery (zoological gardens), animals-for-research...can be categorised. These representations do influence the subjective impact of a possible suffering inflicted to animals. For example, the use of farm animals for testing is better tolerated than the use of pets.

Some of these uses are the subject of a high public concern, and of questions about what they mean for animals. They're animal experimentation and animal intensive breeding, two fields students of scientific disciplines will precisely be confronted with. But the project to bring students to increase their awareness of animals-relationship's ethics is in conflict with the more general framework of scientific studies, in which science is supposed to be free from any value (Rollin, 1990).

3.2 Students

As in most of handbooks, ethical questions triggered by animal experimentation are generally absent from the basic training of French students. The initiative to treat and discuss these issues with students is left to teachers. However, a recent survey found that most young people were uneasy about animal experimentation, or thought that it should be banned. This feeling is reinforced by the fact that if students feel genuinely concerned about the use of animals, they find themselves caught up in pointless protests that seem to go on without any visible results.

In western countries, the use of animals for advancing scientific knowledge, for understanding disease, for developing new medicines, or for testing the safety of chemicals, is highly controversial in public and scientific opinions. While some of the traditional ways animals were used in experiments are on decline, new areas are opening up. Experiments involving genetic engineering are using an increasing number of animals and sometimes, human genes are introduced into an animal. These transgenic animals may then develop diseases that are very similar to those of humans.

An interesting point has been raised concerning the impact of the use of animals in medical training. It has been reported that Canadian neurologists who chose to spend a year of their training experimenting on animals, had so hardened themselves to animal suffering that they were less likely to assess suffering in their patients for quite a while after returning to clinical work (Howell, 1983). The generalised use of animals in students' training without appropriate ethical courses should be then questioned. Students should feel uncomfortable in the absence of a specific teaching on the question of the human-animal relationship in general or about ethical implications of the procedural use of animals. Their behaviour and feelings about their practices should depend on their own (naïve?) position about these issues, this later being inevitably influenced by cultural beliefs (Ronecker,

1999). From a work of Banks and Flora (1977), we recently conducted a study on students' attitude toward animals. Eighty-five undergraduate students were given a list of 30 different species and were first asked to evaluate the capacity of each animal species to feel suffering on a 10-point scale. Subsequently, they were asked to attribute a global "intelligence" score to these species on a 10-point scale as well. After the survey, a discussion was initiated about the responses that were given. From the results and the discussion, it appeared that: (1) suffering is generally regarded as an aptitude strongly correlated with the total level of intelligence attributed to animals (Spearman rank order correlation: $\rho = .791$; $p < 0.001$). (2) If the "intelligence" score's attribution roughly followed a phylogenetic ordering, some species deviate from this linear rule: lower scores were attributed to breeding species (e.g. hens, pigs, sheep), while pets and wild species (e.g. cats, dogs, wolves, foxes) endowed with positive attributed representations, were not only credited for more extended abilities, but also more concern for their welfare. Students then appear much concerned about procedures involving the 'higher' animals like chimpanzees. These animals have highly developed skills and display emotional behaviour looking similar to humans. In the same way, they are disturbed by the thought of cats, dogs and rabbits being used in experiments. Finally, the least valuated or attractive animals, as well as animals which are not kept as pets (i.e. affectively endowed) aroused less concern in students.

From this pioneering study, the investigation should be interestingly enlarged to other countries. Indeed, what are the common standpoints between students? Do the possible differences toward animals follow the tradition that opposes northern to southern cultures?

4. The conflicting place of ethics

4.1 Related to the use of farm animals

Since the early nineties and under the influence of public opinion, rearing conditions and some husbandry practices are controlled by animal protection rules. Today, philosophers and scientists agree with the idea that animals must receive an ethical treatment. The study of rearing conditions and of their relationships with animal welfare is one application of this project. Bernard Denis (2001) suggests the respect of welfare conditions, as defined by scientists, as a first step towards an ethical response. The next one could be the resorting to alternative products coming from selected farms for their rearing conditions.

It means a proposal for individual ethical responses of 'conviction', but a collective responsible ethical response. The consumers' interests, like those of producers, are supposed to work out for the animal a "level of welfare", that is "what is acceptable in terms of "cost" for it's adaptation" (Signoret, 1994) to husbandry conditions. Then, 'animal welfare' becomes a political object in a framework where ethics is about economical transactions and welfare conditions are the result of social negotiations.

It is of importance that teachers also point out to students the limits and contradictions of a social choice such as a contractual ethics. It may end in minor changes in factory farms' rearing conditions. Actually, the acknowledgement of scientific studies is based on the value accorded to proof: to succeed in demonstrating that physiological alterations exist without changes in productivity parameters; in proving that one practice leads to pain or chronic stress; in identifying 'objective' signs of distress or behavioural troubles and assessing whether they could restrict farm animal's adjustment to existing husbandry conditions. Furthermore, an assessment of welfare by proof shrugs off possible negative subjective states that we could not even neither notice nor graduate. While pathologists objectively list diseases, while physiologists identify somatic deteriorations or multifactorial pathologies and ethologists measure abnormal behaviours, no ethical thinking is engaged about increasing industrial pressures on farm animals, such as a higher number of animals, higher prolificity and productivity.

Although it is now merely admitted that the status of 'object' is unsatisfactory to describe animals, the choice of a responsible ethical response still gives them an instrumental value, the one of a negotiated object. In contradiction with what is admitted by scientists: they are quite ready to credit animals for being sensitive, they recognize their ability of suffering, the existence of behavioural needs and of mental states. For example, INRA has been developing a main study program called "AGRI-BEA", which purpose is among others to assess and compare positive and negative emotions accessible to farm animals.

Taking a collective decision of adopting ethical principles would be more consistent with the values attributed to animals and would at last enable us to change our connections with them. For example, our society could give its opinion on the respect of physical integrity (even whether we do not have some objective proofs about pain), or the respect of the species specific behaviours (without waiting for having proved that a lack of achievement of a particular behaviour is a psychological alteration)...

4.2 Related to the use of animals for experimentation

Why the ethical issue of animal using in teaching and experiments hasn't been approached head-on in the courses of biology? The use of animals in experimentation is related to the fact that biology recognizes and admits a relationship between man and animals. However, if biology accepts this idea, it does not put all the relationships on the same foot. Can one consider in the same way amoeba, snails, frogs and chimpanzees? If all species differ from each other, one current species is as well adapted as another. Then the hierarchy is rather a hierarchy of closeness to humans, which does not confer any priority for ethical concerns.

Since at first sight animal experimentation is little treated or even not at all, whereas it is an integral part of biology, how could one consider an opening to this question in a course of biology? Some fundamental questions could be tackled, such as: What do animal experimentation represents? Which impact does it have on medicine and on our knowledge? What are the limits of the explanatory models drawn from animal experimentation? What is their validity? What has animal experimentation brought to medicine and pharmaceutical industry? Are we able to justify experiments on animals when they have no obvious benefit to human health? Do alternative methods exist? Are they effective? Are we ready to give up with animal experimentation? Can one make experimentation "more bearable" for the animal? How? Etc.

The University, among others, has responsibilities in these matters (Bernard, 1986). The first one is due to the general community: to ensure that when research is carried out by its own staff and when research students involve animal subjects, there will be a proper regard for the ethical and moral principles involved. The second one is due to its members: to advise them about the ethical or legal issues that might arise in projects under expectation and to be ready to protect them from unwarranted criticisms or attempts at interference. The public institution provides ineffective assistance to these difficulties.

Local ethics' committees are rare and when they exist, their role is often ineffective because of local rules of functioning and resistances. Indeed, we tend to defend what we are accustomed to. One of the origins of such a resistance is a Bernardian legacy, the primacy of animal experimentation over theory and practice of biomedical research (Dupouey, 1997). Importing chemistry and physics' methods, Claude Bernard shaped a mechanistic physiological paradigm that has reigned for most of the twentieth century and is still resisting, overemphasizing laboratory investigation and conceptualising animal experimentation as the centrepiece of biomedical research. This paradigm is particularly strong in France, its predominance being reinforced by the lack of connection between scientific and philosophical knowledge. Its influence has occulted the understanding that evolution represents a continuum and that what once had been a clear physical and mental distinction between humans and animals has become much fuzzier.

Another parameter is the tendency to rationalize. Rationalization allows us to find out some reasons in order to explain our actions. For instance, dissections are rationalized as "hands-on" experience for students. Rationalization can be used with its worst aspects when economics is implied. Cattle breeders justify their activities by calling themselves "environmentalists," and hunters by calling themselves "conservationists." Animal experimenters, when criticized, tend to defend their work by resorting to reassuring images, such as afflicted children. This sort of thinking leads to the use of prejudice (e.g., rat vs. baby), rather than morally relevant criteria, as a basis for ethical decisions.

If we have responsibilities to both humans and animals, we have to face the moral dilemmas of animal experimentation. The problem is that the use of animal experiments to help caring for people and to alleviate human suffering, clash with our duty to care for animals. That's all very well to be able to say 'yes' to the question "Can we do this?" But we also need to pause and ask another question "Should we do this?" The reason is that doing something just because we have the ability, does not necessarily mean that doing it is right. As members of public opinion and sources of influence on it and as future users of animals, students must be formed to the ethical questioning about their practice: must the experiment cause suffering? Can the experiment give a clear answer? Is it finding new data?

5. Conclusion

Scientific manuals reflect what decision makers consider as significant to be learned. While most of animal production manuals do not deal with ethics, few deal with farm animals, whose welfare and emotions are worth to be studied, but above all, without changing the intensive logic of production, though it is implied in the troubles inventoried. Biology manuals mention animals as models of human knowledge or pathology, but shrug off questions about what we should do or not with them. This lack of proper treatment of relevant ethical issues does not fit with the increasing demand of students. This results in either abolishing ethical sensitivity and judgement in students, or inducing rejection of some disciplines.

Whatever the case, consequences are largely counter-productive regarding the scientific learning process.

Such an observation reflects the research's world unrest. Because searchers are caught up in a vice-like between firstly, the assumption that non-human animals have some moral worth, which generates potent questions about the morality of the practice, and secondly, biotechnologies reducing living beings to marketable things. While new transgenic beings or more productive / prolific farm animals emerge, prospects such as new markets or numerous publications tend to weaken (to sweep away?) all ethical concerns or objections.

References

Amengaud, F., 2001. L'anthropomorphisme: vraie question ou faux débat? Séminaire INRA, Ivry, Janvier 2001.

Banks, W.P. and Flora, F., 1977. Semantic and perceptual processes in symbolic comparisons. Journal of Experimental Psychology: Human Perception and Performance; 3, p. 278-290.

Bernard, J., 1986. De la biologie à l'éthique: nouveaux pouvoirs de la sciences et nouveaux devoirs de l'Homme. Buchet/Chastel, Paris.

Bockaert, J. 1998. Livre blanc sur l'expérimentation animale. CNRS éditions.

Burgat, F., 1997. Animal, mon prochain, Ed. Odile Jacob.

Burgat, F., 1999. L'homme et l'animal: un débat de société, INRA Eds., Paris.

Chapouthier, G., 1990. Au bon vouloir de l'homme, l'animal. Denoël, Paris.

Denis, B., 1996. Le statut éthique de l'animal: conceptions anciennes et nouvelles, Université de Liège.

Denis, B., 2001. Des pistes de réponses éthiques. In L'animal et l'éthique en élevage, Ethnozootechnie, hors-série nr 2, Société d'ethnozootechnie, p. 57-69.

Dupouey, P., 1997. Epistémologie de la biologie: la connaissance du vivant. Nathan Université, Paris.

Gonthier, T., 1999. L'homme et l'animal. La philosophie antique. Collection "Philosophies". PUF, Paris.

Howell, D.A., 1983. British Medical Journal, 11, p. 1894.

Leroy A.M. (1892-1978), cité par Guyomarc'h R., 2002. Enseignement pratique de la zootechnie et préparation aux métiers de l'élevage. In Ethnozootechnie n°68, Elevage et enseignement pratique de la zootechnie. Société d'ethnozootechnie, Clermont-Ferrand.

Quéré F., 1991. L'éthique et la vie. Points, Odile Jacob.

Rollin E., 1990. "Biomedical scientists have been trained under the ideology that suggests that science is, and ought to be "value-free". Animal welfare, animal rights and agriculture, J. of Anim. Sci., 68, p. 3456-3461.

Ronecker, J.P., 1999. Le symbolisme animal: mythes, croyances, légendes, archétypes, folklore, imaginaire... Dangles, Paris.

Sanson, A., 1882, cited by B. Denis, 2001: "the zootechnical problem consist of ... well running the construction of animals machines, of exactly fit them to physical and economical conditions in which their exploitation takes place... There is the essential nature of any industrial problem".

Signoret, J.P., 1994. Comportement et adaptation des animaux domestiques aux contraintes de l'élevage: bases techniques du bien-être animal, INRA Eds, Paris.

Teaching animal bioethics: pedagogic objectives

Michael J. Reiss
Institute of Education, University of London, 20 Bedford Way, London WC1H 0AL, United Kingdom

Abstract

This chapter begins by examining various reasons why students on agricultural and veterinary higher education courses should study ethics. Such teaching might heighten the ethical sensitivity of students, might increase their ethical knowledge, might improve their ethical judgements or might make them morally better people. I then go on to consider what sort of ethics such students should study, examining the possibilities of fundamental issues of ethics, ethics that arises especially in agriculture and veterinary science, appropriate professional codes of ethics, or the ethics of particular local relevance. Finally I look some possible ways in which ethics might be taught to agricultural and veterinary higher education students, examining advantages and disadvantages of each approach.

1. Introduction

There was a time when in most countries students in higher education taking applied science courses would not formally have been taught anything about ethics. But that is changing. It is not just agricultural and veterinary higher education students who learn about ethics; medical students now regularly learn about medical ethics, engineers learn about environmental ethics, and so on. This broadening of educational aims and content reflects a growing societal expectation that graduates in applied science disciplines should know something of ethics in their field rather than merely being expected to 'pick it up' during the course of their professional lives.

2. Why should students on agricultural and veterinary higher education courses study ethics?

But what precisely might be the aims of teaching ethics to students on agricultural and veterinary higher education courses? Based on Davis (1999), at least four can be suggested (Reiss, 1999a).

First, such teaching might heighten the ethical *sensitivity* of participants. For example, students who have never thought about whether certain breeds of dogs should have their tails removed ('docked') or whether there is a right age at which calves should be removed from their mothers might be encouraged to think about such issues. Such thinking can result in students becoming more aware and thus more sensitive. It is not unusual, as a result, to find students saying 'I hadn't thought of that before'.

Secondly, such teaching might increase the ethical *knowledge* of students. The arguments in favour of this aim are much the same as the arguments in favour of teaching any knowledge - in part that such knowledge is intrinsically worth possessing, in part that possession of such knowledge has useful consequences. For example, appropriate teaching about the issue of rights might help students to distinguish between legal and moral rights, to understand something of the connections between rights and duties and to be able to identify fallacies in arguments for or against the notion of animal rights.

There is in agricultural and veterinary ethics a tremendous amount of relevant knowledge for students to learn. This is true of any interdisciplinary subject but seems particularly true when commercial animals are involved. Consider, for example, the issue of whether we should be concerned, on welfare grounds, about fish farms. Relevant considerations, in addition to moral philosophy, include the possibilities of pain detection by fish and fish consciousness (Kaiser, 1997; Rose, 2002). To understand these well requires knowledge of animal behaviour, of neurophysiology, of psychology and of epistemology. If students are to think through the consequences of permitting or prohibiting fish farms then they also need knowledge of such matters as disease transmission (as farmed fish generally have higher rates of parasites and infectious diseases), pollution (from the various chemicals used to treat the parasites and infectious diseases) and employment in rural economies (as there are a lot of people employed in some countries on such fish farms).

Thirdly, such teaching might improve the ethical *judgement* of students. As Davis, writing about students at university, puts it:

"The course might, that is, try to increase the likelihood that students who apply what they know about ethics to a decision they recognize as ethical will get the right answer. All university courses teach judgment of one sort of another. Most find that discussing how to apply general principles helps students to apply those principles better; many also find that giving students practice in applying them helps too. Cases are an opportunity to exercise judgement. The student who has had to decide how to resolve an ethics case is better equipped to decide a case of that kind than one who has never thought about the subject." (Davis, 1999, p. 164-165).

Fourthly, and perhaps most ambitiously, such teaching of ethics might make students *better people* in the sense of making them more virtuous or otherwise more likely to implement normatively right choices. For example, a unit on ethics for student veterinarians might lead the students to reflect more on the possibilities open to them, leading them to be less pressured by the views of others and so resulting in improved animal welfare. There is, within the field of moral education, a substantial literature both on ways of teaching people to 'be good' and on evaluating how efficacious such attempts are (e.g. Wilson, 1990; Carr, 1991; Noddings, 1992). Here it suffices to note that while care needs to be taken to distinguish between moral education and moral indoctrination, there is considerable evidence that moral education programmes can achieve intended and appropriate results (e.g. Straughan, 1988; Bebeau, Rest and Narvaez, 1999).

3. What sort of ethics should students on agricultural and veterinary higher education courses study?

What sorts of ethics should be taught to students on agricultural and veterinary higher education courses? I suggest the following - but note that this is not a suggestion as to how the teaching should be conducted (for which see the last section below):
- Ethics that connects to fundamental issues of ethics - i.e. issues not specific to agriculture and veterinary studies. For example, the fundamental approaches to ethics (consequentialism, Kantian ethics, virtue ethics, the contribution of religious traditions, feminist approaches, etc.).
- Ethics that arises especially in agriculture and veterinary science (use of animals, relations to the natural environment, duties to future generations).
- Appropriate professional codes of ethics.

- The ethics of relevant local issues (whether fish farming, pig husbandry, GM crops, selective breeding of farm animals or whatever).

3.1 Ethics that connects to fundamental issues of ethics

Ethics is the branch of philosophy concerned with how we should decide what is morally wrong and what is morally right. We all have to make moral decisions daily on matters great or (more often) small about what is the right thing to do: Should I continue to talk to someone for their benefit or make my excuse and leave to do something else? Should I give money to a particular charity appeal? Should I stick absolutely to the speed limit or drive 10% above it if I'm sure it's safe to do so? We may give much thought, little thought or practically no thought at all to such questions. Ethics is a specific discipline which tries to probe the reasoning behind our moral life, particularly by critically examining and analysing the thinking which is or could be used to justify our moral choices and actions in particular situations (Reiss, 2003).

3.1.1 The way ethics is done

One can be most confident about the validity and worth of an ethical conclusion if three criteria are met (Reiss, 1999b). First, if the arguments that lead to the particular conclusion are convincingly supported by reason. Secondly, if the arguments are conducted within a well established ethical framework. Thirdly, if a reasonable degree of consensus exists about the validity of the conclusions, arising from a process of genuine debate.

It might be supposed that reason alone is sufficient for one to be confident about an ethical conclusion. However, there are problems in relying on reason alone when thinking ethically. In particular, there still does not exist a single universally accepted framework within which ethical questions can be decided by reason (O'Neill, 1996). Indeed, it is unlikely that such a single universally accepted framework will exist in the foreseeable future, if ever. This is not to say that reason is unnecessary but to acknowledge that reason alone is insufficient. For instance, reason cannot decide between an ethical system which looks only at the consequences of actions and one which considers whether certain actions are right or wrong in themselves, whatever their consequences. Then feminists and others have cautioned against too great an emphasis upon reason. Much of ethics still boils done to views about right and wrong informed more about what seems 'reasonable' than what follows from logical reasoning.

3.1.2 Is it enough to look at consequences?

The simplest approach to deciding whether an action would be right or wrong is to look at what its consequences would be. No-one supposes that we can ignore the consequences of an action before deciding whether or not it is right. Even when complete agreement exists about a moral question, consequences will have been considered. The deeper question is not whether we need to take consequences into account when making ethical decisions but whether that is all that we need to do. Are there certain actions that are morally required - such as telling the truth - whatever their consequences? Are there other actions - such as betraying confidences - that are wrong irrespective of their consequences?

Consequentialists, including utilitarians, believe that consequences alone are sufficient to let one decide the rightness or otherwise of a course of action. Utilitarianism begins with the assumption that most actions lead to pleasure (typically understood, at least for humans, as happiness) and/or displeasure. In a situation in which there are alternative courses of action, the desirable (i.e. right) action is the one which leads to the greatest net increase in pleasure (i.e. excess of pleasure over displeasure, where displeasure means the opposite of pleasure, that is, hurt, hurt or suffering).

Utilitarianism now exists in various forms. For example, preference utilitarians argue for a subjective understanding of pleasure in terms of an individual's own conception of his/her well-being. After all, if I like to spend my Saturday evenings picking my nose, who are you to say that there are other more pleasurable ways in which I could spend my time?

There are two great strengths of utilitarianism. First, it provides a single ethical framework in which, in principle, any moral question may be answered. It doesn't matter whether we are talking about the legislation of cannabis, the age of consent, the patenting of DNA, bull fighting or the genetic engineering of farm animals; a utilitarian perspective exists. Secondly, utilitarianism takes pleasure and happiness seriously. The general public may sometimes suspect that ethics is all about telling people what not to do. Utilitarians proclaim the positive message that people should simply do what maximises the total amount of pleasure in the world.

However, there are difficulties with utilitarianism as the sole arbiter in ethical decision making. For one thing, there is the question as to how pleasure can be measured. Is pleasure to be equated with well-being, happiness or the fulfilment of choice? And, anyway, what are its units? How can we compare different types of pleasure, for example sexual and aesthetic? Then, is it always the case that two units of pleasure should outweigh one unit of displeasure? Suppose two people each need a single kidney. Should one person (with two kidneys) be killed so that two may live (each with one kidney)?

Utilitarians (e.g. Singer, 1993) claim to provide answers to all such objections. For example, rule-based utilitarianism accepts that the best course of action is often served by following certain rules - such as 'Tell the truth', for example. Then, a deeper analysis of the kidney example suggests that if society really did allow one person to be killed so that two others could live, many of us might spend so much of our time going around afraid of being hit over the head at any moment that the sum total of human happiness would be less than if we outlawed such practices.

3.1.3 Intrinsic ethical principles

The major alternative to utilitarianism is when certain actions are considered right and others wrong in themselves, i.e. intrinsically, regardless of the consequences. There are a number of possible intrinsic ethical principles. Currently, and in the West, perhaps the most important such principles are thought to be those of autonomy and justice.

People act autonomously if they are able to make their own informed decisions and then put them into practice. At a common sense level, the principle of autonomy is why people need to have access to relevant information, for example before consenting to a medical procedure such as a surgical operation.

There has been a strong move towards the notion of increased autonomy in many countries in recent decades. Until recently, for example, most doctors saw their role as simply providing the best medical care for their patients. If a doctor thought, for instance, that a patient would find it upsetting to be told that they had cancer, they generally did not tell them. Nowadays, any doctor withholding such information might find themselves sued. Society increasingly feels that the important decisions should be made not by the doctors but by the patients (or their close relatives in the case of children or adults unable to make their own decisions).

Of course, such autonomy comes at a cost. It takes a doctor time to explain what the various alternative courses of action are - time that could be spent treating other patients. In addition, some doctors feel de-skilled, while some patients would simply rather their doctor made the best decision on their behalf. Overall, though, the movement towards greater patient autonomy seems unlikely to go away in the near future. However, autonomy is not a universal good. Someone can autonomously choose to be totally selfish. If society grants people the right to be autonomous, society may also expect people to act responsibly, taking account of the effects of their autonomous decisions on others.

Autonomy is concerned with an individual's rights. Justice is construed more broadly, Essentially, justice is about fair treatment and the fair distribution of resources and opportunities. Considerable disagreement exists about what precisely counts as fair treatment and a fair distribution of resources. For example, some people accept that an unequal distribution of certain resources (e.g. educational opportunities) may be fair provided certain other criteria are satisfied (e.g. the educational opportunities are purchased with money earned or inherited). At the other extreme, it can be argued that we should all be completely altruistic. However, as Nietzsche pointed out, it is surely impossible to argue that people should (let alone believe that they will) treat absolute strangers as they treat their children or spouses. Perhaps it is rational for us all to be egoists, at least to some extent.

3.1.4 Feminist ethics and virtue ethics

Feminist ethics is one of the many products of feminism, chief among the tenets of which is the belief that women have been and still are being denied equality with men, both intentionally and unintentionally. This inequality operates both on an individual level (e.g. discrimination in favour of a male candidate over an equally good female candidate for a senior job) and at a societal level (e.g. poor access to state child care makes it extremely difficult for women in certain careers to return to full-time work after having a child).

Feminist ethics, in the words of Rosemary Tong "is an attempt to revise, reformulate, or rethink those aspects of traditional Western ethics that depreciate or devalue women's moral experience" (Tong, 1998, p. 261). Feminists philosophers fault traditional Western ethics for showing little concern for women's as opposed to men's interests and rights. There has, for example, been a lot more written about when wars are just than about who should care for the elderly. Then there was the discovery that some of the best known and most widely used scales of moral development tend to favour men rather than women because the scoring system favours the rational use of the mind with the application of impartial, universal rules over more holistic judgements aimed at preserving significant relationships between people (Gilligan, 1982). In addition, there is the feminist argument that moral philosophers have tended to privilege such 'masculine' traits as autonomy and independence over 'feminine' ones such as caring, striving for community, valuing emotions and accepting the body.

Virtue ethics holds that what is of central moral significance is the motives and characters of individuals rather than what they actually 'do'. The emphasis is therefore more on those personal traits that are fairly stable over time and which define the moral nature of a person. Think, for example, of the virtues we might desire in someone, whether a friend, an employer, a politician or a farmer. We might hope that they (and we) would be honest, caring, thoughtful, loyal, humane, truthful, courageous, reliable and so on.

Of course, as Aristotle pointed out almost two and a half thousand years ago, any virtue can be taken to excess. Loyalty to one's friends is generally a good thing but it is better to report your friend to the police if you have reasonable cause to think that he or she is about to murder someone.

In practice, working out precisely what the virtuous thing to do in a situation is can be difficult. Consider euthanasia. Is it more caring absolutely to forbid euthanasia or to permit it in certain circumstances? And what exactly are the virtues we would wish to see exercised in animal biotechnology? Despite such difficulties - difficulties which attend every ethical set of principles - there seems little doubt that the world would be a better place if we were all even a bit more virtuous.

3.2 Ethics that arises especially in agriculture and veterinary science

Traditionally, ethics has concentrated mainly upon actions that take place between people at one point in time. In recent decades, however, moral philosophy has widened its scope in two important ways. First, intergenerational issues are recognised as being of importance (e.g. Cooper and Palmer, 1995). Secondly, interspecific issues are now increasingly taken into account (e.g. Rachels, 1991). These issues go to the heart of 'Who is my neighbour?'.

Interspecific issues are of obvious importance in agriculture and veterinary science. Put at its starkest, is it sufficient only to consider humans or do other species need also to be taken into account? Consider, for example, the use of new practices (such as the use of growth promoters or embryo transfer) to increase the productivity of farm animals. An increasing number of people feel that the effects of such new practices on the farm animals need to be considered as at least part of the ethical equation before reaching a conclusion. This is not, of course, necessarily to accept that the interests of non-humans are equal to those of humans. While some people do argue that this is the case, others accept that while non-humans have interests these are generally less morally significant than those of humans.

Accepting that interspecific issues need to be considered leads one to ask 'How?'. Need we only consider animal suffering? For example, would it be right to produce, whether by conventional breeding or modern biotechnology, a pig unable to detect pain and unresponsive to other pigs? Such a pig would not be able to suffer and its use might well lead to significant productivity gains: it might, for example, be possible to keep it at very high stocking densities. Someone arguing that such a course of action would be wrong would not be able to argue thus on the grounds of animal suffering. Other criteria would have to be invoked. It might be argued that such a course of action would be disrespectful to pigs or that it would involve treating them only as means to human ends and not, even to a limited extent, as ends in themselves.

3.3 Appropriate professional codes of ethics

The insufficiency of reason is a strong argument for conducting debates within well established ethical frameworks, when this is possible. Professional codes of ethics can be useful in this regard. Traditionally, the ethical frameworks most widely accepted in most cultures arose within systems of religious belief. Consider, for example, the questions 'Is it wrong to lie? If so, why?'. There was a time when the great majority of people in many countries would have accepted the answer 'Yes. Because scripture forbids it'. Nowadays, though, not everyone accepts scripture(s) as a source of authority. Another problem, of particular relevance when considering the ethics of biotechnology, is that while the various scriptures of the world's religions have a great deal to say about such issues as theft, killing people and sexual behaviour, they say rather less that can directly be applied to the debates that surround many of today's ethical issues, particularly those involving modern biotechnology. A further issue is that we live in an increasingly plural society. Within any one western country there is no longer a single shared set of moral values. Instead there is a degree of moral fragmentation: one cluster of people has this set of ethical views; another has that.

Nevertheless, there is still great value in taking seriously the various traditions - religious and otherwise - that have given rise to ethical conclusions. People do not live their lives in isolation: they grow up within particular moral traditions. Even if we end up departing somewhat from the values we received from our families and those around us as we grew up, none of us derives our moral beliefs from first principles, *ex nihilo*, as it were. In the particular case of moral questions concerning animals, a tradition of ethical reasoning is already beginning to accumulate. For example, most member states of the European Union and many other industrialised countries have official committees or other bodies looking into the ethical issues that surround at least some aspects of our use of animals. The tradition of ethical reasoning in this field is nothing like as long established as, for example, the traditions surrounding such age-old questions as war, capital punishment and freedom of speech. Nevertheless, there are the beginnings of such traditions and similar questions are being debated in many countries across the globe. This may lead to accepted codes of ethics for veterinarians and those in agriculture.

3.4 The ethics of relevant local issues

As a teacher it is good to keep in mind both universal questions (such as those considered above about the difference between utilitarianism and feminist ethics) and local ones. At the very least it is more motivating for students to debate genuine issues of immediate relevance. Furthermore, local issues often serve to 'ground' ethical theory in actualities.

Given the difficulties in relying solely on any one particular ethical tradition, we are forced to consider the approach of consensus (Moreno, 1995) when considering ethical questions, particularly when trying to discern ways forward on the ground, at local level. It is true that consensus does not solve everything. After all, what does one do when consensus cannot be arrived at? Nor can one be certain that consensus always arrives at the right answer - a consensus once existed that women should not have the vote.

Nonetheless, there are good reasons both in principle and in practice for searching for consensus. Such consensus should be based on reason and genuine debate and take into account long established practices of ethical reasoning. At the same time, it should be open to criticism, refutation and the possibility of change. Finally, consensus should not be equated with majority voting. Consideration needs to be given to the interests of minorities, particularly if they are especially affected by the outcomes, and to those - such as young children, the mentally infirm and non-humans - unable to participate in the decision-making process. At the same time it needs to be born in mind that while a consensus may eventually emerge, there is an interim period when what is more important is simply to engage in valid debate in which the participants respect one another and seek for truth through dialogue (cf. Habermas, 1983).

4. How might ethics be taught to students on agricultural and veterinary higher education courses?

Here I can only briefly look at a range of pedagogical strategies, considering some of the advantages and disadvantages of each. Some of the other chapters in this book consider certain of these strategies in more depth.

4.1 Teaching of fundamental ethical approaches

Teaching ethics by going through such fundamental ethical approaches as consequentialism, deontology, virtue ethics, feminist ethics and so forth provides a rigorous and valid grounding. But the approach can appear abstract and may give too optimistic a view of the ease of making ethical decisions in reality. If it is used - and this approach can be used as part of a course of applied ethics - it is particularly important, in my view, for the approach to be even-handed. There are some consequentialists who, both in their writing and in their talks, seem more evangelical about their position than are many religious people about their faith. Similarly, in some quarters the principalist approach has reached the level of a mantra. Equally, virtues are culturally laden. Consider, for example 'honour'; what is the right way for me to behave if you insult me? The answer varies greatly from culture to culture.

4.2 Studying case studies

Case studies can be highly motivating for students. They are seen to be 'relevant' (the highest accolade for some students) and allow students to contribute their own views and to discuss the views of others, whether of their peers, their lecturer(s) or academics in the field. They have considerable flexibility, taking as little as 20 minutes or occupying months of study.

However, some care is needed. Too much background information or too complicated a dilemma may be overwhelming; too superficial an introduction to a case study and students may not engage; too many case studies and students may fail to see connections between cases or get bored. Teachers may need to help students 'debrief' at the end of a case study, so that more general lessons can be drawn out and learnt - even if the lesson is only that sometimes general lessons can't be learnt!

4.3 Role plays

Role plays, though rarely used in teaching about ethics, can be memorable and allow for a lived experience rather than students just engaging in talk. They can also increase empathy so that students see more deeply how others may perceive a situation. Indeed, it can be worth encouraging (but not forcing) students to take on a role different from that which they would occupy themselves. My own view is that it's not a particularly good idea to get students to role play being non-humans (e.g. farm animals); a better idea is to get some of them to act the role of someone who argues that animals have rights, shouldn't be kept in captivity, eaten, or whatever.

However, role plays do make particular demands on both lecturers and students, If you have never run one before, you might want either to get a colleague who has to help you or to start students off not on a full-blooded role play but on a sort of open-ended discussion where all you do is ensure that two or three different views are being articulated. Role play can polarise attitudes and it is always a good idea to 'de-role' at the end of one, so that participants come out of role and get the chance to say anything they want to now that they are again 'themselves'.

4.4 Imitation of lecturers

In a higher education course, students inevitably get to know something of the views of their lecturers and the degree to which their lecturers' actions are consonant with these views. The extent to which students gain such knowledge depends on the structure of the course. If all your teaching is delivered in 50 minute lectures to groups of over one hundred students they will learn far less about you than if you take them on outings or a residential field trip.

Imitation of lecturers is an apprenticeship model in which lecturers are seen as role models. It is a form of embodied learning and is likely to happen to a certain extent in any event. However, awareness of it can make especial demands on lecturers. Similarly, students learn from the whole ethos and structure of an institution as well as from individual lecturers. What sorts of relationships between staff and students are encouraged and which forbidden? What provision is made for students' diets (e.g. vegetarian, vegan, kosher, halal)? Are students allowed to keep pets on campus? And so on.

4.5 Students act authentically by changing their own actions during the course

Finally, do students get the chance to learn authentically by changing their own actions during their course? This is a type of enacted learning which involves getting into the habit of being good through the manifestation of agency. Of course, it requires courses to provide opportunities for students to make such authentic decisions, and so makes demands on course administrators, lecturers and technicians - as well as on the students! What opportunities do students get to choose the subject matter of their project work and other assignments? Do they have any control over the use of animals in their establishment, for example, which species are kept, how they are kept and whether and under what circumstances they are killed?

5. Conclusion

There are many ways of teaching animal bioethics and many ways of conceptualising such teaching. Lecturers can be seen as experts (fine if we lecturers behave as though we have some answers, not so good if we give the impression we have either all of them or none) or travel guides (which means that we have some familiarity with the journey but are open to going a different way and seeing things anew and from different perspectives). Equally, there are many ways of learning animal bioethics. The best learning builds on what a person already knows, has a degree of authenticity, allows the learner a degree of choice, requires the active processing of knowledge and uses a variety of modes, e.g. linguistic, visual and olfactory - try a visit to an abattoir for holistic learning about animal slaughter.

References

Bebeau, M.J., Rest, J.R. and Narvaez, D., 1999. Beyond the promise: a perspective on research in moral education, Educational Researcher, 28(4), p. 18-26.

Carr, D., 1991. Educating the virtues: an essay on the philosophical psychology of moral development and education, London: Routledge.

Cooper, D.E. and Palmer, J.A., 1995. (Eds) Just environments: intergenerational, international and interspecies Issues, London Routledge.

Davis, M., 1999. Ethics and the university, London Routledge.

Gilligan, C., 1982. In a different voice, Cambridge, MA Harvard University Press.

Habermas, J., 1983. Moralbewusstsein und Kommunikatives Handeln, Frankfurt am Main, Suhrkamp Verlag.

Moreno, J.D., 1995. Deciding together: bioethics and moral consensus, Oxford, Oxford University Press.

Kaiser, M., 1997. Fish-farming and the precautionary principle: context and values in environmental science for policy, Foundations of Science, 2, p. 307-341.

Noddings, N., 1992. The challenge to schools: an alternative approach to education, New York, Teachers College Press.

O'Neill, O., 1996. Towards justice and virtue: a constructive account of practical reasoning, Cambridge, Cambridge University Press.

Rachels, J., 1991. Created from animals: the moral implications of Darwinism, Oxford, Oxford University Press.

Reiss, M.J., 1999a. Teaching ethics in science, Studies in Science Education, 34, p. 115-140.

Reiss, M.J., 1999b. Bioethics, Journal of Commercial Biotechnology, 5, p. 287-293.

Reiss, M.J., 2003. How we reach ethical conclusions, in Key issues in bioethics: a guide for teachers, Levinson, R. and Reiss, M. J. (Eds), London, RoutledgeFalmer, p. 14-23.

Rose, J.D., 2002. The neurobehavioural nature of fishes and the question of awareness and pain, Reviews in Fisheries Science, 10, p. 1-38.

Singer, P., 1993. Practical ethics, 2nd edn, Cambridge, Cambridge University Press.

Straughan, R., 1988. Can we teach children to be good? Basic issues in moral, personal and social education, 2nd edn, Milton Keynes, Open University Press.

Tong, R., 1998. Feminist ethics, in Encyclopedia of applied ethics, volume 2, Chadwick, R. (Ed.), San Diego, Academic Press, p. 261-268.

Wilson, J., 1990. A new introduction to moral education, London, Cassell.

Teaching professional ethics: more than moral cognition alone

Vincent Pompe
Department of Animal Management, Van Hall Institute, Leeuwarden, The Netherlands

Abstract

Professionals nowadays cannot act solely on economic and technical principles that guide them to an efficient and effective product or service. They must show private and collective responsibility towards the well-being of living creatures, whether these be humans, animals or nature. The purpose of teaching ethics is to cultivate the moral dimension of the profession in the student. To reach this objective, education should not exclusively concentrate on the transfer of knowledge, but on the broader aspects of moral attitude, such as moral self-structuring, role-taking and ethical problem-solving. With this, tutoring students learn to deal with conflicts between self-interest and public interest, between personal impulse and community requirements. The question is whether the educational task belongs solely to the domain of moral philosophy or whether it needs a substantial amount of moral psychology too.

In this paper, I will demonstrate the importance of moral psychology in education, by explaining the complex fabric of moral behaviour. Insight into the relevant principles, guidelines and codes of ethics must be supplemented with the modification of attitudes, in particular communicative virtues. Moral education for professionals must be communication-oriented with a value-oriented background. I will advocate a moderate shift from ethics as philosophy to ethics as communication. Furthermore, I will dwell on some consequences for an (animal) ethics curriculum if one expands moral education beyond cognition.

1. From justification to implementation

In today's society one can detect an increasing interest in ethics within the professions. This interest is caused by an overall change in the field of professionalism from an industrial to a post-industrial era. Society does not accept that businesses constrain themselves to a perspective that is solely aimed at economic survival by increasing the efficiency of production and services. The public is sensitive to environmental pollution, animal welfare, destruction of natural beauty and resources, violation of human rights and exploitation of Third World (child) labour, especially when these are presented as collateral damage of a business enterprise. The exclusive maximisation of profit through production and sales therefore shifts to a multi-operational task in which moral and social concern is essential for realising business objectives. In a post-industrial era, the instrumental praxis to improve one's economic or strategic position is extended with an ethical dimension to maintain or advance society within a healthy environment and with natural diversity. This era asks for a professional attitude that possesses moral awareness of one's moral abilities, in order to take responsibility towards the well-being of humans, animals and nature. One can clearly see post-industrial qualities in the great variety of codes of ethics that many professional associations have implemented.

The justification of ethics in professional education is as such not a major issue. Professional and academic universities prove their identity to some degree by incorporating societal requirements into the curriculum. The post-industrial paradigm must therefore have some influence on university courses. The aim of education can therefore not solely be to teach instrumental knowledge and skills but also to provide students with insight into moral values and norms and teach them an ability to apply ethics in their line of work.

The problem of ethics in education is not its justification but its implementation. Many curricula have incorporated the subject of ethics, but that alone does not guarantee that society can profit from a substantial moral gain. The result of moral education is highly influenced by the didactical outline and approach. Teaching ethics by transferring knowledge and skills, just as any other subject, such as, biology, sociology and economics, will only result in an explanation of the principles and rules of conduct. It teaches students 'what they ought to do' or 'how they have to act' in a particular situation. This form of teaching ethics is far from adequate, because professional ethics is not a synonym for 'teaching etiquettes' or training social desired behaviours. Knowledge and understanding of ethical principles is necessary but not sufficient for a moral praxis. Moral education is not only about knowledge but also about acknowledgement, not simply about consciousness but also about conscientiousness. In order to teach students an ability to perform a 'full' ethical praxis, one must concentrate on moral attitudes, i.e. personality, will-power and determination. Teaching ethics therefore involves both philosophical features and social-psychological aspects.

In this chapter the psychological process of a moral conduct and the sources that influence the process will be further explained. From the psychological analysis, conclusions can be drawn to improve the implementation of ethics in professional and academic education, in particular animal ethics.

2. The fabric of moral behaviour

The study of moral behaviour was part of philosophy up to the 20th century, although it was not a major component. Only a small number of philosophers paid attention to this subject: Socrates and Plato by linking behaviour with knowledge, Aristotle by reducing it to a person's character and Rousseau and Kant by emphasising self-determination as the centre for morality. In the 20th century sociology became more interested in the moral development of children. This issue was split between two perspectives: the moral socialisation of Emile Durkheim and the moral cognition paradigm of Jean Piaget. In Durkheim's view moral development was a matter of learning, accepting and internalising the norms of one's culture and behaving according to those norms. Piaget turned Durkheim's view upside down by rejecting the claim that society determines the development of behaviour and replacing it with a theory that centres on the individual, having cognition as a thinking process that represents the meaningful reality he constructs (cf. Carr, 1991).

The most discussed theory of moral behaviour is that of Lawrence Kohlberg. In the line of Piaget, Kolhberg underlines the position of the individual in the determination of what is morally right and wrong. The individual interprets a situation, derives meaning from social events and makes a judgement as to the best behaviour. Kohlberg distinguished six stages of moral problem-solving that are logically sequenced, starting from non-moral-rational to full-moral-rational behaviour. The development of morality begins at a pre-conventional level that consists of the stages of obedience and instrumental egoism. At a higher conventional level, moral problems are solved through interpersonal agreement or by orientating towards law and duty. The highest level is the post-conventional one in which moral values and principles are not defined by the authority of a person or group but by rational autonomy. At this level Kohlberg distinguishes the stage of consensus-built social contract and the stage at which problems are solved through orientation towards universal ethical principles with justice as its primate. This theory has been criticised, especially regarding the logical sequence of the stages. Nevertheless, Kolhberg's theory is still useful as a heuristic model for moral education, for instance by regarding the stages as conceptions of how to organise cooperation (cf. Munsey, 1980).

Kohlberg's introduction to moral psychology is further elaborated by James Rest (1984, 1994). Rest claims that there is more to moral development than the maturity of moral judgement. Morality can be divided into cognition, affect and behaviour. This division gives us insight into what psychologically is supposed to happen when moral behaviour actually takes place, or to formulate it in a negative sense: 'what fails when morality is not performed'. Rest developed a theory that is known as the 'Four component model of moral behaviour', because morality can be analysed concerning its sensitivity, judgement, motivation and character.

A person that shows moral behaviour must have some *moral sensitivity* towards a being or party that can be damaged or favoured. He must be aware of how his action affects the interests of others, e.g. discrimination against foreigners who are seeking asylum, the abuse of animals to increase profit or the damage of nature by wasting chemicals. At the same time that person must also be aware of possible lines of action he can take and what the effect of each action would be on others.

Once a person is aware of the social needs and his possibilities to act, his *moral judgement* must decide which line of action is morally the best within the situation. An example is the dilemma between animal welfare and environmental values. We like to see cows grazing in the meadows, because it supports animal welfare but the manure causes environmental problems such as acidity. A person must assess the consequences of his action, or reason from which principle or intention he is going to act. By doing so, moral judgement prevents a person from acting out of revenge or purely out of selfishness.

Sensitivity and judgement alone are not sufficient to generate moral conduct. The component of *moral motivation* must be activated too. Situations are often complex in the sense that different interests and values may play roles that sometimes can be diametrically opposed to each other. A person can be aware that something must be done to stop the suffering, but action does not occur because another interest seems to be more important. Farmers exploiting animals in industrial conditions are aware of the harm they inflict on the animals, but they feel constrained by the free economic market that dictates efficiency of production in order to secure some profit. It is the moral motivation that generates the choice out of the set of moral and non-moral values a person wants to pursue. It puts and reasons priorities.

Prioritising a value will not automatically generate behaviour. One must have a *moral character*, i.e. ego strength, perseverance, toughness and sometimes courage. Moral decision-making can sometimes lead to psychological, social or economic pressure. Can a farmer take a loss of prosperity when he changes from intensive cattle husbandry to extensive ecological farming? The moral character is necessary to lead moral intention ultimately to moral behaviour.

The psychological components are the ingredients of distinct processes that together generate moral behaviour. However, according to Bredemeier and Shields (1994) moral behaviour is more complex than that. One must not only look at the process of moral behaviour but also at the different sources that influence conduct, such as the social context, a person's competencies and a person's ego-processing.

Moral behaviour is mostly expressed within a *social context*. The goals and values a social or corporate environment pursues influences the individual's moral sensitivity. The moral atmosphere in a company that strives for environmental health by reducing chemical waste is different from the one in a firm that sets libertarian targets to maximise profit at almost all costs. Working in one of those atmospheres will either strongly intensify or weaken one's environmental or libertarian commitment. The social structure also gives signals about what conduct is favoured given the situation. This may sometimes be difficult, because society itself is a reactor in which polar values are trying to come to a balance. For example, animal values such as welfare must often compete with human values such as health. Even if we have a benevolent attitude towards animals, strained relations with other types of values remain, such as conformity, security, power, hedonism and self-direction. The social environment might limit or enhance one's behaviour by its power structure, such as gender, age and merit, that determines someone's status or role in the moral issue.

A *person's competencies* are another major source that influences the four processes of moral behaviour. The knowledge and skills a person has acquired affects his own conduct in several ways. Competencies determine the perspective one takes towards a moral case and influence the information one gathers to form an opinion. Professional knowledge and skills shape one's beliefs and attitudes and with that the process of reason. It forms a person's judgement of what is right in the given context, of which moral obligation is most binding and of which line of action is most important. Competencies also determine one's capacity to act autonomously and to solve social problems, such as delaying gratification and preserving a challenging task. The processes of moral behaviour are highly influenced by the sort of professional or academic education one has had; the way one is scholarly nurtured. An education that is solely focused on solving technical problems will bleach one's moral abilities in contrast with a broader education that incorporates social dimensions into the curriculum.

Ego-processing does not only refer to one's nurture but also to one's nature. It is one's psychological profile; the mental qualities one possesses and expresses in one's personality. Ego-processing refers to different mental abilities. One is the intraceptive function that constitutes the capacity of one's engagement with one's own thoughts, feelings and intuition in response to what is happening both internally and in one's environment. Autistic and empathic persons have in this sense contrary intraceptions. Cognitive and regulative functions are other aspects of ego-processing. The first function refers to one's intellectual and logical capacities, while the latter denotes the faculty of coping with emotional conflicts and moral dilemmas. A person might have a gifted brain to understand the very complex dimensions of a moral problem, but at the same time he might be tormented by the impact the problem has on his conscience. His cognitive function is working well, but his regulative function is causing him problems and prevents him from acting morally or from acting at all. Determination is the fourth ability of ego-processing. It refers to mental capacities to concentrate and to set aside disturbing or attractive thoughts and feelings that might negatively influence the task at hand.

The elaboration of Bredemeier and Shields shows the full complexity of moral behaviour. Combining the four components of behaviour with the three sources of influences, Bredemeier and Shields generate a 'Twelve component model of moral action' (see Table 1).

According to Rest and to Bredemeier and Shields is it difficult to give a scientific validation of a logical or empirical sequence in the component of moral action and its sources of influence. It is not strictly the case that sensitivity must precede judgement; judgement must have taken place in order to set the motivation etc. It is more likely that the components interact with each other when moral behaviour is generated. However, at least for heuristic reasons one can construct some order. The components of sensitivity and judgement, whether or not in a dialectic process, can be regarded as a foundation for motivation and character. Regarding the sources, the classification into social context, competencies and ego-processing can be projected on a scale from external to internal influences with the professional skills as an intermediate.

With the twelve components of moral behaviour, one can perceive on the one hand insight into the deficiency that causes moral failure and on the other an outline for the development of educational programs.

Table 1. Twelve component model of moral action.

Sources	Processes			
	Sensitivity Awareness of needs, affects and lines of actions	**Judgment** Justification of the best line of action	**Motivation** Prioritising moral and non-moral values	**Character** Perseverance and withstanding pressure
Contextual Social environment	**Social goals** The community values	**Prevailed principles** The moral atmosphere	**Domain cues** Hints about most suitable conduct	**Power structure** The establishment
Personal competencies Professional knowledge and skills	**Role-taking** Perspective and information gathering	**Reasoning** Attitude and moral analysis	**Self-structure** Setting moral priorities	**Autonomy and social skills** Controlling the social problem
Ego processing Psychological profile	**Intraceptive/evolvement** Autistic and empathic response	**Cognitive processes** Intellectual and logical capacities	**Regulating processes** Coping with emotional conflicts	**Determination** Concentration and willpower

Note: the degree of shading represents the difficulty of teaching; see next section.

3. The difficulty of teaching ethics

The 'Twelve component model of moral action' shows the psychological complexity of moral behaviour. Moral behaviour entails more than knowledge and skills. Ultimately it is a matter of personality, will-power and character. This insight has some consequences for the development and implementation of a moral curriculum. In the moral psychological perspective, teaching ethics might not be as simple as it looks. It is plausible that, within an educational setting, external objectives such as, knowledge and understanding of social affairs are easier to reach than internal objectives such as a modification of someone's psychological profile. With this classification one can form a graduation in the difficulty of teaching ethics, as represented with the degree of shading in Table 1. Besides, the difficulty can be connected with the level of ethics such as basic, extended and full level.

The basic level of moral education in the profession is to concentrate on the social environment and therefore to teach the current social values as well as the moral principles that prevail in the community to pursue these values. The relevance of professional ethics is to make students aware of the different needs and interests that society and its diverse sections pursue as well as the different lines of action it takes to 'solve' the moral conflicts. Students gain some insight into the moral issues, such as animal use and welfare, environment and biodiversity, human health, prosperity and justice. Basic understanding of ethical theories on the issues will give some insight into moral judgement and justification, but one can broaden the topic by giving students some insight into the social priority of non-moral values over moral values, such as economic prosperity over animal welfare. Moreover, knowledge of the current establishment by explaining the power structure, formed by politics, technology and economy, will enlighten the sociological dimension of ethics. The basic level of education aims to teach student the 'know-that' and 'know-how' of moral behaviour. It transfers values as a source to approach and control moral problems. Since value-transfer does not require much input from students, lectures and seminars are suitable forms of education for this topic. The shortcoming of this level of education might be that students are not encouraged to critically reflect on their own moral judgement, especially when their judgement differs from traditional convictions (cf. van der Ven, 1998, ch.4).

The basic level can be extended by concentrating on the development of professional moral skills, such as role-taking, reasoning and self-structuring. The characteristic of this educational level is not as much to transfer values but to clarify values. Value clarification means that education is not about the value as such, but about the process of valuing. The process of clarification can be enriched by developing analytical, hermeneutical and communicative skills (cf. van der Ven, 1998, 34ff, ch6). Communication skills are particularly important, because morality intrinsically has a social quality, and openness is a social requirement. Case studies, peer groups, debates and essay writing are good ways of developing analytical and communicative skills. Students can be asked to defend 'publicly' a line of action from a certain perspective that is not their own. They can analyse the strengths and weaknesses of the opposite theories and determine their own structure of moral and non-moral values and present their prioritisation. To assist this level of education, several measuring instruments have been developed to assess the student's moral performances (Bebau, 1994). The Ethical Sensitivity Test (EST) assesses ability to recognize the ethical issues hidden within a problem the professional encounters. The Defining Issue Test (DIT) measures the extent of moral reasoning and judgement. To assess motivation one can use the Professional Role Orientation Inventory (PROI) that gives some insight into the student's commitment to professional values over personal values. The student's moral character on the competencies level can be assessed with the Professional Problem Solving Index, which measures the student's ability to solve complex moral cases autonomously. Extending the teaching of social values and principles with developing competencies to clarify values, to communicate about values and to solve problems, mature students in their own responsibility towards the well-being of others, whether humans, animals or nature.

For a full moral education one must concentrate, additionally, on the student's personality. This implies that education aims to modify a student's ego-processing in order to make it more capable of acting morally. At this level one must focus on the student's emotional, cognitive and regulative processes as well as on his determination. The main problem with modifying a personality is that it takes a long term to realise. It seems obvious that this full form of moral education is not possible within professional and academic courses. Ethics in the professional curricula is an auxiliary subject. It must supplement the core subjects of the profession, not replace them. However, education can initiate and stimulate ego-processing modifications. One can emphasise the importance of making self-judgments, to predict one's own behaviour; self-evaluations, to judge one's conduct afterwards; and self-responses, by assessing one's ability to perform a moral action (cf. van der Ven, 1998, 55ff). One can easily attach these tasks to practical assignments and trainee periods.

The 'Twelve component model of moral action' shows some consequences of the development and implementation of a moral curriculum. By focusing on moral psychology, teaching ethics becomes a complicated task.

4. Taking teaching animal ethics seriously

Despite all the difficulties any moral education faces, teaching animal ethics has a further complicating factor: the diversity of thoughts and theories. A pluralistic collection of pro- and anti-animal perspectives, of anthropocentric and zoocentric theories, complicates the issue. In this diversity the questions are: to what extent must one describe the social goals regarding human well-being and animal welfare equally; must one be anthropocentric or zoocentric with the philosophical explanation of ethical theories; which principles should be taught, what virtue should be cultivated; can one advocate Peter Singer's sentientism or Peter Carruthers' radicalism? Surely, contemporary animal ethics cannot be monolithic. Pluralism is the state of affair within many moral issues and in particularly the animal one.

The diversity of a moral issue might block the direction of education, because it scatters the subject matter and broadens the options to choose. This has some consequences for teaching. Classical ethics, such as giving an outline of the animal issue, has to be supplemented with moral competencies on how to cope with moral diversity. From that view, value-oriented education has to switch from ethical knowledge to communicative attitude, as Rice and Burbules (1992) advocate. Through teaching communicative virtues, students can cultivate abilities such as patience, tolerance for alternative points of view, respect for differences, willingness to listen thoughtfully and attentively, openness to giving and receiving criticism, and sincerity in their self-expression. Integrating communicative virtues into a regular ethics course can be the spin-off for students to clarify moral priorities and to give insight into their self-structure. It also stimulates the regulating process regarding how to handle emotional conflicts and moral dilemmas. This kind of value-communication cultivates the professional's task to justify his use of animals publicly with the awareness of anthropocentric and zoocentric stereotyping and fallacies.

Noting the complexity of moral behaviour, the sources that influence behaviour and the diversity of the animal issue, an educational shift from value-oriented to a more communication-oriented approach is required. To take this shift seriously, the curriculum, as the educational guide, must meet some desirable conditions (cf. van der Ven, 1985, V).

The social source of influence should not only be taught, but must be actually there. The culture of the faculty is a context in which students mould their moral abilities. This context ought to be an example of professional morality, in order to stimulate students. The faculty's culture can be disseminated by an ethical code or mission statement. Lectures can link their subject with its (potential) moral dimensions.

Professional competencies can, in general, be better taught across the curriculum than as a separate subject. With the limited number of student hours, the gains of a one-time subject, such as, moral role-taking, reasoning, prioritising and problem solving, are significantly less than spreading it out over the curriculum. In short, one time forty hours is much less than four times ten hours of education. Hence, a sensible scheme is to present students with an introductory course of ethics, in which they learn the fundamentals, and then give them several assignments throughout the curriculum in which they apply and improve their expertise and develop their moral professionalism. Other specialist lectures and teachers can in such a scheme easily assist the ethicist, because moral education is more about skills than knowledge, more about form than subject matter. The ethicist is mainly responsible for the outline of moral education in the introductory course. Other teachers can be in charge of implementing ethics in their subject, with the ethicist as an advisor. At this point the moral culture of a faculty that demonstrates multidisciplinary cooperation is significant for the teaching of moral competencies to young professionals.

Although ego-processing is difficult to teach, it can be an implied part of the education. Curricula can influence personality. Moral education might entail social-psychological understanding of moral behaviour as a step to self-insight. This insight, which is an explicit part of the moral assignments, can be further developed by self-judgement, self-evaluation and self-response. Paying attention to self analysis will give the student more insight into the strength and weakness of his morality.

Society demands for professionalism, wich entails an awareness of the contemporary normative issues and skills to judge, justify and act. Curricula that meet these requirement must invest in two fields of moral education. One consists of explaining relevant principles, guidelines and codes of ethics. The other, modifying attitudes by focusing on communication-oriented education with a value-oriented background. Then, ethics as philosophy and ethics as communication will be merged.

References

Bebau, M., 1994. Influencing the Moral dimensions of Dental Practice. in Rest and Narvaez, 1994.

Bredemeier, B. and Shields, D., 1994. Applied ethics and Moral Reasoning in Sport, in Rest and Narvaez, 1994.

Carr, D., 1991. Educating the Virtues. London, Routledge.

Munsey, B., 1990, Moral Development, Moral Education, and Kohlberg, Birmingham, Alabama.

Rest, J.R., 1984. The major components of morality. In W. Kurtines and J. Gewirtz (eds), Morality moral behaviour and moral development. New York, Wiley.

Rest J.R and Narvaez D., 1994, Moral Development in the Profession: Psychology and Applied Ethics, LEA, New Jersey.

Rice S. and Burbules, N.C., 1992. "Communicative Virtues and Educational Relations" Philosophy of Education, p. 34-44.

van der Ven, J.A., 1985. Vorming in waarden en normen. Kok Agora, Kampen.

van der Ven, J.A., 1998. Formation of the Moral Self. Eerdmans, William B. Publishing Company.

Teaching animal welfare to veterinary students

Xavier Manteca[1], Donald M Broom[2], Ute Knierim[3], Jaume Fatjó[1], Linda Keeling[4] and Antonio Velarde[5]
[1]School of Veterinary Science, Universitat Autònoma de Barcelona, Bellaterra, Spain, [2]Dept. of Clinical Veterinary Medicine, University of Cambridge, Cambridge, United Kingdom, [3]Department of Farm Animal Behaviour and Husbandry, Faculty of Organic Agricultural Sciences, University of Kassel, Germany, [4]Department of Animal Environment and Health, Section of Animal Welfare, Swedish University of Agricultural Sciences, Skara, Sweden, [5]Institute for Food and Agricultural Research and Technology (IRTA), Meat Technology Centre, Monells, Spain

Abstract

As public concern over the welfare of animals has increased in many European countries and elsewhere, Veterinary Schools have to include animal welfare as a subject in their curricula. Although some schools have already done so, many others have not. Animal welfare is a relatively new subject and as a consequence there is a lack of knowledge among the academic community as to what contents should be included in a general course on animal welfare aimed at veterinary students. This chapter is intended to help those that have the responsibility to start such a course. We hope that it may be useful as well to those who simply want to compare the courses they teach with other courses taught elsewhere. The specific objectives of this chapter are: (1) to provide a syllabus of the main contents to be included in a general course on animal welfare for veterinary students, (2) to explain why those contents should be included and how they could be taught, and (3) to provide a set of useful references and sources of information for those responsible for teaching animal welfare to veterinary students.

This chapter is largely based on the authors' experience. We hope, nevertheless, that in the long-term it will contribute to build a common set of guidelines for teaching animal welfare to veterinary students in Europe, so that in the future all European Veterinary Schools share at least the main elements of their curricula as it relates to animal welfare.

The chapter is divided into six sections. The first section covers the basic principles of animal welfare, whereas the following five sections deal with animal welfare in each of the main fields in which veterinarians are professionally involved: farm animal housing and husbandry, transport and slaughter of farm animals, companion animals, laboratory animals and wildlife.

I. Basic principles

We suggest that there are three main issues to be dealt with in this section: the concept of welfare, the relationship between welfare and concepts such as needs, feelings, stress, health and pain, and the assessment of welfare.

I.I Concept of welfare

The term welfare requires strict definition if it is to be used effectively and consistently. A clearly defined concept of welfare is needed for use in precise scientific measurements, in legal documents and in public statements or discussion. It is therefore important that students are exposed to a discussion of the concept of welfare. There are many different definitions of animal welfare and the chapter by Duncan and Fraser (1997) provides an useful summary of the different approaches that have been used to define and study welfare. One of the most widely used and useful definitions is that by Broom (1986), according to which the welfare of an individual is its state as regards its attempts to cope with its environment. This definition refers to a characteristic of the individual at the time. The origin of the concept is how well the individual is faring or travelling through life and the definition refers to its state at a particular time (for further discussion, see Broom, 1991a,b, 1996, 2001c, Broom and Johnson, 1993). The concept refers to the state of the individual on a scale from very good to very poor and includes its feelings and its health.

I.2 Welfare and needs, feelings, stress, health and pain

The concept of animal welfare is related to the concepts of needs, feelings, stress, health and pain. The environment is appropriate if it allows the animal to satisfy its needs. Animals have a range of functional systems controlling body temperature, nutritional state, social interactions etc. (Broom, 1981). Together, these functional systems allow the individual to control its interactions with its environment and hence to keep each aspect of its state within a tolerable range. The allocation of time and resources to different physiological or behavioural activities, either within a functional system or between systems, is controlled by motivational mechanisms. When an animal is actually or potentially homeostatically maladjusted, or when it must carry out an action because of some environmental situation, we say that it has a need. A need can therefore be defined as a requirement, which is part of the basic biology of an animal, to obtain a particular resource or respond to a particular environmental or bodily stimulus (Broom 2001a). As pointed out by Broom (1997), these include needs for particular resources and needs to carry out actions whose function is to obtain an objective (Toates and Jensen, 1991; Broom, 1996). Needs can be identified by studies of motivation and by assessing the welfare of individuals whose needs are not satisfied (Hughes and Duncan, 1988a,b; Dawkins, 1990; Broom and Johnson, 1993). The question of the importance of different needs is a good subject for discussion with students.

The feelings of an animal are an extremely important part of its welfare (Broom 1991b). Information can be obtained about feelings using preference studies and other information giving indirect information about feelings can be obtained from studies of physiological and behavioural responses of animals. Feelings are aspects of an individual's biology which must have evolved to help in survival (Broom 1998), just as aspects of anatomy, physiology and behaviour have evolved. They are used in order to maximise its fitness, often by helping it to cope with its environment. It is also possible, as with any other aspect of the biology of an individual, that some feelings do not confer any advantage on the animal but are epiphenomena of neural activity (Broom and Johnson, 1993).

The word stress should be used for that part of poor welfare which involves failure to cope. A definition of stress as just a stimulation or an event which elicits adrenal cortex activity is of no scientific or practical value (Mason 1971, Broom 2001c). A precise criterion for what is adverse for an animal is difficult to find but one indicator is whether there is, or is likely to be, an effect on biological fitness. Stress can be defined as an environmental effect on an individual which over-taxes its control systems and reduces its fitness or seems likely to do so (Broom and Johnson, 1993, see also Broom 1983, 2001c, Fraser and Broom 1997). Depending on the year on which the course is taught, students may or may not be familiar with the physiological changes associated with stress, and this should be taken into account when allocating time to deal with this in the course.

The word "health", like "welfare", can be qualified by "good" or "poor" and varies over a range. However, health refers to the state of body systems, including those in the brain, which combat pathogens, tissue damage or physiological disorder. Health may be defined as *"an animal's state as regards its attempts to cope with pathology"* (Broom 2000). In this statement, animals include humans. The meaning of pathology is discussed at length by Broom and Kirkden (in press). Welfare is a broader term than health, covering all aspects of coping with the environment and taking account of a wider range of feelings and other coping mechanisms than those which affect health, especially at the positive end of the scale. Health is a part of welfare and hence disease always has some adverse effect on welfare (Broom and Corke 2002).

The pain system and responses to pain are part of the repertoire used by animals, including man, to help them to cope with adversity during life. Pain is clearly an important part of welfare. Pain is defined here, following the International Association for the Study of Pain, as an aversive sensation and a feeling associated with actual or potential tissue damage (Broom 2001b).

1.3 Welfare assessment

If animal welfare is to be compared in different situations or evaluated in a specific situation, it must be assessed in an objective way. The assessment of welfare should be quite separate from any ethical judgement but, once an assessment is completed, it should provide information which can be used to take decisions about the ethics of a situation. It is important that the students realise that welfare assessment requires the combination of several measures, some of which appear in Table 1. Most indicators will help to pinpoint the state of the animal wherever it is on the scale from very good to very poor. Some measures are most relevant to short-term problems, such as those associated with human handling or a brief period of adverse

Table 1. Measures of welfare (after Broom, 2000).

- Physiological indicators of pleasure
- Behavioural indicators of pleasure
- Extent to which strongly preferred behaviours can be shown
- Variety of normal behaviours shown or suppressed
- Extent to which normal physiological processes and anatomical development are possible.
- Extent of behavioural aversion shown
- Physiological attempts to cope
- Immunosuppression
- Disease prevalence
- Behavioural attempts to cope
- Behaviour pathology
- Brain changes, e.g. those indicating self narcotisation
- Body damage prevalence
- Reduced ability to grow or breed
- Reduced life expectancy

physical conditions, whereas others are more appropriate to long-term problems (for a detailed discussion of measures of welfare, see Broom 1988a, Broom and Johnson 1993, and Fraser and Broom 2002).

Signs of poor welfare can be conveniently divided into the following categories:

- Physiological measurements. For instance increased heart-rate, adrenal activity, adrenal activity following ACTH challenge, or reduced immunological response following a challenge, can all indicate that welfare is poorer than in individuals which do not show such changes. The impaired immune system function and some of the physiological changes can indicate what has been termed a pre-pathological state (Moberg, 1985).
- Behavioural measures are also of particular value in welfare assessment. The fact that an animal avoids an object or event strongly gives information about its feelings and hence about its welfare. The stronger the avoidance the worse the welfare whilst the object is present or the event is occurring. An individual which is completely unable to adopt a preferred lying posture despite repeated attempts will be assessed as having poorer welfare than one which can adopt the preferred posture. Other abnormal behaviour such as stereotypies, self mutilation, tail-biting in pigs, feather-pecking in hens or excessively aggressive behaviour indicates that the perpetrator's welfare is poor. In some of these physiological and behavioural measures it is clear that the individual is trying to cope with adversity and the extent of the attempts to cope can be measured. In other cases, however, some responses are solely pathological and the individual is failing to cope. In either case the measure indicates poor welfare.
- Disease, injury, movement difficulties and growth abnormality all indicate poor welfare. If two housing systems are compared in a carefully controlled experiment and the incidence of any of the above is significantly increased in one of them, the welfare of the animals is worse in that system. The welfare of any diseased animal is worse than that of an animal which is not diseased.
- Indicators of good welfare which we can use are obtained by studies demonstrating positive preferences by animals. Early studies of this kind include that by Hughes and Black (1973) showing that hens given a choice of different kinds of floor to stand on did not choose what biologists had expected them to choose. We should try to assess the specific functioning of the brain when welfare is good in humans and other animals. Good welfare in general, and a positive status in each of the various coping systems, should have effects which are a part of a positive reinforcement system, just as poor welfare is associated with various negative reinforcers.

An issue that merits discussion is that in all welfare assessment it is necessary to take account of individual variation in attempts to cope with adversity and in the effects which adversity has on the animal (Koolhaas *et al.,* 1983, Cronin and Wiepkema, 1984, von Holst, 1986, Broom, 1987, Benus, 1988). After informing students about different measures of welfare it is useful to discuss with them how to resolve issues where different measures give different information and also the extent to which different individuals use different strategies.

Finally, each assessment of welfare will pertain to a single individual and to a particular time range. In the overall assessment of the impact of a condition or treatment on an individual, a very brief period of a certain degree of good or poor welfare is not the same as a prolonged period. However, a simple multiplicative function of maximum degree and duration is often not sufficient. If there is a net effect of poor welfare and everything is plotted against time, the best overall assessment of welfare is the area under the curve thus produced (Broom 2001c).

2. Farm animal housing and husbandry

2.1 General issues

Farming animals means that they are fed, cared for and given shelter, but housing and management may restrict their performance of natural behaviour and that breeding may lead to traits that are for example associated with pain, ill-health or behavioural restriction. They may be subjected to painful or fear eliciting procedures, and they are usually finally killed by man. The ultimate reason for the farmer to keep the animals is to make his or her living from farming. Other aspects such as working quality and health standards of the farmer, and issues of environmental and consumer protection in society come also into view when tackling the ethical question of which, if any, limitations to animal welfare are acceptable. The basic question about the ethical acceptability of farming and killing animals per se can be treated, for example, based on texts of Sandøe *et al.* (1997) or Comstock (2002).

In general, it should be an aim to enable the students to clearly distinguish between animal welfare and other arguments, such as economics. Both should be based on facts with scientific foundation. The students should further understand that the final weighing of the different arguments against each other involves a subjective decision that is influenced by the basic attitude, including ethics, of the decision-maker. We believe that this intellectual process is especially important with regard to farm animals, as discussions of farm animal welfare often tend to focus on economical aspects. Furthermore, some students will have an agricultural background and may be 'home blind' or want to stay with certain husbandry practices because of tradition. One didactic approach can be to have for example a panel discussion, after sufficient information has been collected, about cases of farming practice that have opposing advantages and disadvantages in the different areas, and to let individual students present and discuss the topic only from the viewpoint of animal welfare, economy, consumer protection etc..

A good understanding of farm animal welfare is essential for those veterinary students that will go into farm animal practice and so are likely to need to deal with welfare problems on farms. In modern veterinary medicine that focuses on prophylaxis and herd health surveillance programmes, knowledge about how to improve animal welfare by good housing and management become increasingly important. Furthermore, quite a number of veterinarians are associated with animal production when they are employed by the industry or when they are working as State veterinarians being responsible for food control including animal welfare.

The basic principles of animal welfare apply to all species, but it is inevitable that many examples even in discussions of principle are taken from animal agriculture (e.g. in Dawkins, 1980; Webster, 1995; Appleby and Hughes, 1997), as, compared with the other categories of species discussed in this chapter, we generally have better knowledge about their welfare and welfare problems. Moreover, many students have some experience of visiting farms and farm animals are relatively easy to access. This therefore provides opportunities for case studies or to involve students in role playing since the facts are often more readily available or the students already have the basic knowledge to contribute in more spontaneous discussions in the classroom.

2.2 Behavioural restriction

A considerable number of animal welfare problems result from a mismatch between the natural behavioural repertoire of the animals and their housing and management. The veterinary students should therefore understand that, in general, animal welfare questions can only be properly tackled on the basis of knowledge of the biology of the species. The range of farm animal species is very wide, from insects such as honey bees, fish kept in aquaculture, birds such as domestic fowl, turkeys or even the relatively undomesticated ostriches to mammals such as pigs, cattle and sheep, but also animals kept for fur production or deer. It will not be possible to cover all possible species used for farming purposes, but we propose that important aspects of behavioural biology and the associated welfare problems should be presented for the more common farm animal species. Here the use of video resources is recommended. The advantage of using video resources is that behavioural processes are more easily understood from a moving image, but also that it is easier for the students to relate to the animals, that they often experience in situations where there is not much opportunity for variable natural behaviour. In our view this is a good impetus for developing a greater respect for the animals. For written information about the natural behaviour of horses, cattle, sheep and goats, pigs, rabbits and domestic fowl the textbook of Jensen (2002) or, in a more general form, Fraser and Broom (1997) can be used as starting points. When basic aspects such as natural behaviour are discussed, it is worthwhile remembering that in most veterinary schools there is relatively little education about poultry, although welfare problems and numbers of affected animals can be considerable in these species. Another aspect is the stage of domestication of the species (Grandin, 1998; Price, 2002) The types of species kept for farming purposes is changing and welfare problems can occur especially in newly domesticated species, partly because their needs may be less known, partly because their capacity to adapt to the new and possibly unfavourable conditions is smaller. Therefore, some knowledge of the effects of domestication on the adaptability of animals should be conveyed, and possibly some ethical aspects of the domestication and selection process be discussed.

One consequence of the mismatch between husbandry conditions and biology of the animal may be abnormal behaviour. This is an example of where the basic principles apply to all species but where it may be easier to teach it in farm animals. Although abnormal behaviour patterns can occur in pets and laboratory animals, a lot of basic research in this topic has been on farm animals (e.g. Lawrence and Rushen, 1993). Students should be aware that rather than being the exception, for some species the type of housing and management results in a very high proportion of animals showing abnormal behaviour patterns, e.g. stereotypies, and this may be regarded by the farmer as normal for that species. The same is for example true for certain physical damage, such as loss of hair or feathers, swellings or injuries, that may be inflicted by repeated collisions with ill-adjusted housing equipment.

More or less specific to farm animals are welfare problems associated with early separation of the mother and young, disruption of social groups or keeping the animals singly or in very large groups. The most severe case is the one of poultry where the eggs are separated from the mother and incubated artificially, and the birds are often kept in extremely large groups. The effects on the behaviour of the animals is only partly understood yet, but certainly of high importance for animal welfare. Useful information about those social issues can be found in Keeling and Gonyou (2001).

2.3 Breeding for high performance

Veterinary students should also be aware of welfare problems associated with high production achieved by intensive directional selection. Problems may include direct effects of rapid growth such as increased risks of leg or cardiac problems in fattening species, especially poultry (Scientific Committee on Animal Health and Animal Welfare, 2000), hunger in the parent generation of the animals kept for fattening purposes (e.g. poultry: Savory and Lariviere, 2000; or pigs: Appleby and Lawrence, 1987; Robert *et al.*, 1997) or increased risks of so-called production diseases such as metabolic problems, lameness and mastitis in dairy cows. The actual incidence of such disorders (except the hunger of the breeders), however, depends on the quality of housing and management. In laying hens another effect of high selection on egg-laying is the killing of about half of all chicks, the male layers, right after hatching, because it is not economical to fatten them. For example in Germany the Animal Welfare Act states that killing of vertebrates is only allowed with an acceptable reason (such as food production, disease control etc.). This practice is therefore often discussed on ethical and legislative grounds.

2.4 Painful or fear-eliciting procedures

Farming practices often include routine procedures that are painful for the animal, such as dehorning, beak trimming, tail docking or castration. Discussion of such procedures should not only weigh the wanted effects against possible negative consequences on behaviour and well-being through pain, sensory or communicative deprivation, but also consider the ethical idea of integrity of species and individuals (see for different views e.g. Rollin, 1995 or Verhoog, 1997). In association with mutilations, but also surgical and other treatments, pain recognition and alleviation is an important issue that should be part of the curriculum. Students should also become aware of the national and possibly European legal requirements regarding permissibility, the necessity for analgesia and methods for the above mentioned procedures. It may also be enlightening to compare standards for different species with each other.

Even procedures that should not lead to significant pain in the animals such as shearing, for example, can be very aversive because they can induce fear. This is one of the areas where the importance of stockmanship and human-animal relationship is becoming increasingly evident (e.g. Hemsworth and Coleman, 1998). By this the stockperson does not only influence production, but also animal welfare.

2.5 Killing

Apart from the general ethical question about killing discussed above, veterinary practitioners will often have to decide on individual farms which treatment to give or what course of action to recommend. Even if a disease or injury is easily treatable, the farmer may decide that the animal is not economically worth the cost of the treatment and so the veterinarian may be asked to put down the animal in any case. With poultry or very young animals the farmer will probably not even call the veterinarian, so by definition the decision is not an easy one for the farmer when veterinary help is requested. The veterinarian will be expected to give advice and this will involve balancing the welfare costs to the animal against the economic worth of the animal to the farmer. This may even involve advising in favour of killing if the suffering to the animal will be high during the course of the treatment, even if the animal is very valuable, such as in the case of a breeding bull or competition horse. However, in other cases, such as killing for diagnostic purposes or for disease control, it is not the welfare of the individual animal concerned that is considered, but the effects on the herd or even the national herd, and economical aspects form an important basis for the decisions. We think that it is very valuable if students have the chance to test and clarify their attitudes regarding such issues before they get in respective situations that need quick and efficient decisions.

The future veterinarian should also be able to give the farmer advice on the humane killing of animals in accordance with animal welfare legislation and be knowledgeable enough to choose the most humane method of killing (see publications form the Humane Slaughter association, http://www.hsa.org.uk; for disease control: Scientific Veterinary Committee, 1997).

2.6 Legislation and other measures to improve animal welfare

Since farm animals represent a very large proportion of the total number of animals in a country, there is usually more legislation on this category than there are on other animals on national as well as on European level (Council of Europe, European Union). Furthermore, unlike veterinarians employed in laboratory animal units, where often strict procedures are in place and training to ensure that the legislation is followed, veterinarians will be visiting farmers who may be unaware or uninformed about legal requirements on even simple points such as stocking density or handling procedures. Thus veterinarians should have a working knowledge of this legislation in their country or where to find it, already when they leave university.

A further aspect is the implementation of animal welfare legislation, for instance in cases of animal neglect or cruelty or other offences with which quite a number of future veterinarians will be confronted. Not only scientific, but also administrative and practical aspects need to be considered such as how to possibly relocate or kill large numbers of animals. We have the experience that it is very interesting for the students to discuss such matters with a veterinarian being in charge of implementation issues, or with lawyers, police or community services to get an idea of how such cases might be handled later in the course of their profession.

Animal welfare standards can also be improved by economical incentives through the public in the case of subsidies for certain types of housing or farming practices, or through the single consumer in the case of quality food schemes. Often veterinarians may also be involved in the setting up of subsidy or food schemes, and in the control of them. An important issue for the control of legal and other standards is how to monitor and assess animal welfare on the individual farm. For instance, a number of papers from a workshop on this topic can be found in Sørensen and Sandøe (2001).

3. Transport and slaughter of farm animals

Transport and slaughter of farm animals are undoubtedly among the most stressful procedures in livestock production. Veterinarians may deal with these issues either as official veterinary surgeons or as consultants on animal welfare.

3.1 Transport and lairage

Transport and lairage are likely to adversely affect animal welfare due to stress, fear or pain. Therefore, we suggest that the students are first exposed to a discussion on the welfare problems encountered by animals during transport and lairage. The reference by the European Union Scientific Committee on Animal Health and Welfare (2002) may be particularly useful. It is also important to discuss how the animals may perceive the stressors encountered during transport and lairage (see Grandin, 1997). The students should bear in mind that not all animals react in the same way to the same stressor. Depending on the previous knowledge of the students, it may be useful, therefore, to allow some time to describe the most important features of the anatomy, physiology and behaviour of the farm species.

The books of Fraser and Broom (1997), and Keeling and Gonyou (2001) are good references on this. Behaviour and stress responses may be different between animals from the same species depending on their genetics, previous experience and breeding conditions (intensive vs. extensive). The references by Grandin and Deesing (1998) and Manteca and Ruiz-de-la-Torre (1996) are very useful.

A second point is how to assess welfare during transport and lairage. Much of this has been covered in the section on basic principles. It may perhaps be useful to include a discussion, even if brief, on the relationship between animal welfare and meat quality in different species. The book by Gregory (1998) is a good reference.

The effect of transport-related stress on the animals's health is increasingly important. Pathogens that do not induce disease in farm animals kept under good conditions may become activated during transport, often due to immunosuppression resulting from stress.

The students should become familiar with the European and national legislation regarding the protection of animals during transport. They should know the minimal requirements to protect welfare during transport. The web page of the European Union provides all the current legislation on welfare during transport.

Finally, students should gain some knowledge of practical measures to improve the welfare of animals during transport. The book by Grandin (2000) provides useful information on this.

3.2 Stunning

The objective of stunning is to render animals unconscious prior to slaughter so that death can occur through bleeding without causing unnecessary pain, suffering and distress. The key point is that the student understands that bleeding does not cause immediate death to the animal, and that unconsciousnsness must last until death ensues. The time to die after sticking depends on the species but also on the sticking methods. Gregory (1998), and Blackmore and Delany (1988) provide useful information about time to brain death after sticking in different species, using different procedures.

Stunning procedures are based on disruption of normal functioning of neurones. It may be useful to allow some time to describe the physiology of the central nervous system as well as the theoretical basis of the unconsciousness induced by the most common stunning methods, including the following:

- Mechanical methods based on concussion and/or physical destruction of brain tissue. Finnie (1997) is a good reference.
- Electrical methods based on the induction to epilepsy. Cook *et al.* (1992, 1995) provides useful information.
- Carbon dioxide stunning based on depression of the central nervous system due to hypercapnic hypoxia and decrease of the intracellular pH. Forslid (1988) is a good reference.

One of the responsibilities of veterinarians is to advise the slaughter personnel on the best stunning system as well as on the welfare requirements of each system. The book by Gregory (1998) provides useful information on this.

Although European legislation requires pre-slaughter stunning, students should know that there is an exemption for animals slaughtered by religious methods. The most important religious methods are the Jewish, called *schechita* and the Muslim, called *halal*. Both faiths state that meat must come from animals that are alive at the time of slaughter. Anil and Sheard (1994), Grandin and Regenstein (1994) and the National Welfare Advisory Committee of New Zealand (NAWAC, 2001) provide useful information on this.

4. Companion animals

Small animal practitioners must frequently face situations where difficult ethical decisions must be made. Some of the main welfare issues that could be addressed in a general course on animal welfare are euthanasia, non-therapeutic surgery, behavioural problems, breed standards, housing of dogs and cats and pain management.

4.1 Euthanasia

The decision of putting an animal down should always be carefully analysed by the veterinarian and the owners as well. The need for euthanasia is easy to understand when the veterinarian is dealing with an animal that experiences severe chronic pain or that is otherwise suffering. Nevertheless, there are certain situations where euthanasia should be applied not for the direct benefit of the animal, but for the benefit of people around it. The small animal practitioner must frequently face other situations where the convenience of euthanasia is at least objectionable, including, for example, pets that are dangerous to people or show otherwise objectionable behaviour, pets whose owners can not afford medical treatment, etc. We suggest that it may be useful to devote some time to discuss these situations with the students. The case of dangerous animals is particularly interesting, as the student must weigh human health and safety against the animal's life. Further, aggression is the most frequent canine behaviour problem referred to specialists (Beaver, 1999). To stimulate the discussion, the students may be asked, for example, to think about possible risk factors that can be useful when deciding if the pet should or should not be put down, including the dog's weight, the intensity of the attacks, the presence in the family of children, elderly or disabled people and whether the aggression seems unpredictable.

4.2 Non-therapeutic surgery

Non-therapeutic surgery is a broad category that includes any surgical procedure that is carried out in the absence of a disease. The list of the commonest procedures includes castration, spaying, declawing, debarking, ear-cropping, tail-docking and teeth-cutting. From a welfare point of view, some of these interventions can be of benefit for animals whereas others can harm them greatly. Some authors believe that all these procedures are in fact mutilations and consequently ethically unacceptable (Young, 1996). However, this is a position not shared by the vast majority of veterinarians. The veterinary student must be trained to differentiate between acceptable and unacceptable non-therapeutic surgical procedures. The factors that can help the veterinarian to decide to perform this procedures can be reviewed in the form of three questions:
• What is the main purpose of the intervention?
• Are there any alternatives to the procedure?
• Is it a painful or a dangerous surgery?
The reasons for such a surgery fail into three broad categories: the control of reproduction, the modification of behaviour (Horwitz et al., 2002) and what can be considered cosmetic surgery.

The prevention of unwanted reproduction can be very beneficial for the entire species, for it helps to control overpopulation. In this respect, most humane societies and veterinary associations around the world endorse campaigns to promote neutering of dogs and cats. The treatment of behaviour problems is also essential, since they can lead to euthanasia and abandonment. According to this, some of these interventions can be understood as truly therapeutic procedures. Regarding cosmetic surgery, it is important to keep in mind that it is never helpful for the animal and has as the only aim to accomplish some arbitrary morphological standard. These points can be discussed using a case-study such as castration, declawing, debarking, ear cropping or tail-docking.

4.3 Behaviour problems

Dogs and cats frequently show behaviours that are either annoying for people, or dangerous for people or themselves. The vast majority of behaviour problems are in fact normal responses that simply bother people. As already mentioned, behaviour problems are one of the main causes of abandonment and euthanasia in companion animal medicine. Also, some behaviour problems involve different degrees of animal stress and anxiety. Thus, the treatment of behaviour problems it is also essential to prevent the animal from suffering. Stress and anxiety can often pass unnoticed to the owners and even to the veterinarian. Separation anxiety can be used as a paradigm of all these facts. It should be stressed that diagnosis and treatment of behaviour problems should be based on the causes of the behaviour and not on its symptoms. Also, the use of positive techniques should always be encouraged. In this sense, commercial electronic devices that can deliver an electric shock to the animal whenever it behaves improperly are a focus of major concern for the majority of veterinary behaviourists. There are many good sources of information on canine and feline behavioural problems, e.g. Horwitz et al. (2002).

4.4 Breed standards

The current concept of a pure dog breed refers to a list of characteristics, both physical and behavioural, that a given dog must fit to be considered member of that particular breed. This relatively new concept started in the United Kingdom back in the nineteenth century and is responsible for the strict selective pressures applied to dogs by professional breeders. Breed standards are often based in pure aesthetic criteria, that can increase the prevalence of physical and behavioural abnormalities. Examples of the former are the high prevalence of respiratory problems in the English Bulldog or the higher rate of elbow dysplasia in Labrador Retrievers. Regarding behaviour problems, examples include stereotypic tail-chasing in Bull Terriers and flank-sucking in Doberman Pinchers. For these reasons, veterinarians must play an active role in the promotion of rational breed standards. Also, veterinary students must understand that euthanasia should never be applied to dogs and cats that do not meet a breed standard, a practice that is still unfortunately found within our profession. The book by Ruvinsky and Simpson (2001) is a good source of information on inherited disorders of pure breed dogs.

4.5 Housing of dogs and cats

There are few studies on the welfare of housed dogs and cats (Hubrecht, 1995; Rochlitz, 2000). Some of the issues that could be considered include pen size, environmental enrichment and the need for social interaction.

4.6 Pain management

Pain management is of obvious importance in a course on animal welfare. Students should be familiar with the most common signs of pain in dogs and cats as well as with methods to alleviate it. Since some of these issues may be dealt with in other subjects of the curriculum, the amount of time devoted to them will depend on the year in which the animal welfare course is taught, among other things.

5. Laboratory animals

5.1 Introduction

The use of live animals in research and for teaching purposes raises a number of ethical questions related to the welfare of animals. Veterinarians may face these questions either as a consequence of their research activity, if they pursue a scientific career, or as a result of being employed in laboratory animal units or as members of an Institutional Animal Care and Use Committee (IACUC). We believe veterinary students should be exposed both to the general principles around which the ethical debate over the use of live animals in research revolves as well as to more "technical" issues. Also, students should become familiar with the legislation on the use of animals for research and teaching, both the European regulations and the laws of their own country. The final aim of teaching, rather than making the students remember a list of facts, would be to provide them with the intellectual tools to understand the problem and develop their own opinion.

Most veterinary students are, or should be, reasonably familiar with the general biology of the most common domestic species, such as dogs, cats, cattle, pigs, horses, sheep and chicken. They may not be so familiar, however, with the general biology of the species most commonly used in biomedical research, i.e. rats and mice. This knowledge, however, may be important to understand some welfare problems related, for example, to the housing of these animals. It may be useful, therefore, to allow some time to describe the most important features of the anatomy, physiology and behaviour of these species. The book by UFAW (1999) is a very good source of information on this.

FELASA (Federation of European Laboratory Animal Science Associations) has elaborated a set of recommendations for the education and training of persons involved in animal experiments (FELASA, 2001). Although these recommendations are intended for very specialised courses, rather than for general courses on animal welfare, they are still worth looking at.

5.2 Basic principles

In 1959, Russell and Burch put forward the so-called *Three Rs Principle*. This is now widely accepted by the scientific community as one of the main guiding principles in the use of live animals in research. Besides, we believe it may provide a very useful framework to discuss some of the important concepts that should be taught in a general course on animal welfare. According to Russell and Burch, three issues must be taken into consideration when using live animals in research or for teaching purposes:

- *Replacement*, i.e. whether the use of live animals can be replaced by an alternative method, such as "*in vitro*" or computer-simulation models. The web pages of the University of California Center for Animal Alternatives (UCCAA), the Norwegian Inventory of Alternatives (NORINA) and the Animal Welfare Information Center (AWIC) -among others- provide useful information on alternative methods that may help to illustrate this concept. Also, the reference by Balls (1994) be very useful

- *Reduction*, i.e. whether the same objectives can be achieved with fewer animals, for example by improving the experimental design. The references by Festing *et al.* (2002) and Festing (1994) are very useful.

- *Refinement*, i.e. whether the amount of suffering caused to the animals used in the experimental procedure can be reduced to a minimum. The possibility to refine an experiment largely revolves around the ability to assess the pain and discomfort caused to the animals and this is dealt with below. The reference by Flecknell (1994) provides a good summary of the issue of refinement.

The key point when teaching the *Three Rs Principle* is that the student understands that laboratory animal welfare is not only about reducing the amount of suffering caused to the animals, but also about considering the possibility to reduce the number of animals used.

A second point of general interest about the ethical acceptability of animal experiments is whether the benefits of the experiment outweigh the suffering of the animals (see Bateson 1986 and Stafleu *et al.*, 1999). In fact, this question should be a main concern for the members of an IACUC and is therefore important that veterinary students are exposed to it. In practical terms, however, it is often very difficult to balance the significance of a particular piece of scientific research against the welfare of the animals. We believe that it is important that veterinary students realize this difficulty and, therefore, become aware of the problems faced by those having the responsibility to decide which experiments are ethically acceptable. To illustrate these difficulties, it may be very useful to use a case-based approach, in which the students are given an experimental procedure and have to think about its benefits and the suffering caused to the animals, and reach a conclusion about its acceptability. The book edited by Orlans *et al.* (1998) provides several good examples of this approach.

In order to balance the benefits of an experiment against the suffering caused to the animals, it is imperative to be able to assess the amount of suffering before actually starting the experiment. One possibility is to use an analogy-based reasoning, so that a procedure that would probably cause pain or distress in a human is considered in principle as being painful or distressful to a laboratory animal. We think it may be very useful to discuss the benefits and shortcomings of such an approach, as it is commonly used in any debate over the ethical acceptability of animal experiments. Both Bateson (1996) and Stafleu *et al.*(1992) are good references on this.

Apart from this analogy-based reasoning, one of the tools commonly used to do this are the so-called "invasiveness scales" or "severity banding", that are currently used by IACUCs in many countries. In 1990, for example, a working party of the Laboratory Animal Science Association provided a system for assessing severity as minimum, intermediate or maximum (Wallace *et al.*, 1990). We also found the reference by Orlans (1996) very useful. Even though a general course on animal welfare may be too short to analyse in detail how these scales have been worked out, it may still be useful that the students know, at the very least, their existence.

5.3 "Technical" issues

An important point to be taught to the students is that laboratory animals may experience suffering in a variety of forms, including not only pain but also stress and behavioural restriction (see section 1 and 2 in this chapter). Also, it is necessary for the student to realize that suffering may occur not only during the experimental procedure, but also afterwards, and that it may be caused as well by the housing system. The book by UFAW (1999) is a very good source of information on housing of laboratory animals. An issue that is related to housing and that should be mentioned in the context of laboratory animal welfare is environmental enrichment. Some references on this topic are given in the next section on wildlife.

As already mentioned, one of the requirements set out by the *Three Rs Principle* is that of refinement. We believe that a discussion on how to refine different experimental procedures is well beyond the scope of a general course on animal welfare. Nevertheless, a discussion on how pain and distress in laboratory animals can be assessed is, in our opinion, one of the most important contents to be included in such a course. Indeed, this knowledge can be applied -with some modifications- to any experimental procedure and recognising pain and distress should be one of the main concerns for any veterinarian involved in laboratory animal care. On this issue, we suggest the paper by Morton and Griffiths (1985) as a very interesting source of ideas. Although published almost 20 years ago, this paper is still widely cited. The authors discuss the main signs of pain and discomfort in laboratory animals, including changes in body weight, appearance, clinical variables and behaviour. Also, they suggest a system to obtain an overall assessment of how much the animal is suffering. Obviously, the methodology suggested by Morton and Griffiths to assess pain and discomfort should be taken only as a starting point, among other reasons because each experimental procedure requires a slightly different protocol. It may be useful, therefore, to present this paper first and then ask the students to develop their own system to assess pain and discomfort for a given experimental procedure. The references by the NRC (1992), by Flecknell (1994) and by Morton (1997), as well as the web page of *Laboratory Animals*, may also be very useful.

Another issue that is related to the assessment of pain and discomfort is that of the so-called "humane end-points". In many procedures, it is a main ethical requirement to establish a set of criteria that will allow the experimenter to identify those animals that are experiencing intense pain and that should therefore be humanely killed to avoid further suffering. As one of the responsibilities of the veterinarians that are responsible for laboratory animal welfare is to help the researchers on this issue, we believe a discussion on "humane end-points" should be included in the course. Some of the references we find most useful are those by the Canadian Council of Animal Care (1998) and by Hendriksen and Morton (1998). This discussion could be enriched with an overview of the different methods of euthanasia and its advantages and disadvantages from an animal welfare point of view (AVMA, 2001; Close *et al.*, 1996, 1997).

Finally, and also on the issue of refinement, analgesia is obviously an important topic. This, however, is usually covered in other subjects of the veterinary curriculum. Thus, depending on the year on which the course on animal welfare is taught, it may be useful to refer to these other subjects.

5.4 Legal issues

Students should be made aware of the main points raised by the Council Directive 86/609/EEC on the protection of animals used for experimental and other scientific purposes. It is important to realise that this directive is in train to be modified, particularly in relation to housing conditions of laboratory animals. Also, they should be exposed to the national legislation of their country. Finally, in countries were IACUC are mandatory or common, it could be useful to allow some time for one member of the University's IACUC to deliver a short presentation on the function and type of work done by the IACUC.

6. Wildlife

6.1 Introduction

Veterinarians are increasingly involved in wildlife medicine, husbandry and conservation and should be familiar with a number of welfare issues related to these professional activities. Veterinarians employed in zoos, captive breeding centres and wildlife rehabilitation centres are responsible for the welfare of the animals and therefore should know how to assess it and how to apply measures to improve it; environmental enrichment is probably the most widely used of such measures. Wildlife conservation in nature often involves the capture, handling and transport of animals; as veterinarians are ideally suited to carry out these operations, they should be aware of their animal welfare implications. Finally, some wildlife species are increasingly popular as companion animals and a basic knowledge of their needs may be useful. It is likely, however, that a detailed discussion on the care of exotic pets is well beyond the scope of a general course on animal welfare, even more so if the course is in the first years of the curriculum. Nevertheless, it may be advisable that the person responsible for teaching animal welfare discuss this particular aspects with those teaching exotic animal medicine.

6.2 Basic principles

We suggest that a course on animal welfare that deals, even if very briefly, with wildlife, should start by considering the biodiversity crisis. This would include a description of the current rate of species extinction as well as a discussion of its main causes. There are many good textbooks on conservation biology that may be useful as a source of information, including those by Caughley and Gunn (1995), Meffe and Carroll (1997), Primack (1998, 2000) and Pullin (2002). The book by Leakey and Lewin (1995) is also interesting. Although habitat destruction is by far the main cause of extinction, there are other causes, some of which are particularly relevant to veterinarians. The trade in wild animals that results from the pet market is one of them and has been identified as an important factor threatening several species, including, for example, many parrots and related species. A discussion on the effects of international trade on wildlife as well as on the *Convention on International Trade of Endangered Species* (CITES), should therefore be included. The reference by Hemley (1994) may be useful. Also, the book by Stattersfield and Capper (2000) on threatened birds may be of interest to illustrate the effects of the pet trade on wild birds; also, it provides a lot of information on the causes of extinction.

Another issue that should be included is related to zoos. Even though zoos have been criticised on animal welfare grounds, it is also true that they have played an important role in conservation. The main elements of the debate over the goodness of zoos can be included in the course. Special emphasis should be given to the widely accepted principle that modern zoos should be devoted to research, conservation and education, and should give a high priority to the welfare of animals. The role of zoos in conservation can be highlighted by giving examples of successful captive breeding programmes. The references by Bostock (1993), Mench and Kreger (1996), Norton *et al.* (1995) and Tudge (1991) are particularly interesting. The booklet published by the *Universities Federation for Animal Welfare* (1988) is also useful.

6.3 Assessing and improving the welfare of captive wild animals

As with other animals, the welfare of captive wild animals should be assessed by combining different types of welfare indicators, such as disease, changes in behaviour and physiological measures of stress. This has been explained elsewhere in this chapter.

As already mentioned, environmental enrichment is one of the most commonly used methods to improve the welfare of captive wild animals and we suggest that a discussion of its basic principles -even if cursory-should be included in the course.

The book edited by Shepherdson *et al.* (1998) is an important reference. Also, Carlstead and Shepherdson (2000) give a good review of the effects of environmental enrichment on the welfare of captive wild animals. The web pages of the *Animal Welfare Information Center* and *The Shape of Enrichment* are also particularly interesting, among many other reasons because they provide examples of environmental enrichment programs for several species. The issue of environmental enrichment can be approached by asking the students to design a program of enrichment for a particular species. They can be required to gather information on the natural history of the species and to review enrichment methods that have been already used in that species, and then to suggest their own enrichment method. It may be also useful to relate this issue with that of how to assess welfare, so that the students might think how to find out whether their enrichment program is improving the welfare of the animals. Environmental enrichment has oftentimes positive effects on welfare, but it is not without problems, and this should at least be mentioned. Haemisch *et al.* (1994) provide a god example of the potentially negative effects of environmental enrichment.

6.4 Welfare issues related to the capture, handling and transport of wild animals

The welfare of wild animals can be put at risk when captured, handled and transported. As veterinarian may have an active role in these operations, a discussion on their welfare implications may be of interest. The references by the American Society of Mammalogists (1998) and by Gaunt and Oring (1999) may be useful. Depending on the background of the students -mainly whether they have already had a basic course on pharmacology-, a discussion on the use of drugs to alleviate stress in wild animals could be included in the course. Ebedes (1992) provides a good overview of this topic. We suggest, however, that it is important to emphasize that using drugs is simply an aid and should never replace good handling.

References

American Society of Mammalogists, 1998. Guidelines for the capture, handling and care of mammals as approved by the American Society of Mammalogists. J. Mammal.79, p. 416-1431.
American Veterinary Medical Association, 2001. 2000 Report of the AVMA Panel on Euthanasia. J.Am.Vet.Med.Assoc., 218, p. 671-696.
Anil, M.H., McKinstry, J.L., Gregory, N.G., Wotton, S.B. and Symonds, H., 1995. Welfare of calves - 2, Increase in vertebral artery blood flow following exsanguination by neck sticking and evaluation of chest sticking as an alternative slaughter method. Meat Science,.41, p. 113-123.

Anil, M.H. and Sheard, P.R., 1994. Welfare implications of religious slaughter. Meat Focus Int. 10, p. 404-405.

Anil, M.H., Whittington, P.E. and McKinstry, J.L., 2000. The effect of the sticking method on the welfare of slaughter pigs. Meat science 55, p. 315-319.

Appleby, M.C. and Hughes, B.O., (eds.), 1997. Animal welfare. CAB International, Wallingford.

Appleby, M.C., and Lawrence, A.B., 1987. Food restriction as a cause of stereotypic behaviour in tethered gilts. Anim. Prod. 45, p. 103-111.

Bager, F., Braggins, T.J., Devine, C.E., Graafhuis, A.E., Mellor, D.J., Tavener, A. and Upsdell, M.P, 1992. Onset of insensibility at slaughter in calves: effects of electroplectic seizure and exsanguination on spontaneous electrocortical activity and indices of cerebral metabolism. Research in Veterinary Science, 52, p. 162-173.

Balls, M., 1994. Replacement of animal procedures: alternatives in research, education and testing. Lab. Anim. 28, p. 193-211.

Bateson, P., 1986. When to experiment on animals. New Scientist 1496, p. 30-32.

Bateson, P., 1991. Assessment of pain in animals. Anim. Behav. 42, p. 827-839.

Benus, I., 1988. Aggression and coping. Differences in behavioural strategies between aggressive and non-aggressive male mice. Ph.D. thesis, University of Groningen.

Blackmore, D. and Delany, M., 1988. Slaughter of Stock. Publication No 118, Veterinary Continuing Education, Massey University, Palmerston North, New Zealand.

Bostock, S. St. C., 1993. Zoos and Animal Rights. London, Rouletdge.

Broom, D.M., 1981. Biology of Behaviour. Cambridge University Press, Cambridge.

Broom, D.M., 1983. The stress concept and ways of assessing the effects of stress in farm animals. Applied Animal Ethology 1, p. 79.

Broom, D.M., 1986. Indicators of poor welfare. British Veterinary Journal 142, p. 524-526.

Broom, D.M., 1987. Applications of neurobiological studies to farm animal welfare. In Biology of Stress in Farm Animals: an Integrated Approach. ed. P.R. Wiepkema and P.W.M. van Adrichem, Current Topics in Veterinary Medicine and Animal Science 42, p. 101-110. Dordrecht: Martinus Nijhoff.

Broom, D.M., 1988. Relationship between welfare and disease susceptibility in farm animals. In Animal Disease - a Welfare Problem, ed. T.E. Gibson, p. 22-29. London: British Veterinary Association Animal Welfare Foundation.

Broom, D.M., 1991. Animal welfare: concepts and measurement. Journal of Animal Science 69, p. 4167-4175.

Broom, D.M., 1991. Assessing welfare and suffering. Behavioural Processes 25, p. 117-123.

Broom, D.M., 1996. Animal welfare defined in terms of attempts to cope with the environment. Acta Agriculturae Scandinavica Section A. Animal Science Supplement, 27, p. 22-28.

Broom, D.M., 1997. Welfare evaluation. Applied Animal Behaviour Science, 54, p. 21-23.

Broom, D.M., 1998. Welfare, stress and the evolution of feelings. Advances in the Study of Behavior, 27, p. 371-403.

Broom, D.M,. 1999. Animal welfare: the concept and the issues. In Attitudes to Animals: Views in Animal Welfare, ed. F.L. Dolins, p. 129-142. Cambridge, Cambridge University Press.

Broom, D.M., 2000. Welfare assessment and problem areas during handling and transport. In Livestock Handling and Transport, 2nd ed, ed. T Grandin, p. 43-61. Wallingford, CAB International.

Broom, D.M., 2001. Coping, stress and welfare. In Coping with Challenge: Welfare in Animals including Humans. Ed. D.M. Broom, p. 1-9. Berlin, Dahlem University Press.

Broom, D.M., 2001. Evolution of pain. In Pain: its nature and management in man and animals, ed. Soulsby, Lord and Morton, D. Royal Society of Medicine International Congress Symposium Series, 246, p. 17-25.

Broom, D.M., 2001. The use of the concept Animal Welfare in European conventions, regulations and directives. Food Chain 2001, p. 148-151, Uppsala, SLU Services.

Broom, D.M. and Corke, M.J., 2002. Effects of disease on farm animal welfare. Acta veterinaria Brno, 71, p. 133-136.

Broom, D.M. and Johnson, K.G., 1993. Stress and Animal Welfare. Dordrecht, Kluwer 211 p.

Broom, D.M. and Kirkden, R.D., (in press). Welfare, stress, behaviour and pathophysiology. In: Veterinary Pathophysiology, ed. R.H. Dunlop and C.-H. Malbert. Ames, Iowa State University Press.

Canadian Council of Animal Care, 1998. Guidelines on choosing an appropiate endpoint in experiments using animals for research, teaching and testing. Canadian Council of Animal Care, Ottawa.

Carlstead, C. and Shepherdson, D., 2000. Alleviating Stress in Zoo Animals with Environmental Enrichment. IN: G.P. Moberg and J.A. Mench (eds.) The Biology of Animal Stress. Basic Principles and Implications for Welfare. Wallingford, CAB International, p. 337-354.

Caughley, G. and Gunn, A., 1995. Conservation Biology in Theory and Practice. Oxford, Blackwell Science.

Close, B., Banister, K., Baumans, V., Bernoth, E., Bromage, N., Bunyan, J., Erhardt, W., Flecknell, P., Gregory, N., Hackbarth, H., Morton, D. and Warwick, C., 1996. Recommendations for euthanasia of experimental animals: Part 1. Lab. Anim. 30: 293-316.

Close, B., Banister, K., Baumans, V., Bernoth, E., Bromage, N., Bunyan, J., Erhardt, W., Flecknell, P., Gregory, N., Hackbarth, H., Morton, D. and Warwick, C., 1997. Recommendations for euthanasia of experimental animals: Part 2. Lab. Anim. 31, p. 1-32.

Comstock, G.L. (ed.), 2002. Life Science Ethics. Iowa State Press, Iowa

Cook, C.J., Devine, C.E., Gilbert, K.V., Smith, D.D. and Maasland, S.A., 1995. The effect of electrical head-only stun duration on electroencephalographic-measured seizure and brain amino acid neurotransmitter release. Meat Science, 40, p. 137-147.

Cook, C.J., Devine, C.E., Tavener, A. and Gilbert, K.V., 1992. Contribution of amino acid transmitters to epileptiform activity and reflex suppression in electrically head stunned sheep. Research in Veterinary Science, 52, p.48-56.

Cronin, G.M. and Wiepkema, P.R., 1984. An analysis of stereotyped behaviours in tethered sows. Annales de Recherches Vétérinaires 15, p. 263-270.

Daly, C.C., Gregory N.G., and Wotton S.B. (1987) Captive bolt stunning of cattle: effects on brain function and role of bolt velocity. British Veterinary Journal, 143(6), p. 574-580.

Dawkins, M.S., 1980. Animal Suffering: The Science of Animal Welfare. Chapman and Hall, London

Dawkins, M.S., 1990. From an animal's point of view: motivation, fitness, and animal welfare. Behavior and Brain Sciences, 13, p. 1-61.

Duncan, I.J.H. and Petherick, J.C., 1991. The implications of cognitive processes for animal welfare. Journal Animal Science 69, p. 5017-5022.

Duncan, I.J.H. and Fraser, D., 1997. Understanding animal welfare, In M C Appleby and B O Hughes (eds) Animal welfare Wallingford, CAB International, p. 19-31.

Ebedes, H., 1992. The use of tranquillizers in wildlife. Department of Agricultural Development, Pretoria.

EU Scientific Committee on Animal Health and Welfare, 2002. The welfare of animals during transport (details for horses, pigs, sheep and cattle).

FELASA, 2001. FELASA recommendations for the education and training of persons involved in animal experiments. Laboratory Animals Ltd Reprinted 2001. London: The Royal Society of Medicine Press.

Festing, M.F.W., 1994. Reduction of animal use: experimental design and quality of experiments. Lab. Anim. 28, p. 212-221.

Festing, M.F.W., Overend, P., Ganes Das, R., Cortina Borja, M. and Berdoy, M., 2002. The Design of Animal Experiments. Reducing the Use of Animals in Research through Better Experimental Design. Laboratory Animal Handbooks NO 14. London, The Royal Society of Medicine Press Limited.

Finnie, J.W., 1997. Traumatic head injury in ruminant livestock. Australian Veterinary Journal, 75 (3), p. 204-208.

Flecknell, P.A., 1994. Refinement of animal use-assessment and alleviation of pain and distress. Lab. Anim. 28, p. 222-231.

Forslid, A., 1987. Transient neocortical, hippocampal and amygdaloid EEG silence induced by one minute inhalation of high concentration CO_2 in swine. Acta Physiologica Scandinava 130, p. 1-10.

Forslid, A. and Augustinsson, O., 1988. Acidosis, hypoxia and stress hormone release in response to one-minute inhalation of 80% CO_2 in swine. Acta Physiologica Scandinava 132, p. 223-230.

Fraser, A.F. and Broom, D.M., 1997. Farm animal behaviour and welfare. 3rd ed., CAB International, Wallingford.

Gaunt, A.S. and Oring, L.W., 1999. Guidelines to the use of wild birds in research, 2nd ed. The Ornithological Council.

Grandin, T., 1997. Assessment of stress during handling and transport. Journal of Animal Science, 75, p. 249-257.

Grandin, T., 2000. Livestock handling and transport. CAB International, London, United Kingdom.

Grandin, T. (ed.), 1998. Genetics and the behavior of domestic animals. Academic Press, San Diego.

Grandin, T. and Deesing, M., 1998. Genetics and behaviour during handling, restraint and herding. In Genetics and the behaviour of domestic animals. Academic Press, California, USA.

Grandin, T. and Regenstein, J.M., 1994. Religious slaughter and animal welfare: a discussion for meat scientists. Meat Focus International, March 1994, p. 115-123.

Gregory, N.G., 1998. Animal Welfare and Meat Science. CAB International, London, UK.

Haemisch, A., Voss, T. and Gartner, K., 1994. Effects of environmental enrichment on aggressive behaviour, dominance hierarchies, and endocrine states in male DBA/2J mice. Physiology and Behavior 56, p. 1041-1048.

Hemley, G. (ed.), International Wildlife Trade. A CITES Sourcebook. Washington, Island Press.

Hemsworth, P.H. and Coleman, G.J., 1998. Human-livestock interactions: The stockperson and the productivity and welfare of intensively farmed animals. CAB International, Wallingford.

Hendriksen C.F. and Morton, D.B., 1998. Humane endpoints in animal experiments for biomedical research. Proceedings of the International Conference, 22-25 November 1998, Zeist.

Hughes, B.0. and Black, A.J., 1973. The preference of domestic hens for different types of battery cage floor. British Poultry Sciences 14, p. 615-619.

Hughes, B.O. and Duncan, I.J.H., 1988a. Behavioural needs: can they be explained in terms of motivational models? Applied Animal Behaviour Science, 20, p. 352-355.

Hughes, B.O. and Duncan, I.J.H., 1988b. The notion of ethological 'need', models of motivation and animal welfare. Animal Behaviour 36, p. 1696-1707.

Jensen, P. (ed.), 2002. The ethology of domestic animals: an introductory text. CAB International, Wallingford

Keeling, L.J. and Gonyou, H.W. (ed.), 2001. Social behaviour in farm animals. CAB International, Wallingford.

Knierim, U. and Jackson, W.T., 1997. Legislation. In: Appleby, M.C. and Hughes, B.O., (eds.): Animal welfare. CAB International, Wallingford.

Koolhaas, J. M., F. Schuurman, and D. S. Fokkema. Social behavior in rats as a model for the psychophysiology of hypertension. In: Biobehavioral Bases of Coronary Heart Disease, edited by T. M. Dembroski, T. H. Schmidt, and G. Blumchen. Basel: Karger, 1983, p. 391-400.

Lawrence, A.B. and Rushen, J., 1993. Stereotypic animal behaviour: Fundamentals and applications to animal welfare.

Leakey, R. and Lewin, R., 1995. The Sixth Extinction. Patterns of Life and the Future of the Humankind. New York: Anchor Books.

Manteca, X and Ruiz-de-la-Torre, J., 1996. Transport of extensively farmed animals. Applied animal behaviour science, 49, p. 89-94.

Mason, J.W., 1971. A re-evaluation of the concept of 'non-specificity' in stress theory. Journal of Psychiatric Research, 8, p. 323-33.

Meffe, G.K. and Carroll, C.R., 1997. Principles of Conservation Biology. Sunderland, Massachusetts, Sinauer.

Mench, J.A. and Kreger, M.D., 1996. Ethical and Welfare Issues Associated with Keeping Wild Mammals in Captivity. In D. G. Kleiman, M. E. Allen, K. V. Thompson and S. Lumpkin (eds.) Wild Mammals in Captivity. Principles and Techniques. Chicago, Chicago University Press., p. 5-15.

Moberg, G.P., 1985. Biological response to stress: key to assessment of animal well-being? In Animal Stress, ed. G.P. Moberg, G.P. p. 27-49. Bethesda, Md: American Physiological Society.

Morton, D.B. and Griffiths, P.H.M., 1985. Guidelines on the recognition of pain, distress and discomfort in experimental animals and an hypothesis for assessment. Vet. Rec. 116, p. 431-436.

Morton, D.B., 1997. Ethical and refinement aspects of animal experimentation. In: Veterinary Vaccinology, p. 763-785 (P.P. Pastoret, J. Blancou, P. Vannier and C. Verschueren, Eds.). Elsevier, Amsterdam.

National Research Council, 1992. Recognition and alleviation of pain and distress in laboratory animals. National Academy Press, Washington.

NAWAC, 2001. Discussion paper on the animal welfare standards to apply when animals are commercially slaughtered in accordance with religious requirements. National animal welfare advisory committee, New Zealand.

Norton, B.G., Hutchins, M., Stevens E.F. and Maple T.L., 1995. Ethics on the Ark. Zoos, Animal Welfare and Wildlife Conservation. Washington: Smithsonian Institution Press.

Orlans, F.B., Beauchamp, T.L., Dresser. R., Morton, D.B. and Gluck, J.P., 1998. The Human Use of Animals. Case Studies in Ethical Choice. Oxford University Press, Nueva York.

Poole, T., 1999. The UFAW handbook on the care and management of laboratory animals. 7th ed., volume I: Terrestrial vertebrates. Blackwell Science, Oxford.

Price, E.O., 2002. Animal domestication and behavior. CAB International, Wallingford.

Primack, R.B., 1998. Essentials of Conservation Biology. Sunderland, Massachusetts: Sinauer.

Primack, R.B., 2000. A Primer of Conservation Biology. Sunderland, Massachusetts: Sinauer.

Pullin, A.S., 2002. Conservation Biology. Cambridge: Cambridge University Press.

Robert, S., Rushen, J. and Farmer, C., 1997. Both energy content and bulk of food affect stereotypic behaviour, heart rate and feeding motivation of female pigs. Appl. Anim. Behav. Sci. 54, p. 161-171.

Rochlitz, I., 2000. Recommendations for the housing and care of domestic cats in laboratories Laboratory Animals 34, p. 1-9.

Rollin, B.E., 1995. Farm animal welfare: social, bioethical and research issues. Iowa State University Press, Ames.

Russell, W.M. and Burch, R.L., 1959. The principles of humane experimental technique. Methuen, Londres.

Sandøe,P., Crisp, R. and Holtug, N., 1997. Ethics. In Appleby, M.C. and Hughes, B.O. (eds.), Animal welfare. CAB International, Wallingford

Savory, C.J. and Lariviere, J.M., 2000. Effects of qualitative and quantitative food restriction treatments on feeding motivational state and general activity level of growing broiler breeders. Applied Animal Behaviour Science 69, p. 135-147.

Scientific Committee on Animal Health and Animal Welfare, 2000. The welfare of chickens kept for meat production (broilers). SANCO.B3/AH/R15/2000, European Commission, Brussels.

Scientific Veterinary Committee, 1997. The killing of animals for disease control purposes. European Commission, Brussels.

Shepherdson, D.J., Mellen, J.D. and Hutchins, M. (eds.), 1998. Second Nature. Environmental Enrichment for Captive Animals. Washington: Smithsonian Institution Press.

Sørensen, J.T. and Sandøe, P. (eds.), 2001. Assessment of animal welfare at farm or group level. Acta Agric. Scand., Sect. A, Animal Sci., Suppl. 30.

Stafleu, F.R., Rivas, E., Rivas, T., Vorstenbosch, J., Heeger, F.R. and Eynen, A.C., 1992. The use of analogous reasoning for assessing discomfort in laboratory animals. Anim. Welfare 1, p. 77-84.

Stafleu, F.R., Tramper, R., Vorstenbosch, J. and Joles, J.A., 1999. The ethical acceptability of animal experiments: a proposal for a system to support decision-making. Lab. Anim. 33, p. 295-303.

Stattersfield, A.J. and Capper, D.R., 2000. Threatened Birds of the World. Barcelona: Lynx Edicions.

Toates, F. and Jensen, P., 1991. Ethological and psychological models of motivation: towards a synthesis. In J.A. Meyer and S. Wilson (Eds) Farm Animals to Animats, MIT Press, Cambridge, p. 194-205.

Tudge, C., 1992. Last Animals at the Zoo. How Mass Extinction Can Be Stopped. Oxford, Oxford University Press.

Universities Federation for Animal Welfare, 1988. Why Zoos? UFAW Courier No. 24. Potters Bar: UFAW.

Verhoog, H., 1997. Intrinsic value and animal welfare. In van Zutphen, L.F.M. and Balls, M. (eds). Animal alternatives, welfare and ethics. Proc. 2nd World congress on alternatives and animal use in life sciences, Utrecht, The Netherlands, 20-24 October 1996, Elsevier, Amsterdam, p. 169-177.

von Holst, D., 1986. Vegetative and somatic components of tree shrews' behaviour. Journal of the Autonomic Nervous System, Supplement. p. 657-670.

Wallace, J., Sanford, J., Smith, M.W. and Spencer, K.V., 1990. The assessment and control of the severity of scientific procedures on laboratory animals. Report of the Laboratory Animal Science association Working Party. Lab. Anim. 24, p. 97-130.

Webster, J., 1995. Animal welfare: A cool eye towards Eden. Blackwell Science, Oxford.

Legal texts

Council of Europe: European convention for the protection of animals kept for farming purposes 1976, supplemented by a protocol of amendment 1992 (http://book.coe.int/gb/cat/liv/htm/l76.htm) with a number of recommendations concerning different farm animal species (available under http://www.admin.ch/bvet). National ministries should also be able to provide copies of those.

European Union

Commission of the European Communities (1986). Council Directive 86/609/EEC of 24 November 1986 on the approximation of laws, regulations and administrative provisions of the Member States regarding the protection of animals used for experimental and other scientific purposes. Official Journal L 358, 18/12/1986, pp. 0001-0028;

Council directive 93/119/EC of 22 December 1993 on the protection of animals at the time of slaughter or killing;

but also regulations on marketing standards, e.g. for eggs, poultry meat and organic products, may effect animal welfare (http://europa.eu.int/eur-lex/en/), see Knierim and Jackson (1997)

National legislation: Some animal welfare legislation can be found on the Internet, e.g.:

Germany: http://www.verbraucherministerium.de/

Animal welfare act in English: http://www.tiho-hannover.de/einricht/tsb/act_transl.pdf

Switzerland: http://www.admin.ch/bvet

United Kingdom: http://www.defra.gov.uk/animalh/welfare/default.htm

Some selected web pages

Courses and general information: http://www.animal-info.net

European legislation on animal welfare:

http://europa.eu.int/comm/food/fs/aw/aw_references_en.html#95-29

Laboratory Animals: http://www.lal.org.uk

University of California Center for Animal Alternatives:

 http://www.vetmed.ucdavis.edu/Animal_Alternatives/main.htm

Animal Welfare information Center: http://www.nal.usda.gov/awic/

Norwegian Inventory of Alternatives: http://oslovet.veths.no

The Shape of Enrichment: http://www.enrichment.org

Some selected video resources

Verhaltensweisen von Rindern (behaviour of cattle) I-III
    (http://www.iwf.de/iwfger/3medien/medien_in.html)

Verhalten beim Hausschwein (behaviour of the domestic pig) I-IV
    (http://combi.agri.ch/lmz/lehrbuch/video.htm - website only in German)

Verhalten beim Haushuhn (behaviour of the domestic fowl) I-III
    (http://combi.agri.ch/lmz/lehrbuch/video.htm - website only in German)

Teaching ethics to agricultural and veterinary students: experiences from Denmark

T. Dich[1], T. Hansen[1], S.B. Christiansen[2], Pernille Kaltoft[3] and P. Sandøe[2]
[1]The Royal Veterinary and Agricultural University, Department of Food Economics, Unit of Learning and Bioethics, Rolighedsvej 26, DK-1958 Frederiksberg C, Denmark, [2]The Royal Veterinary and Agricultural University, Centre for Bioethics and Risk Assessment, Groennegaardsvej 8, DK-1870 Frederiksberg C, Denmark, [3]National Environmental Research Institute, Department of Policy Analysis, Frederiksborgvej 399, P.O. Box 358, DK-4000 Roskilde, Denmark

Abstract

Over a ten year period interdisciplinary courses in bioethics have been developed at the Royal Veterinary and Agricultural University in Copenhagen. The first course was a voluntary course mainly directed at agronomy students. Recently, two obligatory courses have been developed, one aimed at veterinary students, the other for students in agronomy, agricultural economy and biotechnology. The aim of this paper is to try to give a systematic account of the experiences from developing these courses. First it is being discussed why there is a need for agriculturalists and veterinarians to understand ethics. Secondly the three courses are presented. Two important ideas go through the development of the courses: i) Ethical theory is taught in tandem with real agricultural and veterinary problems and actual issues within agricultural science and veterinary medicine. ii) Students work on projects in which they have to view an issue within agricultural science, agricultural economy or veterinary medicine from at least two ethical perspectives. Thirdly a number of factors which seem to be important for the success of the courses are singled out.

1. Introduction

In 2000 the Danish education minister and university rectors agreed that all undergraduate programmes of university education in Denmark should include an introductory philosophy course. Unlike a previous philosophy course (offered in Denmark until 1971), this course will be tailored to suit individual study programmes so that connections are made with research within the relevant fields of study. The only requirement placed on the content of courses is that they must enable the students to view their subject from a broader perspective. Thus a course may view the relevant field of study from, for example, a historical, an ethical, an epistemological or sociological perspective.

But how can a successful introduction to philosophical subjects be combined with subjects, such as agricultural science and veterinary medicine, which are mainly taught as natural sciences? And why is it important to combine subjects in this way?

Over the last decade, the Royal Veterinary and Agricultural University in Denmark has offered courses on ethics to students in agricultural science and veterinary medicine. In what follows, we describe our own experience of developing and running these courses. We begin by discussing the societal need for agricultural and veterinary students to be educated in ethical issues. We then examine three courses dealing with philosophical subjects: two presently being offered in, respectively, agricultural science and veterinary medicine, and one that is being developed for students of agricultural science, agricultural economics and biotechnology. The paper concludes with a discussion of the conditions under which these kinds of course and other courses dealing with philosophical subjects are most likely to be successful in an area of study based on natural science.

2. The societal need for agriculturalists and veterinarians to understand ethics

Modern, highly productive and intensive agriculture has been publicly criticised for a number of years. Intensive food production systems are criticised for their negative environmental impact (e.g. nitrogen leaching and low levels of biodiversity as a consequence of pesticide use and monoculture), poor standards of animal welfare, unsafe products (e.g. BSE and salmonella), and potentially hazardous technologies (e.g. genetic modification). Traditionally agriculturalists and veterinarians have been loyal to farmers and agriculture, but there is a growing feeling that they should understand and accord respect to the perspective of civil society. Today's farmer, for example, may want advice on production methods that are not only profitable in the short run but also acceptable to society at large.

It is also true that some of the technical solutions to problems in farming offered by agricultural science and veterinary medicine have provoked negative responses. Prominent examples are the use of pesticides in crop cultivation and the preventative use of antibiotics in animal production. Thus a 'value gap' has developed between experts and lay people.

This value gap is especially noticeable in the case of gene technology. At present, Europeans are very critical of genetic engineering when the techniques are employed in relation to food (Durant *et al.*, 1996). Members of the scientific community have a much more positive attitude (Meyer and Sandøe 2001), however, and it is likely that a similarly positive attitude is taken by those in related professions, such as agricultural science and veterinary medicine.

A society in which public trust in intensive agriculture and industrialised food production, and indeed in experts in general, is being eroded presents a challenge to agricultural scientists and veterinarians. Developments in this trust relationship will inevitably affect students in their professional careers - e.g. as consultants, veterinary practitioners or researchers. The students therefore need to obtain a certain level of expertise in the issues involved in public scepticism. This expertise will involve the ability to acknowledge, and reflect on, a range of ethical perspectives on agricultural production and veterinary work.

The expanding role of animals in human society, e.g. from mainly being part of agriculture to becoming part of the family, has resulted in the development of distinct branches of veterinary medicine. Both between branches (e.g. large animal and small animal practice) and within the same branch different norms regulating the treatment of animals and their owners apply. Developments in the treatment of companion animals have provoked new ethical questions: How far are we willing to go in the treatment of sick animals? Is treatment with chemotherapy or organ transplantation acceptable? How is the value of a dog's life to be weighed against the costs of the treatment, the side effects of the treatment and the feelings of the owner?

Animals are also used for scientific purposes such as research on human disease. Reservations about the use of experimental animals are frequently expressed in society today. Often the public are sceptical and question whether the expected benefits of research can justify the suffering imposed on the animals.

In the course of their work, people with professional involvement in agriculture and veterinary medicine will often be expected to give 'expert opinions' and advice. Already, both veterinarians and agricultural scientists are asked to make judgements and decisions about ethical issues in their everyday working lives. If professionals are to cope with these requests, their training must acquaint them with a range of relevant discourses (on risk, the environment, safety, sustainability, animal welfare etc.), and with the values that these discourses embody. Without such acquaintance, it will be impossible for the professionals of the future to meet the expectations of society and perform their role in a satisfactory and satisfying way.

3. The experience of running a course in bioethics for students of agricultural science and agricultural economics

The course grew out of an existing module in environmental science and ethics established in the mid-1990s. Gradually the environmental science part of the course evolved into a branch of science, which needed to be viewed in an ethical perspective. From 1999 the course was renamed and officially converted into a course on ethics and agriculture.

Until January 2002, the course took place over an intensive three-week period (giving each student 6 ECTS), and each year 50 students have participated. The main objective was to provide the students with the ability to analyse ethical and other value-related controversies about agriculture, whether those controversies are between lay people and experts or involve experts only.

The challenge, in a course like this, is to combine knowledge of ethics (different traditions and conflicting theories), knowledge of agro-scientific subjects, and knowledge of the public debate (different stakeholders, different perceptions of risk etc.). During the first week, lectures, plenary discussions and small group sessions introduced ethical theory and various controversies relating to agriculture. A range of ethical theories were presented in their historical context, and different schools within animal ethics and environmental ethics were presented as well. A number of guest lecturers were invited, primarily from within the university, to present relevant cases.

The two first days of the course were devoted to a general introduction to ethics: we focused on the need for ethical reflection as a professional competence and looked at a few theoretical schools within ethics. Throughout both days, examples with direct relevance to agriculture were provided. The third day was devoted to animal welfare. It included a number of lectures by ethologists and lectures on animal ethics. The theme of the fourth day was the controversy over pesticide use in agriculture and the ideas of organic farming. During the day, the students also attended lectures on environmental ethics. On the fifth day, students looked at genetic engineering from different perspectives. A lecture was given by a scientific expert on plant biotechnology, and a lecture on public opinion and associated ethical positions was given by a sociologist. Discussions about the relation between biotechnology and world food supply, about patenting and EU and national regulation were raised. On the sixth day there was a presentation of the use of economic analysis to prioritise environmental policies and on the ethical assumptions underlying such analysis. The extent to which lecturers themselves included ethics in their presentations varied, but the coordinators, both of whom have a philosophy background, aim to bring relevant ethical points into discussions between the students and the guest lectures.

After the first, very intensive week, which was spent mostly in the classroom, the students were divided into groups (see Figure 1). They then choose a subject for project work from among prepared options. The options included: 'Is organic farming a good strategy in a world with food scarcity?', 'Animal welfare in pig or poultry production', 'Sustainable use of pesticides?', 'The use of economic incentives to reduce use of pesticides or artificial fertilisers'. To help the students to remain focused on interdisciplinary analysis, the coordinators gathered relevant literature. Each theme was anchored by a supervisor from the agro-scientific subject area of the project, and each group was assigned an ethics supervisor.

The project work was considered an essential pedagogical feature of the course, because it forced the students to apply ethical analysis to subject matter that they usually encounter as students in the agricultural university. This is why the course actively supported the interdisciplinary approach by having supervisors of both science and ethics assigned to each group.

At the end of the three-week period, two days were dedicated to a seminar at which all students attending the course were present to learn from the work of the other groups. Each group gave an oral presentation of their written report and was

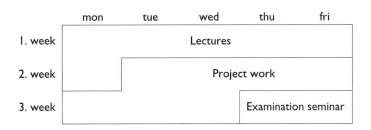

Figure 1. Structure of the bioethics course.

examined by the two supervisors. After the presentation and examination, the project group engaged in discussion with the rest of the students.

The course was non-compulsory and open to all students at the university, and those attending it did not need any special pre-qualifications. In practice most of the participants have been students of agricultural science or agricultural economics. However, students at all levels - from first year to final year - have attended. In most years there were a few forestry or landscape students in each class, and from time to time other students from other universities joined the course as well.

The pedagogical benefit of having a mixed group of students is, roughly, that it broadens the horizons of participants. Older students bring specialised agro-scientific knowledge to the course. Students with a farm background (those who have been raised on farms or were farmers themselves or have worked on farms) bring practical knowledge of crops, animals, farming systems and technologies. None of the students have formal qualifications in ethics, values or society, and it is a matter of individuality what they bring with them. Sometimes it seems that the youngest students were most comfortable with the kind of broad cultural/political discussion that is characteristic in Danish high school education. This perhaps gave them an advantage in interdisciplinary analytical work, probably because they were less specialised in their way of thinking.

The course was always thoroughly evaluated at its conclusion. The students gave a written evaluation in the form of replies to a questionnaire, and oral discussion of the course also took place. The evaluation was used to develop the course from year to year, e.g. by replacing lectures or supervisors to ensure the best outcome, adding opponent groups for the final examination seminar and changing textbooks. Evaluations to date, over four years, show that the students believe that the course is a relevant part of their education (see Figure 2). Efforts to combine ethical theory and real agricultural problems were also evaluated. The evaluations show continuous improvement in integration here (see Figure 3).

The course was held for the last time in 2002. It was closed not as a result of lack of success - it was actually one of the most popular optional courses offered to agronomy students - but because a compulsory course with broadly similar aims will replace it. The plans for this new course are described later in this chapter.

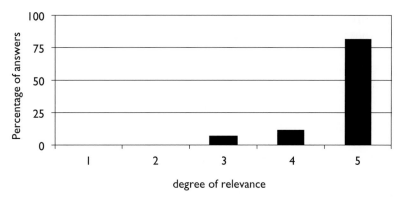

Figure 2. *Evaluation of course relevance. Students were asked: "Do you find this course relevant as a part of your education?" Score 5 was given for highest degree of relevance. The data are averaged over four years (1999-2002), covering answers from 149 students.*

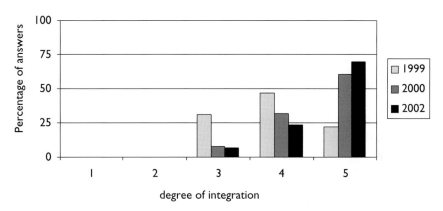

Figure 3: *Evaluation of the integration of ethics and agricultural science issues. Students were asked: "Is ethics being related to agricultural science issues in a satisfactory way?" Score 5 was given for most satisfactory integration. It can be seen that perceived integration improves between 1999 and 2002 (no data on this issue are available for 2001). The figure covers answers from 114 students.*

4. The experience of running a course in animal ethics for veterinary students

The course was first held in autumn 2001 as a compulsory part of the veterinary study programme (first year). It runs for 13 weeks, giving 6 ECTS, and is organised as a mixture of lectures (2 hours x 8 weeks), exercises (2 hours x 13 weeks) and group based project work (2-4 hours x 5 weeks) (see Figure 4). The aim of this arrangement is to combine both theory and practice on the one hand, and teacher directed and student directed learning on the other, so that the students' ability to handle ethical questions within the veterinary profession is raised.

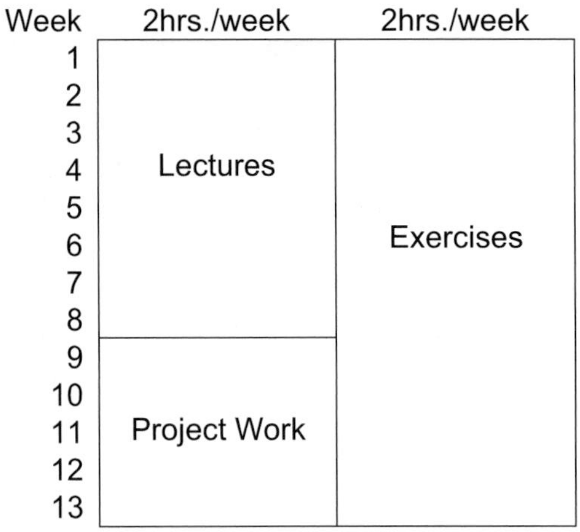

Week	2hrs./week	2hrs./week
1		
2		
3		
4	Lectures	
5		
6		Exercises
7		
8		
9		
10		
11	Project Work	
12		
13		

Figure 4. Structure of the animal ethics course

The immediate purpose of the course is to introduce different ethical positions to the students, and to give them the opportunity to work with and reflect upon ethical questions raised by the veterinary profession. The expectation is that, by the end of the course, students will have acquired skills and knowledge relating to:
• Identification and analysis of ethical questions raised by animal use
• Different tasks and roles in the veterinary profession and connected ethical questions
• Different scientific positions, in a historical perspective, and perceptions of animals within different scientific traditions
• Different perceptions of disease and health, viewed in a veterinary context
• Planning and working with a problem-based project in groups
• Oral and written scientific presentation

The lectures are arranged as 'double lectures' - each one combining the perspectives of a veterinarian and a philosopher. Different topics are dealt with each week, including: 'What is an animal?', 'Production animals', 'Transport and slaughtering',

'Exotic and wild animals', 'Companion animals', 'Food safety and risk assessment', and 'Animal experiments'.

In each lecture, a guest lecturer with a relevant, typical veterinary background first introduces some actual ethical dilemmas within his or her field of work. After this, a lecturer with a philosophical background elaborates on the dilemmas from a more explicitly ethical/philosophical angle. Students then have the opportunity to raise questions and debate with the two lecturers.

At the weekly theme-based exercises students reflect on 'real world' situations from the veterinary profession. The 'real world' situation is presented via description of a small case, role-play, a film, a questionnaire, a CD-ROM and so on. Questions to be discussed are then set out. The students are taught in classes of approximately 30 and work in groups of 4-6. After the group discussion, they summarise their discussion and conclusions for the class. Finally the teacher/facilitator picks-up the overall conclusions of the discussions and locates them in a broader ethical perspective.

The project work begins at the end of the eight-week period, and the last five weeks of the course are fully dedicated to it. The students work in groups of 4-6. Under guidance of the facilitator, they select an ethical dilemma or question within the seven veterinary themes and analyse it from at least two ethical viewpoints. Examples of project themes are: 'The use of laboratory animals for medical research', 'Leg amputation on dogs', 'Protection of the panda' and 'Pre-slaughter handling of pigs'. The students have access to various sources of information, including literature and people with knowledge of the chosen topic (resource persons), and it is up to them to make best use of these sources. They are introduced to literature searching at the University Library and attend lectures on written and oral communication.

Reports are presented orally at a final seminar for the same group of students as were together for exercises. All project groups have an opponent group, and this group provides them with oral feedback after their presentation. This, together with the facilitator's comments, forms the basis for the final evaluation of the project and the overall assessment of the student (pass/fail). Some groups are asked to prepare a 'postscript' using the facilitator's comments before they can pass the course.

During its first year, the students evaluated the course each week, halfway through and at its end (Christiansen *et al.* 2002). The evaluation was undertaken orally as well as through questionnaires assessing different aspects of the course by grades

1-5 (5 for maximum score). This rather rigorous monitoring was opted for in order to develop next year's course and to modify teaching during the present course. It was found that the students considered ethics important in a veterinary context. Many students explained that they achieved insights into ethical issues at an early stage in their courses. This meant that they were able to reflect on ethical issues that they would normally not be exposed to until later in their studies, and that they were equipped in a timely way with tools to handle ethical dilemmas. Through the discussions, the students developed their own personal opinions and arguments, but they also derived a real insight into other ways of viewing veterinary problems - something they felt would be useful in their professional career. A high proportion of the students (98%) felt better prepared to handle ethical questions after completing the course (see Figure 5).

The project work is considered by both students and teachers to be a very good learning process. It helps the students to develop their understanding of ethics, and more generally it is an effective *tour d'horizon* of a range of ethical perspectives on veterinary questions (see Figure 6). It also helps the students to develop critical thinking skills, and increase their moral imagination to address ethical dilemmas, which may have no clear solution or exact answer. Allowing the students to choose an ethical issue to scrutinize serves to strengthen the engagement of the students in their project work. There is also a social benefit in that the group work gives first year students an opportunity to get to know each other and thereby makes their early study easier.

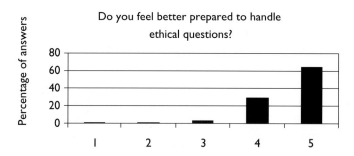

Figure 5. View of the students regarding their ability to handle ethical questions after participating in the course. Score 5 was given for the highest level of feeling prepared to handle ethical questions. The figure covers answers from 92 students (Christiansen et al. 2002).

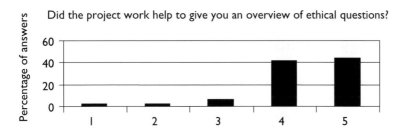

Figure 6. The students' view of the contribution of the project work. Score 5 was given for finding the project work most helpful. The figure covers answers from 92 students (Christiansen et al. 2002).

At the beginning of the course, many students feel that they lack the necessary professional background to discuss the ethical questions raised during the exercises. They acknowledge that this is taken into account in the teaching and in the teaching material, but they still think that they would benefit more from the course later in their studies. Others, and an increasing number as the course proceeds, find that the course is best taken early on, because it provides them with tools to handle ethical dilemmas they will meet later in their studies. Some have proposed a follow-up course at the end of their studies. This issue is yet to be discussed with the study board (Christiansen *et al.* 2002).

In general the students react to the course very well. They find it exciting, inspiring and relevant to them (see Figure 7).

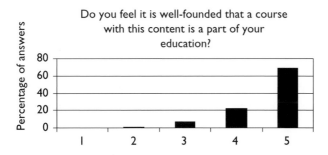

Figure 7. The students' view of the relevance of the course. Score 5 was given for highest degree of relevance. The figure covers answers from 91 students (Christiansen et al. 2002).

5. A new course: Introductory philosophy of science for agricultural science, agricultural economics, and biotechnology students

In spring 2003 a new compulsory course is offered to first year agricultural science, agricultural economics, and biotechnology students. It builds upon the experience of the two courses described above. There are two versions of the course. For the biotechnology students, it has exactly the same mix of exercises and project work as the veterinary course (6 ECTS) (see Figure 1). In the version for agricultural science and agricultural economics students (9 ECTS), a project work module has been added (see Figure 8). In fact, in this version of the course, more time will be spent on lectures and exercises on philosophical themes as time to work on the project will fall outside the course schedule itself; and topics such as sociology and theory of knowledge will also figure more prominently. Both versions of the course run for 13 weeks.

The first eight themes to be dealt with in the lectures, which are the same in both versions of the course, have an ethical perspective: 'Utilisation of the land: what is possible and what is acceptable?', 'The sustainability concept: the ethical dilemma', 'Plants of the future: what is the role of gene technology?', 'The use of gene technology in agriculture: the opinion of the lay people and NGOs', 'The ethical foundation for nature- and landscape management', 'Risk and scientific uncertainty:

Week	2hrs./week	2hrs./week	2hrs./week
1			
2			
3			
4			
5			
6			
7			
8	Lectures	Exercises	Project Work
9			
10			
11			
12			
13			

Figure 8. The structure of the introductory philosophy of science course (for students in agricultural science and agricultural economy)

what about things we don't know?', 'The global distribution of food' and 'Animal welfare: the ethological and epidemiological perspective'. The three last lectures focus on philosophy of science: 'Quantitative and qualitative science', 'What is science?' and 'Reductionism and holism'.

The lectures here are also arranged as 'double lectures': scientists from the agricultural science, biotechnology, economics and sociology will appear as guest lecturers together with a lecturer with a philosophical background. In the ensuing exercises, the students will have the chance to work on and discuss different questions and discussion points relating to the theme. To give an example, students might be asked to identify and discuss the different values behind various sustainability objectives in relation to 'real world' situations. In order to make the debate relevant to the students, examples will be adjusted their field of study.

For the project work, groups will be able to choose a problem or dilemma from within the eight ethical themes. Again they will have access to an ethics advisor and a resource person/facilitator with expertise in the chosen area. As the course has been organised for about 150 students divided into seven classes and subdivided into (approximately 30) groups of 4-6 persons, it requires quite a few resource personnel. These resource persons, who are mainly scientific staff at the university, have so far agreed to work on a voluntary basis.

6. What is needed to set up interdisciplinary ethics courses?

A number of factors have contributed to the success of the two courses that have already been run (whether the third will become a success remains to be seen):
- For a number of years, interdisciplinary research has been continuing in many of the subjects covered by the course - research, moreover, that involves an ethical perspective. This means that research-based knowledge of the ethical issues presented in the course is available.
- A Chair in Bioethics has been created at the university, securing a key person who is both in charge of teaching and doing research in the field of bioethics. This initiative sends a signal to students, staff at the university and institutions from which guest lectures are recruited that the subject is of importance.
- A number of high-profile teachers of veterinary medicine and the disciplines that go into the core curriculum of agricultural science and agricultural economy have been involved. These teachers give lectures and can help with supervision of the projects. The fact that these teachers, who serve as role models to the students, take ethical issues seriously sends an important signal to the students.

- The courses are systematically evaluated in relation to a set of explicit goals. Whenever some component in a course - be it a theme, a lecture or an exercise - does not work, as it should, appropriate changes are made.
- A group of four teachers from the field of learning and methodology have taken on the task of running the exercises and the project work. Their commitment to the idea of using ethics as a perspective from which students can synthesize ideas and knowledge is of great importance. It is therefore crucial that the further education of these teachers is supplemented by courses in ethics and other subjects of philosophy.

There may be other ways to realise the goals of the courses presented here. However, it is important to underline the two key ideas. First, ethical theory must be taught in tandem with real agricultural and veterinary problems and actual issues within agricultural science and veterinary medicine. Using real examples highlights the relevance of ethics to the chosen field of study. Secondly, students must work on projects forcing them to view an issue within agricultural science, agricultural economy or veterinary medicine from at least two ethical perspectives. This way, the project work facilitates and stimulates active and independent reflection on ethical issues.

References

Christiansen, S.B., Dich, T., Hansen, T., and Sandøe, P., 2002. Evaluering af kursus 103018, Veterinærmedicinsk videnskabsteori, efterår 2001. Den Kgl. Veterinær- og Landbohøjskole.

Durant, J., Bauer, M.W. and Gaskell, G. (eds.), 1996: Biotechnology in the public sphere - A European source book. Science Museum, London.

Meyer, G. and Sandøe, P., 2001. Dialogue on biotechnology in relation to plants. Project Report 1, Centre for Bioethics and Risk Assessment, Copenhagen.

What is your conception of ethics?
A questionnaire for engineer and science students

Laurent Rollet[1] and Michel Marie[2]
[1]*ENSGSI, INPL-Nancy, 8, rue Bastien Lepage, 54000 Nancy, France*
[2]*Sciences Animales, ENSAIA, INPL-Nancy, B.P. 172, 54505 Vandoeuvre cedex, France*

Abstract

In order to help engineers and scientific students to construct their own ethical thinking, a questionnaire based on sentences extracted from philosophical works belonging to different schools of thoughts has been designed. This paper describes the development of this test (named "What is your conception of ethics?"), and its final refinement. Sentences from 31 authors, representative of ethics of Good, Duty, Utilitarianism, or Modern ethics have been used to build a set of 45 quotations to which students had to give a quick and spontaneous appreciation. The general degree of interest for ethical issues was also recorded, and this exercise lead to the drawing of an individual ethical cartography. A more efficient form of the questionnaire has been finally elaborated on the basis of 30 quotations from 25 authors. Multi-factorial analyses of the sets of data gave detailed information on the types of responses and on the weight of individual quotations in the definition of the profiles. They showed also a link between Good and Duty ethics and positive interest on one hand, and between Utilitarianism and Modern conceptions on the other hand. In our conditions, 40 % of the 260 students involved in the testing of the method were shown to be clearly determined for one particular trend, and Duty ethics appeared to be the most popular. This method was found to be attractive, and viewed as a good basis for further discussions and reinforcement of the moral philosophy framework developed during other teaching activities.

1. Introduction

The advancement of science and technology constitutes a legitimate source of wonder, but it also raises many questions. It is thus not very surprising to notice an increasing concern of the general public for ethical issues. From this point of view, engineer schools and science faculties often try to widen the horizons of thought of their students by providing them occasions of an ethical reflection. Since these students will be the principal actors of scientific and technological development, it is of the utmost importance to create spaces of ethical training for them, the major stake being the elaboration of a personal thinking on the ethical implications of their future scientific practices.

To achieve such a goal two educational strategies might be adopted. The first strategy could be to rely on a problem-based-learning approach. Knowing that the main thing is to train technicians, engineers or researchers (and not philosophers), one can legitimately suppose that this ethical training should be short, effective, centred on practical issues and shouldn't give too many details about philosophical questions. The second strategy could consist in a philosophy-based-learning methodology. In this perspective, the main presupposition would be that it is almost impossible to propose a teaching in ethics which wouldn't have any philosophical content and which would completely overlook the history of moral doctrines. Of course, the two educational approaches are not contradictory and it is likely that most of the existing bioethics trainings are based on a mixed strategy (both applied and theoretical). The major stake is to elaborate an applied teaching of ethics (medical ethics, animal bioethics, etc.) while taking into account the deepness and richness of philosophical reflection.

Since several years, the École Nationale Supérieure d'Agronomie et des Industries Alimentaires of Nancy proposes to its students a training in bioethics. This optional teaching is devoted to practical issues (animal bioethics), but its aim is also to provide an introduction to the major historical trends of moral philosophy. It consists of several lectures, but our objective is also to involve students in a personal and spontaneous reflection, i. e. helping them to establish connections between their personal moral values and several essential notions, which constitute the core of ethical thinking.

In this perspective, we tried to elaborate a new educational device. We created an 'ethical questionnaire' entitled "What is your conception of ethics?". This questionnaire is composed of a set of philosophical quotations devoted to ethical issues. For each single quotation, students are asked to give their own personal evaluation, using a numerical scale interpreted in terms of agreement or disagreement. The whole questionnaire is organised according to a subjacent philosophical structure that corresponds to several representative trends in moral philosophy. Through this questionnaire we have several objectives:

1. To give to each single student an 'ethical diagnosis' on the basis of his/her spontaneous answers.
2. To provide an evaluation of the students' interest for ethical issues.
3. To show the students that ethical issues are in connection with some philosophical problems and that they cannot be solved by a simple decision device[89].
4. To provide an introduction to moral philosophy and to lay the foundations of discussions on contemporary (bio)ethical issues.
5. To arouse a real and individual reflection on ethical questions.

This paper aims at giving an outlook of the building and use of this philosophical questionnaire, and will be largely opened to methodological reflections. In a first part, we'll give a formal description of the first version of our questionnaire and an explanation of our philosophical presuppositions concerning its functioning. In a second part, we'll provide further details about its practical functioning and the way it was used in several educational situations; this will lead us to provide some statistical information on the students' answers. Finally, we'll conclude with some methodological remarks concerning this ethical tool and we'll describe its final revised version.

2. Description of the questionnaire

As far as we know, this questionnaire is unique in the field of ethics education in France (at least in engineer schools and faculties of science). A first version of this questionnaire was elaborated and used in 2000 by Alain Giré and Laurent Rollet at the Institut National des Sciences Appliquées de Lyon (Giré 2001). Nevertheless, the questionnaire described here is very different in its form and functioning and was elaborated in 2002 at the Institut National Polytechnique de Lorraine. It was first intended as an educational device for science and engineers students and it is the result of a collective thinking concerning ethics and philosophy teaching for non-specialists. Since it is a relatively new device, it has only been used during two years and it must therefore be considered as a work in construction.

2.1 The corpus of authors

Our questionnaire has been formally conceived as a set of 45 quotations dealing with ethical issues and organized according to a subjacent structure. A large part of the quotations is philosophical, in the sense that it comes from well-identified classical philosophical works. Nevertheless, it is not a simple collection of definitions about ethics or morality; apart from several definitions, one can also find paradoxical formulations, aphorisms, critical and provocative judgements. Moreover, our corpus of authors is very large since it includes philosophers, novelists, psychoanalysts, and so on (Table 1).

Our basic methodological principle concerning this corpus was to give all quotations without any information concerning the names of the authors or their cultural background (occidental versus oriental philosophy, for instance), the books they come from or the time they were formulated. This means that all quotations were placed on an equal footing. We neither altered or simplified them nor changed any of their technical terms. Moreover, as far as possible, we tried to select quotations that were in accordance with their authors' philosophical doctrines[90].

Table 1. The corpus of authors.

1	Aristotle	17	Helge Krog
2	Marcus Aurelius	18	Jacques Lacan
3	Yves Barel	19	Emmanuel Levinas
4	Jeremy Bentham	20	Niccolà Machiavelli
5	Henri Bergson	21	Karl Marx
6	Ambrose Bierce	22	John Stuart Mill
7	Albert Camus	23	Friedrich Nietzsche
8	Chang-Hung	24	Max Planck
9	Alexis De Tocqueville	25	Karl Popper
10	René Descartes	26	Svâmi Prajnânpad
11	Alain Etchegoyen	27	John Rawls
12	Jürgen Habermas	28	Jean-Paul Sartre
13	Ernest Hemingway	29	Seneca
14	David Hume	30	Baruch Spinoza
15	Hans Jonas	31	Max Weber
16	Immanuel Kant		

[90] Nevertheless, it was not always possible because some quotations sometimes bear multiple interpretations. Furthermore, it was quite difficult to link some of the aphorisms and paradoxical quotations with the eventual philosophical doctrines of their authors (this problem holds for instance for Hemingway, Bierce or Krog).

2.2 The evaluation scale

The main thing concerning this questionnaire is that it is intended as a spontaneous device. The evaluation of each quotation must be done quite rapidly. This means that the agreement or disagreement concerning these ethical sentences is a matter of a spontaneous judgement. Our purpose is to reveal the students that their 'naïve', personal and subjective judgements can be linked with an "objective" and structured philosophical corpus.

Students are asked to mark each judgement with a figure between 0 and 100 (Table 2). Nevertheless they have a total freedom concerning the degree of precision of the grades (it can be 10, 20, 32, 78 etc.).

The questionnaire was designed to offer the students two essential pieces of information: on the one hand, an indication of their degree of interest for ethical issues; on the other hand, a 'philosophical cartography' of their ethical positions according to their answers. In this perspective, we made a distinction between two kinds of quotations: a first set of 11 quotations was selected according to its ability to express a positive or a negative position concerning ethical reflection; a second set of 34 quotations was chosen for its capacity to express explicit philosophical theses concerning ethics.

Table 2. The evaluation scale.

0	50	100
I totally disagree	Indecision	I totally agree

2.3 Evaluation of the degree of interest for ethical issues

The first set contains quotations such as:
- "Have I done something for the general interest? Well, then, I have had my reward. Let this always be present to thy mind, and never stop doing such good". (Marcus Aurelius, *Meditations*). [Q1 in Table 4]
- "Moral: Conforming to a local and mutable standard of right. Having the quality of general expediency. Immoral: Inexpedient". (Ambrose Bierce, *The Devil's Dictionary*). [Q22]

The first quotation is rather positive and implies a kind of voluntarism concerning ethics, whereas the second is more negative and quite cynical. On this basis, positive quotations were given a positive coefficient (1) and negative quotations a negative one (-1).

The evaluation of the students' interest for ethics is the result of the addition of the marks proposed to the 11 quotations. However, our reasoning for this evaluation is the following: a good mark for a positive judgement (in example 78) must imply some sort of interest for ethics while a good mark for a negative judgment implies probably a kind of scepticism or a lack of interest; therefore, it must be counted as a negative figure (for instance -78) and subtracted from the general amount. Finally, the result of the addition of the 11 marks gives a figure, which aims at measuring the intensity of the students' interest, on an intensity scale ranging from -400 to +700.

Of course, this indication doesn't pretend to be objective and the purpose is not to express a judgement on the students. It is rather conceived as a trick for a thorough discussion.

2.4 Ethical 'cartography'

The second set of 34 quotations is far more essential and constitutes the core of the original questionnaire. Each quotation expresses a specific ethical position that can be linked with a philosophical trend. We thus had to choose an ethical classification so as to organise the whole set. Many classifications could be envisaged, some of them being very technical and complex. Considering our educational purpose (ethical training for non-philosophers) we decided to adopt a relatively simple and classical structure. We thus organised our 34 quotations according to 4 major philosophical trends (some sentences referring to two trends):
1. Ethics of Good: 7 quotations;
2. Ethics of Duty: 7 quotations;
3. Utilitarianism: 7 quotations;
4. Contemporary Ethics: 16 quotations.

Currents (1), (2) and (3) correspond to a classical and traditional classification of moral reflection that can be found in many philosophical handbooks, such as Armand Cuvillier's, *Cours de Philosophie* (Cuvillier 1954). The fourth current is conceived as an aggregation of various modern positions concerning ethics; it is therefore represented by a larger amount of quotations[91].

---

[91] For reasons of concision and in order to avoid mistranslations, we will not present the whole set of quotations but only a selection. On the other hand, one will find at the end of this article a complete presentation of the final version of the questionnaire.

Ethics of Good[92] is based on the idea that Good is a value towards which any person should tend. Indeed, this philosophical trend represents an ideal of wisdom and virtue but it is also opened to an eudemonist conception of life in the sense that the pursuit of Good is an essential element of happiness. Ethics of Good finds its purest representations in ancient moral traditions such as Plato's or Aristotle's. Nevertheless, Spinoza, Malebranche, or even utilitarian philosophers such as John Stuart Mill developed similar conceptions. Basically, one can distinguish two kinds of the Ethics of Good, and the difference depends of the definition given to the notion of Good. On the one hand, many philosophers would consider Good as a metaphysical entity, as an absolute idea and put forward the necessity of a spiritual elevation in order to reach it (Aristotle, Spinoza). On the other hand, it is also possible to propose an empirical definition of Good in terms of concrete and sensible reality; in such a perspective, Ethics of Good could be linked with Epicurean philosophy, as well as with utilitarianism. Here are some of the quotations of our questionnaire concerned with this first moral trend:

- "Every art and every inquiry, and similarly every action and pursuit, is thought to aim at some good; and for this reason the good has rightly been declared to be that at which all things aim". (Aristotle, *The Nichomachean Ethics*). [Q4]
- "Blessedness is not the reward of virtue, but virtue itself; neither do we rejoice therein, because we control our lusts, but, contrariwise, because we rejoice therein, we are able to control our lusts". (Spinoza, *Ethics*). [Q13]
- "For it's not enough to have a good mind. The main thing is to apply it well. The greatest souls are capable of the greatest vices as well as the greatest virtues, and those who proceed only very slowly, if they always stay on the right road, are capable of advancing a great deal further than those who rush along and wander away from it". (Descartes, *Discourse on Method*). [Q3]

Ethics of Duty can be seen as a representation of morality in which the notion of Duty is clearly separated from the notion of Good. Duty can then be considered as a pure obligation, without any justification. This kind of ethics is often leashed together with the idea of transcendence, i. e. the idea of rigid authority independent from human will. It is possible to distinguish several conceptions of this Ethics of Duty. First, one could consider duty as an arbitrary divine commandment; this was for instance the case of several Middle-Age theologians who claimed that God is the one and only source of the essence of Good and that man is totally obliged to model his conduct to these commandments. Secondly, one can define Duty as the obedience to a tradition. Social order, as well as moral order, is thus seen as a constraint that man has to accept. There are of course various forms of this ethics

---

[92] This philosophical trend can be clearly linked with Virtue Ethics.

of tradition but most of them are authoritarian, pessimistic and anti-individualist[93]. Thirdly, the most important kind of Ethics of Duty is represented by Kant's moral philosophy. It is based on a rationalist interpretation of the notion of duty, in the sense that it can be considered as the transcendence of reason for man. The law of reason is thus a categorical imperative. Kant's moral doctrine claims that the only moral intention is the intention that implies an action in conformity with duty. It is a formal conception of morality which finds its source in three principles: a principle of universality, a principle of respect of human dignity and a principle of autonomy. Our questionnaire contains several quotations devoted to the Ethics of Duty, but most of them are clearly linked with a Kantian conception of duty. Kant's formulations are therefore very present, as well as quotations of Hans Jonas, who is a contemporary representative of this deontological ethics:

- "Hence also morality is not properly the doctrine how we should make ourselves happy, but how we should become worthy of happiness". (Kant, *Critique of Practical Reason*). [Q6]
- "Act according to a maxim which can be adopted at the same time as a universal law". (Kant, *Groundworks for the Metaphysics of Morals*). [Q10]
- "Man is the only being that can have a responsibility; this power leads to the duty. The capacity of responsibility rests on man's ontological faculty to choose, knowingly or deliberately, between alternatives of the action". (Hans Jonas, *The Imperatieve of Responsibility*).[94] [Q17]

From the ethical point of view, utilitarianism is a theory that melts its principles of justice and research of happiness not on an ideal standard but on a real standard (resulting from observation and experience). The utilitarian philosopher claims that the general sum of satisfaction (or average utility) is the source of justice. Things that benefit the greatest number of people, and which increase the total balance of satisfaction for a group or a given community can be considered as right. Utilitarianism doesn't claim that the notion of Good must meet that of utility but it is rather presented in the form of a theory of the common good, founded on a liberal conception of community life. Together with Jeremy Bentham (1748-1832), John Stuart Mill (1800-1873) is one of the great figures of utilitarianism. Bentham thus claimed: "The nature placed the humanity under the authority of two sovereign masters, the pain and the pleasure. Them alone indicate what we must make and determine what we will make. [...] The principle of utility recognizes this vassalage

---

[93] It is for instance the case of Joseph de Maistre's and Louis de Bonald conceptions. The latter thus claimed that "in society, there are no rights but only duties".

[94] Our own translation. The sentence in French is the following: "L'homme est le seul être connu qui puisse avoir une responsabilité ; le pouvoir même entraîne avec lui le devoir. La capacité de responsabilité repose sur la faculté ontologique de l'homme à choisir, sciemment ou délibérément, entre des alternatives de l'action."

and room to the foundation of the system whose goal is to build the factory of happiness to the means of the reason and the law". Utilitarianism is very much appreciated in the Anglo-Saxon world and its schemes are very present in the resolution of the ethical problems raised by scientific and technical development: indeed, ethics committees generally try to solve dilemmas by building utility calculi of the type Cost / Benefit. Utilitarianism is thus well represented in the field of bioethics. Here are some of the quotations that we chose in order to present this philosophical trend:

- "The said truth is that it is the greatest happiness of the greatest number that is the measure of right and wrong". (Bentham, *A Fragment on Government*). [Q11]
- "Actions are right in proportion as they tend to promote happiness, wrong as they tend to produce the reverse of happiness. [...] Happiness which forms the utilitarian standard of what is right in conduct, is not the agent's own happiness, but that of all concerned". (J. S. Mill, *Utilitarianism*). [Q21]
- "What you judge good afterwards is moral, what you judge badly afterwards is immoral" (Hemingway)[95]. [Q23]

Finally, some quotations of the questionnaire refer to modern or post-modern ethical positions. We deliberately chose to propose this fourth category in order to counterbalance the classical representation of ethics proposed in the first 3 trends. This modern class is quite heterogeneous but its aim is to propose several points of view that could not be clearly linked with the Ethics of Good, Ethics of Duty or Utilitarianism. Most of the 16 quotations of this category are at odds with traditional ethics and can be ordered according two major orientations.

The first orientation relates to a set of moral conceptions that one could qualify as existentialist. They put the concept of subjectivity (and inter-subjectivity) forward and they insist largely on the concepts of autonomy and free will. They put the stress on the openness of ethical choices and on the capacity of each individual to create his own values and to become actor of his own life. Various authors are represented in this first category (Sartre, Levinas, Barel, Etchegoyen) and some of them can be regarded as existentialists only in a broad interpretation of the term (Bergson for instance). Here are several representative quotations:

---

[95] Our own translation. Its French version is the following: "Ce qui est moral c'est ce que vous jugez bon après, ce qui est immoral c'est ce que vous jugez mal après".

- "Nature informs us by a precise sign that our destination is reached. This sign is joy. I say joy, I do not say pleasure. Joy always announces that life has succeeded, that it gained ground, and wherever there's joy, there's creation". (Bergson, *La conscience et la vie*)[96]. [Q32]
- "Man is nothing else than his project. He only exists insofar he fulfils himself. He is thus nothing else than the whole set of his actions, nothing else than his life". (Sartre, *L' existentialisme est un humanisme*)[97]. [Q27]

The second orientation relates to sceptical, critical - or even cynical - positions towards morality. These various conceptions insist on the alienating power of moral prescriptions and on the fact that they restrain the expression of passion, desire or vitality. They thus put forward the idea of a natural morality and a conception of ethics that takes psychoanalytic approaches into account. Among the various authors represented here, we'll mention Nietzsche, Marx or Lacan.

- "Laws, ethics, religion are in my eyes as many bourgeois prejudices behind which hide as many bourgeois interests". (Marx)[98]. [Q16]
- "Anti-natural morality - that is, almost every morality which has so far been taught, revered, and preached - turns, conversely, against the instincts of life: it is condemnation of these instincts, now secret, now outspoken and impudent". (Nietzsche, *Twilight of the Idols*). [Q14]
- "The only ethical fault is to renounce desire". (Lacan)[99]. [Q18]

Besides these major trends, one can also find quotations from other prominent authors. They aim at presenting such ideas as the ethics of discussion (Jürgen Habermas [Q35]), the principle of equity in the context of a liberal thinking (John Rawls [Q21, Q30]) or the principle of responsibility as an answer to the problems of technological civilisation (Hans Jonas [Q14, Q16]).

To conclude, this classification doesn't assert to be totally exhaustive. However, we tried to choose a structure able to offer the students a general classification of moral philosophy.

---

[96] Our own translation. The French quotation is: "La nature nous avertit par un signe précis que notre destination est atteinte. Ce signe est la joie. Je dis la joie, je ne dis pas le plaisir. La joie annonce toujours que la vie a réussi, qu'elle a gagné du terrain, et partout où il y a de la joie, il y a création."

[97] Our own translation. The French quotation is: "L'homme n'est rien d'autre que son projet, il n'existe que dans la mesure où il se réalise, il n'est donc rien d'autre que l'ensemble de ses actes, rien d'autre que sa vie."

[98] Our own translation. The French quotation is: "Les lois, la morale, la religion sont à mes yeux autant de préjugés bourgeois derrière lesquels se cachent autant d'intérêts bourgeois."

[99] Our own translation. The French quotation is: "La seul faute éthique est de céder sur son désir."

2.5 Concluding remarks on the functioning of the questionnaire

During the development of this questionnaire, the major difficulty was to choose the suitable quotations for each trend. This implied to solve at least three problems. First, we had to assemble a large collection of quotations from various sources. Secondly, we had to insure the philosophical coherence of the corpus so that our ethical diagnosis was acceptable (which doesn't mean, as we shall see later, that the students' positions are always coherent). Thirdly, since the answers to the questionnaire must be spontaneous, we had to estimate the degree of complexity of each single quotation in order to avoid misinterpretations.

The functioning of our evaluation grid is therefore the following (see Table 3). It consists of a six columns table. If a student disagreed with quotation 1 and proposed a grade of 26, this figure is multiplied by the coefficient corresponding to the suitable philosophical trend and the result is written in the appropriate cell. If the student globally agreed with quotation 2 and proposed a mark of 67, this figure is multiplied once by coefficient 0,33 (Duty) and twice by coefficient 0,67 (Utilitarianism), since the quotation bears a pluralist interpretation. The filling up of the table can be made quite quickly following this simple rule. At the end of the table, students are asked to add up the notes of each column and to divide the result by a specific figure, which corresponds to the number of quotations in each philosophical trend.

The final figures (in our example 32, 61, 45, 50) represent the global repartition of the student's answers according to each philosophical trend. Finally, it is possible to propose a 'cartography' of his ethical position with the help of a radar graph (see Figure 5 for examples).

Table 3. Overview of the evaluation grid.

Number of the quotation	Mark	Good	Duty	Utilitarianism	Modern
Quotation 1	26	1			
		26			
Quotation 2	67		0,33	0,67	
			22,11	44,89	
Quotation 3	80				1
					80
...					
Total		224	427	315	800
Divide by		7	7	7	16
Final mark (example)		32	61	45	50

3. Use of the questionnaire in the classroom

3.1 Modalities of use

The original questionnaire has been submitted to a total of 220 students belonging to schools of engineers located in Nancy (France), in nine different sessions. Students were asked to express their approval/disapproval to each of the 45 quotations in a time generally comprised between 60 and 75 minutes. When understanding difficulties arose, comments were done in order to avoid misinterpretation, but taking care not to influence the responses. At the end of this phase, students were asked to report the scores on a second sheet (evaluation grid) disclosing the coefficients to be applied, and to compute the sum for each column, obtaining a final value to be reported on an ethical profile and on a scale of interest.

Immediate exploitation consisted in the identification of the schools of thought represented in the classroom and in comments, for each of these ones, of the most representative sentences.

3.2 Responses to the original questionnaire

The worst noted quotation, with a mean score of only 15.4 / 100, belongs to the ethics of Good [Q39]; 80% of the responses being inferior or equal to 20 (Figure 1A). Surprisingly, two essential sentences, one from Aristotle [Q4, score: 28] and one from J. Bentham [Q11, score: 35] are also badly noted, with, nevertheless, 23 % and 38 % of the responses greater than 50 (Figure 1 B - C).

On the other hand, the best ranked quotations (Figure 1, D - G) are sentences from H. Jonas [Q19 and Q17, scores: 70], J. Rawls [Q28, score: 73] and Aristotle [Q35, score: 80]; this last one, with 87 % of the notes greater than 60 and 55 % greater than 90, is in fact not discriminating.

Numerous quotations received a score close to 50 (Figure 1 H - J); in some cases, this denotes an indeterminate position, with scores centred around 50 [Q1, with 45 % of the responses lying between 40 and 60], when, in other situations [such as Q31 and 9], both favourable, negative and neutral responses are observed, which reveals a diversity of opinions.

Mean scores for each school of thought were 58 for Duty, 53.5 for Modern, 52.5 for Utilitarianism and only 47.7 for Good. On a scale from -400 to +700, mean value for interest was 160, with 67 % of the scores lying between +100 and +300, and 50 % between +100 and +226.

A principal components analysis has been conducted on this set of data in order to identify the quotations contributing the most to each trend, and to evaluate the consistency of the students' positions. For ethics of Good, Q9 and Q13 are well correlated ($r = 0.346$), and Q2 and Q9 show a still significant but much lower correlation ($r = 0.158$); all these quotations refer explicitly to virtue. Surprisingly, a classical sentence from Aristotle [Q4] referring to Good is not (or negatively) correlated to these first three. For ethics of Duty, three sentences from I. Kant [Q 6, 10 and 15] are only moderately correlated (r between 0.145 and 0.174), and sentence Q17, from H. Jonas, is slightly linked to Q10. Higher correlation (between 0.200 and 0.266) is observed with sentences associated to other schools of thought, or with ethical interest. Considering Utilitarianism, the strongest link is observed between one sentence from D. Hume [Q5] and one from J.S. Mill [Q21, $r = 0.169$]. Such contradictory observations denote some lack of coherence between the students' responses. We tried to reduce these discrepancies, as far as they could be due to the methodology itself, by refining this first version of the questionnaire.

3.3 Elaboration of the final questionnaire

In order to facilitate the interpretation of the results of the test, and to enhance its ergonomics, this first set of observations has been taken into account for the elaboration of a final version, based on only 30 quotations: ambiguous sentences have been suppressed, each classical school of thought has been represented by the four most typical and discriminative sentences, the modern trends by twelve, and eight quotations have been selected in order to detect positive (4) or negative (4) interest in ethics (two of which are also associated to modern trends). This new questionnaire is presented in Table 4.

Forty more students have been confronted to this new form, so all together 260 responses have been analysed using this 30 quotations grid.

3.4 Responses to the final version of the questionnaire

The distribution of scores for each school of thought is presented on Figure 2; the best mean is observed for the ethics of Duty (57.5), with some individual responses reaching 95 or 98; then come Modern (55.5), Utilitarian (49.1) and Good (45.4) trends. Interest score (Figure 3) is distributed between -295 and +285, with a mean of 45.4.

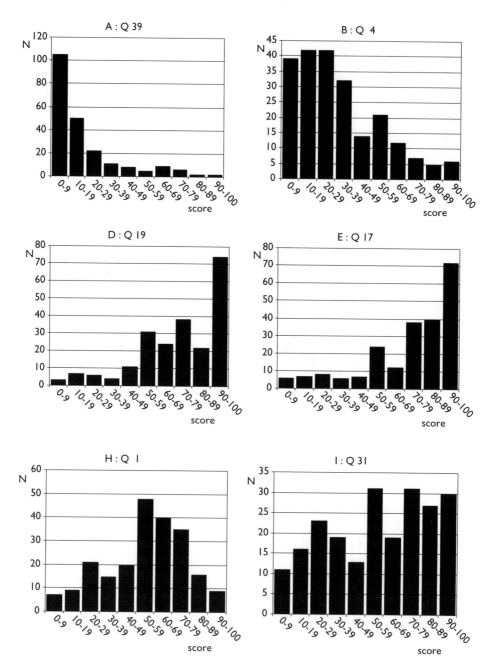

Figure 1. Observed responses for quotations with low (A-C), high (D-G) or medium (H-J) scores (N=220).

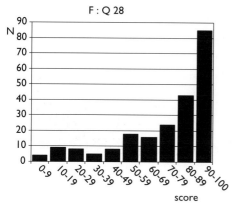

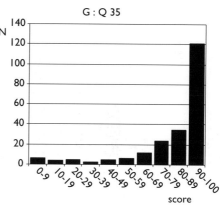

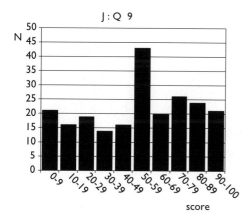

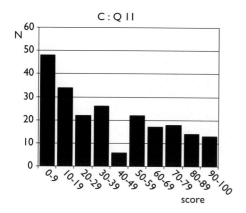

Table 4. Final set of quotations.

No	Quotation	Good	Duty	Util.	Mod.	Int.
1	Have I done something for the general interest? Well, then, I have had my reward. Let this always be present to thy mind, and never stop doing such good]. **Marcus Aurelius**, *Meditations*					I
2	A honest person doesn't care about the lack of respect towards her statute; she only cares about the lack of respect towards her virtue. She isn't ashamed if her salary is low; she's ashamed only if her wisdom is limited. **Chang Hung**	I				
3	For it's not enough to have a good mind. The main thing is to apply it well. The greatest souls are capable of the greatest vices as well as the greatest virtues, and those who proceed only very slowly, if they always stay on the right road, are capable of advancing a great deal further than those who rush along and wander away from it. **René Descartes**, *Discourse on Method.*					I
4	Every art and every inquiry, and similarly every action and pursuit, is thought to aim at some good; and for this reason the good has rightly been declared to be that at which all things aim. **Aristotle**, *The Nichomachean Ethics.*	I				
5	We must renounce the theory, which accounts for every moral sentiment by the principle of self-love. We must adopt a more public affection, and allow, that the interests of society are not, even on their own account, entirely indifferent to us. Usefulness is only a tendency to a certain end; and it is a contradiction in terms, that anything pleases as means to an end, where the end itself no wise affects us. **David Hume**, *An Enquiry Concerning the Principles of Morals.*					I
6	Hence also morality is not properly the doctrine how we should make ourselves happy, but how we should become worthy of happiness. **Immanuel Kant,** *Critique of Practical Reason.*		I			
7	There doesn't exist any ethics in the world that can neglect this: in order to reach 'good' ends we are most of the time obliged to take into account, first morally (or at the very least dangerous) dishonest means, and second the possibility or eventuality of annoying consequences. No ethics in the world is neither able to tell us when and how a morally good end justifies the means and the morally dangerous consequences. **Max Weber**, *Science as a vocation, politics as a vocation.*				I	
8	If I had to write a book on ethics, it would consist of 99 white pages and on the last one I would write: Love is enough! **Albert Camus**.				I	

Table 4. Continued.

No	Quotation	Good	Duty	Util.	Mod.	Int.
9	It is thus an error to ask the reason why one seeks virtue. Because it is to want something that is above the supreme. You ask what I seek in virtue? Virtue itself, because there's nothing better, because it's itself its own price. **Seneca**, *About Happy Life.*	I				
10	Act according to a maxim which can be adopted at the same time as a universal law. **Immanuel Kant**, *Groundworks for the Metaphysics of Morals.*		I			
11	The said truth is that it is the greatest happiness of the greatest number that is the measure of right and wrong. **Jeremy Bentham**, *A Fragment on Government.*			I		
12	Morality is the best tool in order to lead humanity by the nose. **Friedrich Nietzsche**.				I	-I
13	Blessedness is not the reward of virtue, but virtue itself ; neither do we rejoice therein, because we control our lusts, but, contrariwise, because we rejoice therein, we are able to control our lusts. **Baruch Spinoza**, *Ethics.*	I				
14	Anti-natural morality - that is, almost every morality which has so far been taught, revered, and preached - turns, conversely, against the instincts of life: it is condemnation of these instincts, now secret, now outspoken and impudent. **Friedrich Nietzsche**, *Twilight of the Idols.*				I	
15	So act as to treat humanity, whether in thine own person or in that of any other, in every case as an end withal, never as means only. **Immanuel Kant,** *Groundworks for the Metaphysics of Morals.*		I			
16	Laws, ethics, religion are in my eyes as many bourgeois prejudices behind which hide as many bourgeois interests. **Karl Marx**				I	-I
17	Man is the only being that can have a responsibility; this power leads to the duty. The capacity of responsibility rests on man's ontological faculty to choose, knowingly or deliberately, between alternatives of the action. **Hans Jonas**, *The Imperative of Responsibility.*		I			
18	The only ethical fault is to renounce desire. **Jacques Lacan**.				I	
19	Act in such way that the effects of your action are compatible with permanence of an authentically human life on Earth. **Hans Jonas**, *The Imperative of Responsibility.*				I	
20	Man makes himself; he is not made beforehand, he makes himself by choosing his morality, and the pressure of circumstances is such that he's obliged to choose one. **Jean-Paul Sartre**, *L' existentialisme est un humanisme.*				I	

Table 4. Continued.

No	Quotation	Good	Duty	Util.	Mod.	Int.
21	Actions are right in proportion as they tend to promote happiness, wrong as they tend to produce the reverse of happiness. [...] Happiness which forms the utilitarian standard of what is right in conduct, is not the agent's own happiness, but that of all concerned. **John Stuart Mill**, *Utilitarianism*.			I		
22	Moral: Conforming to a local and mutable standard of right. Having the quality of general expediency. Immoral: Inexpedient. **Ambrose Bierce**, *The Devil's Dictionnary*.					-1
23	What you judge good afterwards is moral, what you judge bad afterwards is immoral. **Ernest Hemingway**.			I		
24	The only standards that can claim to be valid are those which could be accepted by all the people concerned, within the framework of a practical discussion. **Jürgen Habermas**, *Remarks on discourse ethics*.				I	
25	I seek to trace the novel features under which despotism may appear in the world. The first thing that strikes the observation is an innumerable multitude of men all equal and alike, incessantly endeavouring to procure the petty and paltry pleasures with which they glut their lives. Each of them, living apart, is as a stranger to the fate of all the rest - his children and his private friends constitute to him the whole of mankind; as for the rest of his fellow-citizens, he is close to them, but he sees them not - he touches them, but he feels them not; he exists but in himself and for himself alone; and if his kindred still remain to him, he may be said at any rate to have lost his country. **Alexis de Tocqueville**, *Democracy in America*.				I	
26	It is necessary to whoever arranges to found a Republic and establisch laws in it, to presuppose that all men are bad and that they will use their malignity of mind every time they have the opportunity. **Niccolò Machiavelli**, *The Discourses upon the First Ten Books of Titus Livy*.					-1
27	Man is nothing else than his project. He only exists insofar he fulfils himself. He is thus nothing else than the whole set of his actions, nothing else than his life. **Jean-Paul Sartre**, *L' existentialisme est un humanisme*.				I	
28	Each person has an equal right to a fully adequate system of equal basic freedoms, which is compatible with the same system of freedoms for all. **John Rawls**, *Political Liberalism*.				I	

Table 4. Continued.

No	Quotation	Good	Duty	Util.	Mod.	Int.
29	If, in every domain, the triumph of the life is creation, don't we have to assume that human life has its raison d'être in a creation, which, unlike that of the artist and the scientist, can continue constantly for each men. The creation of oneself by oneself, the expansion of personality, which draws much from little, something of nothing and unceasingly add to the richness of the world. **Henri Bergson**, *La Consciousness et la vie.*				I	
30	The search for truth and the idea of its approximation are some other ethical principles; as well as the idea of intellectual probity and fallibility, which lead us to the criticism of oneself and tolerance. **Karl Popper**.					I

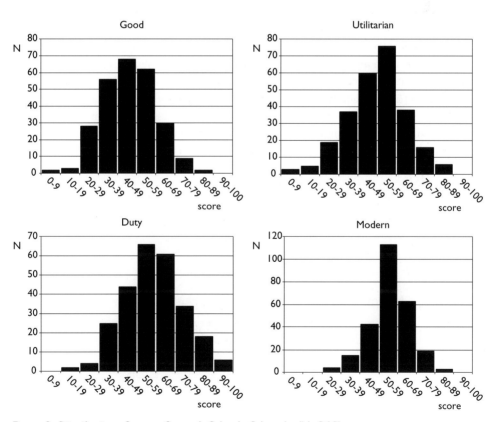

Figure 2. Distribution of scores for each School of thought (N=260).

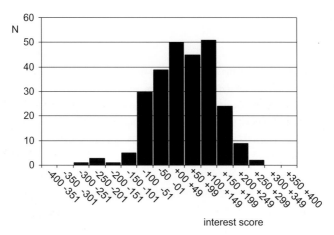

Figure 3. Distribution of interest scores (N=260).

The determination of each student for one specific trend has been considered as established when the score of one trend was equal or superior to 60 %, and greater by at least 10 % than any other score. In these conditions, 158 students (61 %) do not show any particular trend, while 63 (24 %) give responses typical of Duty, 21 (8 %) of Modern, 13 (5 %) of Utilitarian, and only 5 (2 %) of Good ethics. The mean ethical profile observed for each of these categories of students is reported in Figure 4, and typical profiles for individuals belonging to these categories are shown in Figure 5.

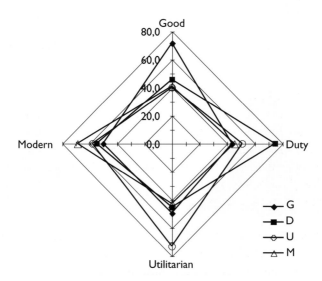

Figure 4. Mean ethical profile for each trend.

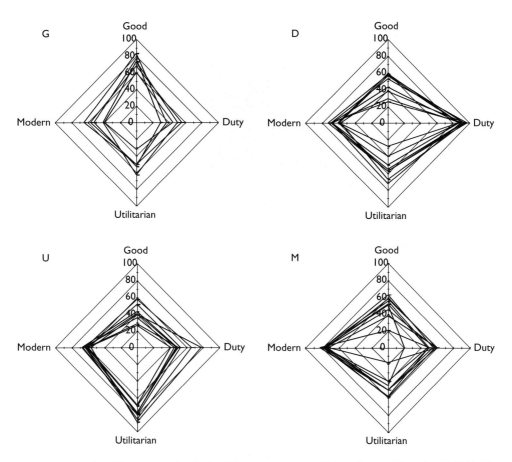

Figure 5. Ethical profile for typical individuals belonging to one of the schools of thought (G, D, U, M).

A second principal components analysis, performed on the second set of data, defined two main axes, representing 60 % of the total variation, and opposing Good and Duty (and also interest) to Utilitarian and Modern trends. In Figure 6A, individual students are represented by points, and the students belonging to a specific category are linked by lines. Furthermore, a hierarchical classification has been performed and lead to 6 groups, reported in Figure 6B. Group 1 expresses a significantly higher ethical interest (mean score: 131) and rates better quotations 1, 5, 19 and 30. Group 2 (upper part of the graph) overvalues Duty (mean score: 71.1), Good (52.5) but also interest (119.6) and quotations 15 and 2. Group 3 (left) corresponds to a low level of determination, with low scores for all main trends, while Group 4 (upper right) overvalues Utilitarianism (68.1) and quotations 11, 21

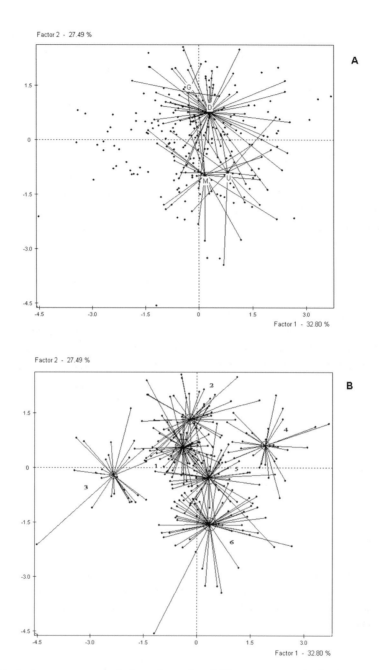

Figure 6. Principal components analysis performed on 260 responses to the final questionnaire. A: position of individuals identified by they adherence to a specific trend (G: Good, N=5; D: Duty, N=63; U: Utilitasianism, N=13; M: Modern, N=21). B: distribution of the 6 groups of responses.

and 23 as well as Duty (76.9) and quotations 6 and 10. On the centre of the graph, Group 5 emphasizes on Good (55.9) and quotations 4, 9 and 13. On the lower part of the graph, Group 6 represents Modern responses (61.5) and quotations 14 and 18, as well as low interest (-55.4) and negative visions of ethics (quotations 12, 16, 22). Such an analysis, performed *a posteriori*, could be of use in editing fine ethical profiles and helping in the interpretation of the responses.

In a classroom context, it takes less than one hour to fill this final version of the questionnaire, and to compute the profile. A debriefing can be done just after (during one hour for example) in order to draw the main conclusions and to take note of the reactions of the group. These quotations can be used later in more detailed sessions, during lectures devoted to philosophical matters, or during ethical case studies analyses.

Students' comments stressed either on the interest of such an exercise in order to introduce the ethical questions, and the fact that it leads to a personal reflection, or on the difficulty to understand some sentences, particularly due to the fact that they are out of their context. It has been suggested to perform a second test, at the end of the ethical teaching module, in order to appreciate the personal evolution of the students; a questionnaire or test based on alternative choices in response to problems derived from real situations could be a good option.

Acknowledgements

We thank Mrs Marie-Françoise Chevrier, and the students involved in the testing, for their help in the elaboration of this method.

References

Cuvillier, A., 1954, *Cours de Philosophie*, 2 volumes, Paris, Armand Colin.

Giré, A., 2001, "Représentation de l'éthique chez de futurs cadres", *in* Association Lyonnaise d'Éthique Économique et Sociale (Ed.), *Regards croisés sur l'éthique : personne, entreprise, société*, Lyon, Éditions Textes et Prétextes, p. 313-326.

An introduction to Problem-Based Learning and its application to an animal bioethics curriculum

Alison J. Hanlon
Faculty of Veterinary Medicine, University College Dublin, Ireland

Abstract

The learning objectives of most third level institutions are to promote deep-learning, and problem-solving abilities, providing students with life-long learning skills, which they can continue to use after graduation. However, research suggests that traditional didactic teaching methods may fail to achieve these learning objectives, and instead support passive and shallow learning. As a consequence, new teaching methods have been developed, to enhance the educational and learning power in a teaching environment.

Problem-based learning (PBL) is a relatively recent teaching strategy, which places the responsibility of learning on the student. Its general aims are to develop critical thinking and communication skills, as well as knowledge acquisition, in a problem-solving context. PBL differs from traditional didactic teaching in several ways. It is a student-centred mode of teaching, where the students determine the learning objectives within the context of a problem. It is problem-based and not necessarily subject based, demonstrating the interdisciplinary nature of most real-life problems; the teacher acts as a facilitator and not as a provider of facts, thus supporting learner autonomy.

There are different PBL models, and, as with other forms of teaching and learning, there are guidelines that need to be adhered to in order to maximise learning potential and prevent shortfalls in PBL. This chapter will address all of the above issues, providing case examples, to give the reader a greater understanding of PBL.

1. Introduction

There is increasing awareness that traditional didactic teaching strategies may not satisfy the learning objectives or aspirations of third level institutions. The conventional format of lecturing supports passive learning, where the lecturer provides the information and the students sit passively, making notes on what the lecturer is saying. The line of communication is normally one-way, from lecturer to student, with the students accepting the information being presented. In general, the lecturer (in accordance with the Faculty's teaching curriculum) determines the learning objectives or content of the lecture. Little or no account may be taken of areas familiar to the students or indeed of different learning styles (Felder, 1993). As a consequence, the student may fail to engage with the material, and not relate it to existing knowledge thus failing to achieve an understanding of the material being presented.

In contrast, PBL gives students autonomy to determine their own learning needs within the context of the case presented. Under the supervision of a facilitator, the students will reason through a problem, explaining terminology or concepts to each other as they arise, or if they are unable to do so, then such terms and concepts will be classified as learning issues. Together with the learning issues, the students develop hypotheses or ideas as to how the problem has arisen, and discuss their plans on how the case can be solved. PBL aims to develop critical thinking in the student, providing them with the necessary communication, information retrieval and reasoning skills necessary for their working life.

For PBL to be an effective form of learning, certain criteria must be fulfilled. These are discussed below.

2. The problem

The effectiveness of PBL is partly determined by the quality of the case, or problem, presented to the students. Research has shown that real life cases stimulate greater interest than contrived problems. Real life cases often have the added benefit of being multidisciplinary in nature. This enables students to integrate their knowledge in an applied context and, conversely, avoids students compartmentalising subjects to a specific course or lecturer (Engel, 1997). Furthermore, PBL allows both horizontal and vertical integration of the curriculum in contrast to subject-based curricula (Summerlee, 1997).

In veterinary sciences, real patient cases should be used, because they are associated with authentic data in the form of laboratory reports, as well as other artefacts such as x- rays. Authentic cases provide a valuable opportunity for preclinical students to learn about clinical aspects of veterinary sciences such as reading x-rays, developing medical vocabulary and having to deal with ethical dilemmas. The cases used should represent common clinical problems that have serious consequences (Bushby, 2000). Authentic cases stimulate student interest, and therefore motivate student learning.

Real-life cases can be tailored to anticipate all possible areas of inquiry, thus maximising the learning benefits. Each case should be based on a clearly defined set of learning objectives, or a curriculum matrix. Each document should end at a decision point, to stimulate discussion and maximise learning issues before students receive more information about the case (Bushby, 2000). Details of the diagnosis should not be provided in the first tutorial, to ensure that students will explore all possibilities including basic anatomy, physiology, pathophysiology and differential diagnosis (Bushby, 2000). It is also important to modulate the amount of information given in each tutorial, so that the students do not feel overwhelmed (Barrows, 1988).

3. The PBL model

The problem can be presented to students in different formats. For example, it can be contained in a single document (often a single page), or a series of documents, which are progressively released. The latter is referred to as *gradual disclosure*. Each new document in gradual disclosure is only distributed once the previous document has been thoroughly analysed. A set number of documents are released in each tutorial. Problems based on gradual disclosure have the capacity to be more complex, and funnel the direction of learning.

The approach taken to PBL is based on a hypothetico-deductive system of analysis. The analytical process has been divided into a number of clearly defined steps. For example:
1. Define the problem.
2. Define what is known.
3. Identify what is not known.
4. Prioritise and assign learning issues.
5. Self-directed study to research the learning issues.
6. Discuss researched learning issues with group members.

7. Apply the researched issues to the problem, and review the hypothesis.
8. Reflect on the learning process (Summerlee, 1997).

Several authors have devised analytical frameworks, which are used in tutorials, to ensure that students conduct a thorough evaluation of the problem. Barrows (1988) advocates structuring the analysis using four headings:

- *Facts* - a list of factual information contained in the case document; and existing knowledge (including researched learning issues).
- *Ideas or hypotheses* - based on the known facts.
- *Learning issues* - items for independent study such as terms, concepts or structures contained in the case document, of which the students have superficial or no knowledge.
- *Plans* - plan of action.

4. The facilitator

Stinson and Milter (1996) reported that facilitators are '*the central variable in effective implementation of PBL.*' The role of the facilitator is multifaceted. One of the primary roles is to support the development of metacognitive skills in the students. Barrows (1988) refers to metacognition as being the '*executive function in thinking*' and '*thinking about thinking*'. Finding and processing information to solve problems can be achieved in various ways. For example some may come from memory and be reapplied to the new context, and the remainder may be found by searching the literature or consulting an expert. The processing and application of the information is referred to as *metacognition*. For students inexperienced at PBL, it may be necessary for the facilitator to mirror the process by asking students the questions that he/she would ask when confronted with a problem. As students become more familiar with the process, and are aware of how to approach a problem, the facilitator will no longer need to prompt metacognitive thinking.

At this point, the priorities of the facilitator's responsibilities change to 'coaching' the students: keeping the process moving, and ensuring that each step is adequately addressed. The facilitator also has to monitor the educational progress of each group member. This can be achieved by encouraging each individual to contribute to group discussions and, by using metacognitive questioning, the facilitator can ascertain whether a student is experiencing difficulties. Learning difficulties should be addressed as part of the PBL tutorial process (Barrows, 1988), for example, assigning specific learning issues to a student experiencing difficulty, to help broaden their understanding of the concept, or encouraging peer coaching, where two students conduct their research together.

The input of the facilitator will therefore change according to the experience of the group of students. With new students, who have no prior experience of PBL, the facilitator will have to model the process, addressing the problem by asking metacognitive questions, and probing the students about knowledge they may already have which is related to the problem. As the students take on progressive responsibility for their learning, the facilitator acts as a coach up to the point where the group can function independently, thus allowing the facilitator's role to fade out.

Academics unfamiliar with the PBL process generally believe that the facilitator should be a content expert in the subjects covered in the tutorials. However, the facilitator should not at any stage provide information to the group during the tutorial session. Morales-Mann and Kaitell (2001) reported that content experts found it more difficult to facilitate than non-expert facilitators, probably because they felt frustration at not being able to provide students with information during the tutorial. In addition, Albanese and Mitchell (1993) state that content experts tended to be more directive in tutorials, which might be detrimental to the PBL process. Barrows (1988) acknowledges that the ideal facilitator would be an expert and a good tutor, but because it is logistically difficult to have a sufficient number of experts, he advocates non-experts who are good tutors. Although being a content expert is unnecessary, it may have disadvantages as well as advantages (Albanese and Mitchell, 1993). One concern of non-expert facilitators is that their students may learn incorrect facts. However, this does not normally occur if the metacognitive process is adhered to. Reassurance can be offered to all facilitators through training or by providing them with a guide to each PBL case.

As with any new teaching strategy, training is paramount. Training seminars for new facilitators are essential, because their role in PBL differs from that of the conventional tutor. Inadequate training of facilitators causes feelings of anxiety and stress, and Morales-Mann and Kaitell (2001) suggest that this may affect student learning. If appropriate training is provided, non-expert academic staff, technical staff, postgraduate students and peers, such as final year students, can facilitate PBL tutorials. However, there may be limitations to using peers as facilitators, Steele *et al.* (2000) reported that they had a tendency to take short-cuts, therefore undermining the PBL process.

5. The tutorial process

With traditional teaching methods, students are often passive, and reluctant to ask or answer questions. The majority may also feel that it is inappropriate to challenge the teachers' ideas or suggest alternatives. In contrast, with PBL, the students have autonomy to direct the learning goals within the boundaries of the curriculum, and so it is essential to create a safe learning environment in which all students feel willing to participate.

The first tutorial normally starts with an ice-breaker, to create a relaxed and familiar learning environment, for example, introductions where each group member, including the facilitator, gives an outline of their professional interests and future aspirations. As well as breaking the ice, introductions can reveal areas of special knowledge or interest, which may be of benefit to the PBL case (Barrows, 1988). Agreeing the ground rules also occurs at the beginning of the first session. The ground rules are determined by the group, for example they may wish to have rules on punctuality, and respect for each other's opinions in tutorial discussions. PBL guidelines are also relevant to the ground rules. For example, it is important to remind students that, in PBL, silence indicates consensus, and that students must be able to summarise their researched learning issues, without reading directly from textbooks.

PBL is structured to develop communication skills and confidence in students by assigning roles within the group. At the beginning of each session, volunteers are invited for the roles of scribe, reader and chairperson. The scribe is responsible for recording the main points discussed by the group on to a flip chart or white board, the reader has to read out case documents to the group, and the chairperson is responsible for managing group discussions. The roles are rotated with each tutorial session.

Once the rules, roles and guidelines have been established at the first session, then the first document is circulated to the group, and is read out by the reader. Thereafter, the analysis of the case begins. Using Barrows' hypothetico-deductive system of analysis (1988), the students have to identify the facts, and develop their hypotheses or ideas, plans, as well as identifying learning issues, all of which are recorded onto a flip chart, or white board, by the scribe. Learning issues may be simple, for example, unfamiliar terminology used in the case document, or more complex, such as a biological structure or system of which the students may either have superficial or no knowledge. Students inexperienced at PBL may be slow to identify learning issues, and so, initially, the facilitator may have to help the students to identify gaps in their knowledge by using metacognitive questioning e.g. asking them to clarify a term, or illustrate a chemical pathway. As students gain experience with PBL, they will explain potential learning issues to each other, often to avoid having to create new learning issues (and consequently more independent study). It is important for the facilitator to probe such explanations, to test their depth of understanding.

After the document has been thoroughly analysed by the group, if the PBL model being used is gradual disclosure, then the group will be asked for the plans they propose in order to try to solve the problem. At this stage, the next document is circulated and the analysis repeated. As each new fact comes to light, the students have to reappraise and amend the list of facts, learning issues, ideas and plans, which have been recorded by the scribe.

After the analysis of the final document, the facilitator asks for a volunteer to summarise the case. This is followed by the assignment of learning issues to group members. As part of the learning process, the facilitator may ask the students about how they will attempt to research their learning issues. The latter offers the opportunity of discussing different resources, and thus broadens the student's awareness of research options.

A self or peer assessment may be conducted either at the end of each PBL tutorial, or at the end of the PBL case. This enables students to reflect on their learning, and provide a critique on their contribution to the case. In my experience, there may be reluctance by group members to conduct a peer review.

In subsequent sessions, the learning issues agreed in the previous tutorial are discussed. Students have to explain how the new knowledge can be applied to the context of the problem, and whether it changes the group's hypothesis or plans. In addition to their research findings, the facilitator asks for a brief critique on the source of the material. Learning issues may continue to arise during the final PBL tutorial, and will have to be researched. In such situations, it is the responsibility of the students to circulate their research to all other members of the group, even though the PBL case has ended.

6. The tutorial group

PBL is designed for small group teaching. Recommendations on the optimal group size range from 5 to 10 students. The composition of the tutorial group changes with each new case. This is to prevent the group dynamic from stagnating, where certain individuals assume a leadership role, and to increase student exposure to different group dynamics.

6.1. Resources

Apart from having a sufficient number of venues, and time within the course curriculum, the main resource implications for PBL are staff to facilitate and prepare cases (*case authors*). As discussed above, academic as well as non-academic staff can be facilitators, as long as appropriate training is provided. Support should be offered, especially to inexperienced facilitators. This can be achieved by providing a *Facilitator's Guide*, holding an introductory workshop on PBL, and/or a Question and Answer session on the PBL case for facilitators prior to the first tutorial. Other possibilities include a shadowing programme, where a new facilitator observes a PBL tutorial or establishing a mentoring system for new facilitators.

Ideally the tutorial room should consist of free-standing chairs and tables. The facilitator should arrange the furniture prior to the tutorial, to optimise group cohesiveness and discussions. For example, remove any unwanted chairs, and place the remaining chairs beside each other in a circle, so that the students are sitting close together and will be able to achieve good eye contact. It is important to have a common record of group discussions, so that the group can reflect on their hypotheses and amend the facts as the case and their knowledge progresses. In most cases a flip chart or white board are used.

In terms of scheduling, allowance should be given for independent study, as well as the tutorials. Students often have difficulty adjusting to PBL (Stinson and Milter, 1996; Summerlee, 1997). Introducing it early in the curriculum may reduce the degree and duration of difficulty experienced by students. There is debate about whether curricula should be exclusively PBL, or mixed with traditional didactic methods. If PBL is being used with conventional teaching, timetables should be scheduled to complement the PBL case. For example with a PBL case on downer cow syndrome in cattle, practicals on gross anatomy of the cow, can be scheduled immediately before the start of or during the PBL case.

6.2. Assessment

Traditional methods of assessment may not be appropriate for PBL, because in contrast to conventional didactic courses, PBL aims to develop problem-solving skills, in addition to knowledge acquisition (Kelly *et al.*, 1997; Norman, 1997). A variety of assessment methods are used for PBL (Swanson *et al.*, 1997). For example, the *'progress test'*, is used at the University of Maastricht (Gijselaers 2001; Verwijnen *et al.*, 1982 cited in Norman, 1997), and is devised to monitor the development of knowledge acquisition and understanding in students over the time course of the curriculum. The same test is administered several times a year and contains a range of material taken from the full spectrum of the curriculum. *'Objective structured clinical examinations'* are used to test the clinical skills and knowledge of medical and dentistry students (Kelly *et al.*, 1997). The Faculty of Veterinary Medicine, at UCD, uses combinations of assessments: open book examination, in the format of a PBL tutorial, multiple choice questions and facilitator evaluation of student contributions to tutorials.

In addition to Faculty assessment, Barrows (1988) advocates using self and peer evaluation as an important 'life skill' at the end of each PBL case. The evaluation is based on the students' views of their own contribution to the process, including a reflection on what they have learnt, as well as assessing the contribution of their peers.

6.3. Using PBL in an animal bioethics curriculum

PBL offers the opportunity to increase the students' awareness of animal bioethics in a dynamic learning environment and to provide essential training to deal with animal welfare cases, as part of the professional development of veterinary undergraduates. At the Faculty of Veterinary Medicine, UCD, animal welfare cases are presented to second year undergraduates. The cases are presented as single documents. They encompass clinical problems, animal husbandry, legislation as well as an important ethical dimension. Students are invited to adopt the role of a government veterinarian who is called to the scene of the animal welfare case. Using the Barrow's four headings (facts, ideas, learning issues and plans), the students analyse each case under the guidance of a facilitator. At the end of the tutorial, learning issues are assigned. In the second tutorial, each group has to present their case to the rest of the class. This is followed by a question and answer session directed by the students, and a brief summary of the 'experts' actions' presented by the facilitator (all cases used are based on real life events).

6.4. Case studies

Two animal welfare cases are presented, with an outline of the analysis conducted by second year veterinary undergraduates. The analysis reflects the prior knowledge and experience of the students.

Case 1. Bovine castration

Four bullocks aged between six and eight months of age were presented with necrotic testes, excessive granulation and fly strike of the wounds at the neck of the scrotum, arising from attempted ring castration two weeks earlier. All four bullocks stood with arched backs, were reluctant to move, appeared depressed, and were inappetent and pyrexic. The animals had been left at grass following application of the rubber rings and had not been inspected for five days. A part-time farmer owned the animals.

Table 1. Bovine castration

Facts	Ideas	Learning Issues	Plans
4 bullocks between 6 -8 months of age that have been ring castrated	Ring castration may not be appropriate in cattle of 6-8 months age.	Regulations on ring castration. Why is ring castration not advocated in older animals? What is best practice for animals aged 6-8 months old?	
Testes are necrotic with excessive granulation		What causes granulation? How should it be treated?	Treat cattle
Fly strike		What causes fly strike? How should it be treated?	Treat cattle for fly strike
Bullocks standing with arched backs and appeared depressed, inappetent and pyrexic	Cattle are in pain	Signs of acute and chronic pain in cattle	Administer analgesics?
Bullocks not inspected for 5 days after castration	Cattle not inspected frequently enough	Regulations on frequency farmer must inspect animals	Advise farmer on best practice; Prosecution on animal welfare grounds should be considered

Case 2 Downer cow

A multiparous Friesian cow arrived at the abattoir in lateral recumbency, after being transported for 60 km. The cow had calved 7 days previously to an oversized

Table 2. Downer cow

Facts	Ideas	Learning Issues	Plans
Multiparous Friesian downer cow		Causes of downer cow syndrome? Prognosis of downer cow syndrome?	
Oversized Charolais calf	Inappropriate feeding management of the cow pre-calving?		Check calving history of cow. Check feeding management of herd. Check Easy Calving Index of Charolais sire.
Calving jack used within 1 hour of appearance of foetal hooves	Incorrect use of calving jack?	Timing of intervention at calving? Guidelines on use of calving jack?	
Calf dead on delivery		Factors affecting morbidity and mortality of calves at calving	Do a post-mortem on the calf
Cow lifted for approx. 2 hours/day	Lifting may cause bruising Vet called too late		
Veterinary attention sought 7 days post-calving		Are there regulations on timing of veterinary intervention?	
Casualty slaughter		Regulations governing casualty slaughter. What are the alternatives?	
Cow certified fit for transport	The cow should not have been transported. Vet may have felt pressurised by farmer to certify cow.	What is the legal definition of 'fit'? What is the certification process?	
Cow arrived at abattoir in lateral recumbency			
The Official Veterinarian at the abattoir euthanased the cow	Biosecurity risks of letting downer cow into abattoir?	Why was the cow euthanased on the trailer and not in the abattoir?	
A post-mortem revealed extensive bruising and pale soft exudative meat	Transport and daily lifting caused bruising.	Causes of pale soft exudative meat?	
The carcass was condemned	Carcass condemned because of extensive bruising and meat failed to set	What are the reasons for condemning a carcass?	

Charolais calf, which was born dead on delivery. The farmer had assisted with the calving within 1 hour of the appearance of the foetal hooves, using a calving jack. The cow collapsed during calving and remained paralysed. Using a sling/front-end loader, the cow was lifted for approximately two hours per day. The cow was kept on a deep bed of straw and milked in the morning and evening. The farmer stated that the cow continued to feed and drink normally during this period.

After 7 days, when the cow showed no sign of recovery, the farmer called in his local veterinary surgeon who examined the cow and recommended casualty slaughter. The cow was certified by the veterinarian as fit for transport, under the proviso that she should be supported and stabilised during transit, using straw bales.

On arrival at the abattoir, the official veterinarian inspected the cow and decided to euthanase it in the trailer. A post-mortem was conducted and revealed extensive bruising. The meat was classified as pale, soft, exudative and subsequently failed to set. As a result the carcass was condemned.

7. Summary

There are a variety of models of PBL (Savin-Baden, 1999). From an educational perspective, PBL has numerous advantages over conventional didactic methods (e.g. Boud and Feletti, 1997). However, if it is used inappropriately, PBL will not lead to robust learning (Stinson and Milter, 1996; White, 1996). Factors such as facilitator training and support, and the quality of the cases, as well as the logistics of schedules are essential to optimise the benefits of PBL.

Acknowledgements

The Royal College of Veterinary Surgeons, kindly granted permission to use a Sample Case Report (Case One) from the Certificate in Animal Welfare Science, Ethics and Law. I gratefully acknowledge Ms P. Barry-Walsh MVB MRCVS Cert Wel, M.Appl.Sc. author of Case Two. The analysis of each case was taken from two tutorial groups in the second year of study of the Veterinary Medicine course, University College Dublin.

References

Albanese, M.A. and Mitchell, S., 1993. Problem-Based Learning: A Review of Literature on its Outcomes and Implementation Issues. Academic Medicine, 68, p. 52-81.

Barrows, H.S., 1988. The Tutorial Process. Springfield, Illinois: Southern Illinois University School of Medicine.

Boud, D. and Feletti, G.E., 1997. Changing Problem-Based Learning. In D. Boud and G. Feletti (Eds.), The Challenge of Problem Based Learning. London, Kogan, p. 1-14.

Bushby, P., 2000. PBL - The Dublin Experience. March 2000. Unpublished newsletter.

Engel, C.E.,1997. Not Just a Method but a Way of Learning. In D. Boud and G. Feletti (Eds.), The Challenge of Problem-based Learning. London: Kogan Page, p. 17-27.

Felder, R.M., 1993. Reaching the Second Tier: Learning and Teaching Styles in College Science Education. J. College Science Teaching, 23 (5), p. 286-290.

Gijselaers, W.H., 2001. PBL: Issues in Learning, Instruction and Curriculum Design. PBL Workshop, Dublin Institute of Technology, September 2001.

Kelly, M., Shanley, D.B, McCartan, B, Toner, M and McCreary, C., 1997. Curricular adaptations towards problem-based learning in dental education. Eur. J. Dent. Educ., 1, p. 108-113.

Morales-Mann, E.T. and Kaitell, C.A., 2001. Problem-Based Learning in a New Canadian Curriculum. J. Advanced Nursing, 33 (1), p. 13-19.

Norman, G.R., 1997. Assessment in Problem-Based Learning. In D. Boud and G. Feletti (Eds.), The Challenge of Problem-Based Learning. London: Kogan Page, p. 263-268.

Norman, G.R. and Schmidt, H.G., 1992. The psychological basis of problem-based learning: a review of the evidence. Academic Medicine, 67, p. 557-565.

Savin-Baden, M., 1999. Problem-Based Learning in Higher Education: Untold Stories. Buckingham: SRHE and OU.

Steele, D.K., Medder, K.D., and Turner, P., 2000. A comparison of learning outcomes and attitudes in student- versus faculty-led problem-based learning: an experimental study. Med. Educ., 34 (1), p. 23-29.

Stinson, J., and Milter, R.G., 1996. Problem-Based Learning in Business Education: Curriculum Design and Implementation Issues. In L. Wilkerson and W.H. Gijselaers (Eds.), Bringing Problem-Based Learning to Higher Education, Theory and Practice. San Francisco, Jossey-Bass.

Summerlee, A.J.S., 1997. Making sense of problem-based learning. Proceedings of EAEVE Veterinary Education Symposium, p. 11-24.

Swanson, D.B., Case, S.M., and Van der Vleuten, C.P.M., 1997. Strategies for Student Assessment. In D. Boud and G. Feletti (Eds.), The Challenge of Problem-Based Learning. London: Kogan Page, p. 269-282.

White, H.B., 1996. Dan Tries Problem-Based Learning: A Case Study. In L. Richlin (Ed.), To Improve the Academy, Volume 15, p. 75-91.

The construction and application of cases in bioethics education

Tjard de Cock Buning[1] and Ellen ter Gast[2]

[1]Institute for innovation and Trausdisciplinary Research, Faculty Earth and Lifescience, Free University Amsterdam, The Netherlands, [2]Department of Philosophy and Science Studies, Radbout University Nijmegen, The Netherlands

Abstract

Cases are important tools when training and examining the analytical skills of students of ethics. A methodological tutorial is given for those who want to design and incorporate 'cases' in their teaching programs. It is shown that different phases in bioethical training (i.e. motivational, analytical and examination phases) demand different types of educational cases. A list of educational (sub)goals is linked to various tools and approaches which can be applied in the classroom. One of the listed goals is that a student should be able to separate his/her private opinion from a generally valid argument. Another is, that he or she should be able to analyse a case from a disinterested perspective (the impartial perspective). Finally, four examples of typical case exercises are described in detail and illustrated by imaginery teacher-student dialogues. The illustrative material of this chapter was collected by the authors over several years and practised at various Master and Bachelor levels in the bio-medical and life sciences.

1. Introduction

'Designed cases are almost real-life situations constructed with a specific educational goal.' This definition implies that not every real-life situation is suitable for educational purposes. The major difference between real-life situations and designed cases is that constructed cases are intelligent reductions of real-life situations by omitting all distractions and pieces of irrelevant information. In order to accomplish that reduction it is important to be aware of the educational goal to be reached. Different types of cases can be used in different phases of the teaching process and their use depends on the different teaching goals.

Cases used in bioethical training usually present a problem, and more specifically a dilemma problem. Dilemmas are forced choices between two or more rejectable or mutually exclusive options. If one opts for one solution, the 'unfairness' of not opting for the other is immediately recognised. But in a dilemma one can never choose both options.

1.1 Example of a dilemma

Tom has a veterinary practice of farm animals. One day he diagnoses disease X in a pig at a large pig farm. Antibiotics will cure the animal. The farmer calculates that the cost of the animal's loss of weight during the antibiotics therapy is greater than the cost of the therapy itself. The therapy is not economic, so he asks the vet to kill the animal. The vet is now in a dilemma. On the one hand there is his intention to heal sick animals (the ethical principle of 'doing well for the animal'), which is against the wish of his client. On the other hand is his financial dependency on his client (the socio-economic principle of 'doing well for himself'). Whatever he decides to do, he will never be fully satisfied by the outcome. If he chooses for the animal (and against the farmer), the farmer might choose another vet or the farmer and his colleagues might even decide to boycott the vet altogether. If the vet chooses to kill the animal, he will secure the bond with his client, but he will feel he has betrayed his own fundamental motivations, ideals and values which he may have had since he started as a student.

Real-life dilemmas are often traumatic for people who feel responsible for the choice. After all, dilemmas force a choice which possesses at the same time the most reasonable arguments against that choice. Dilemmas sometimes condemn people to a lifelong uncertainty about the rightness of a particular choice. In such situations people are also very vulnerable to the arguments of those who are in favour of the other option. Vulnerable, because one experiences these arguments as 'good arguments' in one's own heart. Greek tragedies, for instance, are constructed around dilemmas and show us the devastating effects of a single choice on entire families and nations.

Students who are trained in ethical skills have to learn to cope with dilemmas but not by applying debating tricks to escape from the problem (techniques like: replying with an hilarious dilemma to ridicule the dilemma; 'rebutting the dilemma' or by stating that one side of the dilemma is not relevant, 'grasping the horns', and thus suggesting that the whole dilemma is malconstructed and irrelevant). Dilemmas are often the core of ethical problems. So, one has to teach the student to deal with dilemmas analytically and rationally. Dilemmas, therefore, demand a well-structured argumentation in order to defend the best option among all negative solutions.

Presenting students with real dilemmas as cases which they must solve is a powerful tool in the teaching of bioethics. There are several ways of presenting a case. Newspaper articles are very suitable because of their challenging style and their topical character. Thematic television documentaries can also be used as they provide conflicting background information. Inviting speakers is an option if the speakers' attitudes demonstrate that they are still struggling with the dilemma of the case. From personal experience they can illustrate the fact that ethical dilemmas are not easy to solve in superficial or formalistic approaches.

2. Practical skills and theoretical background

Teachers should be aware that untrained students often reply with a personal opinion to a case. They declare it 'evident' that solution A or B is the right one. However, in dilemma situations there are no 'evidently right solutions'. In ethical decision making it is the structure of the argument that counts: the absence of suggestive and rhetorical tricks, the logical consistency of the argument and the balance of arguments for and against. This implies that students should be thoroughly taught about the various pitfalls in arguments (fallacies, rhetoric), before they enter a substantial analysis of the arguments for and against in a case.

So, in addition to solving cases students need to be trained in argumentation and they have to learn some ethical theories. Theoretical knowledge and practical skills go hand in hand in the teaching process. The focus of this chapter is on the practical use of cases. Nevertheless some remarks about teaching theory need to be made. Cases can serve as powerful tools to illustrate the importance of argumentation in ethical problem solving. The art and theory of argumentation, however, are not discussed during but before or after the discussion of the case. The same goes for ethical theory. The training of practical skills (problem solving) and the acquisition of theoretical knowledge are separate elements in the teaching process.

Usually the teacher selects a specific case for a specific goal. After discussing the case the teacher reflects upon the theoretical background. In the beginning, much emphasis is put on argumentation theory. Later on in the process more emphasis is put on ethical theory, principles, norms and values.

In this chapter we only discuss the use of cases in bioethical teaching. Guidelines to avoid fallacies and other pitfalls for proper argumentation and ethical theory are discussed elsewhere in this book.

3. Phases in the educational process

Three different phases can be distinguished in the educational process: the motivational phase, the analytical phase and the examination. For each phase different types of cases are used.

The first phase in the teaching process can be described as the *motivational phase*. Biology and veterinary students need to be motivated to think about the ethical aspects of their work. Very often this is new to students. Therefore it is important to select a case that is close to the students' experience and at the same time provocative. In the motivational phase cases are used as a tool to sensitise students for the theoretical (hidden) structures that lie behind the case.

For example, students have to choose between an experiment with ten monkeys and an experiment with one hundred rats. In the discussion that follows the students' vote for either option the underlying principles are made clear.

3.1 Classroom procedure

The teacher starts with a short description of the experiment. In this case the experiment involves the testing of a new antidepressant. Animals are made depressed in order to obtain an animal model for human depression. Both the monkey and the rat model are presumed scientifically suitable to induce physiological and behavioural signs of depression (e.g. the animals show little interest anymore in the environment, are apathetic, and have no motivation to escape from mild electric shocks). After a short discussion about the methods and goals of the experiment the teacher presents the dilemma. *'You, a researcher in biomedical sciences, have to choose to conduct the experiment either with one hundred rats or with ten monkeys. What do you do?'* Students have to raise their hands in favour of option A or option B (in our experience the distribution is usually fifty/fifty).

Then the teacher asks the students individually to give the motivation for their choice. Typical answers underlying option A are: *'I choose the one hundred rats because monkeys have more sensation of pain. Monkeys look more like man. Monkeys have a better memory of pain and stress induced upon them.'*

Typical answers that motivate option B are: *'I choose for the monkey because in this way only ten animals are used instead of one hundred. Monkeys live longer and therefore in relation to the total life span the amount of stress is relatively smaller. Monkeys look more like human beings so the experimental data are more reliable.'*

After discussing the different arguments the teacher asks the students what the underlying principle in each choice is. In this phase students are challenged to formulate principles and separate these principles from their first intuition and the 'hard' facts.

The guiding principles in favour of option A are: always choose the animal with the lowest sense of pain or the lowest level of consciousness, or, always choose an animal that is lowest on the evolutionary ladder. The guiding principles in favour of option B are: always choose the experiment with the smallest number of animals, and, always choose the animal that gives the most reliable experimental output. The teacher writes these principles on the blackboard. In the next round the teacher asks the students about the facts: *'What do you know about the sense of pain and the level of consciousness in rats and monkeys? What is the most reliable experimental procedure? What is the difference between fact, estimation and intuition?'*

Three different columns emerge on the blackboard: 'intuition', 'underlying principles' and 'supporting facts'. The teacher explains that in ethical dilemmas a systematic analysis of intuition, principles and facts is necessary (see also the chapter on 'reflective equilibrium').

In the *analytical phase* cases are used as exercises to train the students' ability to apply given theoretical tools, concepts or theories. These cases might be more complicated and rich in the sense that several relevant ethical aspects, very often in contradiction to one another, play a role. Emphasis is put on the skills needed to describe these ethical aspects and to integrate them in a justifiable and sound argument in favour of one of the options. These cases usually give rise to controversies about relevant facts, the relevant moral principles, values, norms and theories. Even more, they give rise to controversies about the rational and the emotional reasoning behind the given options.

A good example of a case that can be used for this purpose is the case of the seals in the Dutch Wadden Sea. For many years the population of seals was threatened with extinction. Many initiatives were taken to save individual seals in order to rescue the population. As a result the seal became a sort of symbol for wildlife protection in Holland. Nowadays the population of seals is healthy and the seals are no longer threatened. The question is, what ought we to do now when seals get sick as a result of a viral infection. Do we let nature take its course and leave the sick animals to die or do we rescue the less fit individuals, which measure could decrease the general fitness of the population?

In this typical ecological case the zoocentric (the life of the individual seal) and the eco-centric (the fitness of the population or the ecosystem) perspectives can be distinguished. Also, a difference can be made between a deontological solution as opposed to a consequential one. Students can be asked whether they look at the long-term consequences or if their decision is based upon a different rule of principle. Furthermore, students are asked to make a distinction between the private, professional and public perspectives in the case under discussion. Do they have an interested or a disinterested and objective perspective?

In the *examination phase,* cases are used to test the students' knowledge and their ability to handle the skills acquired during the training. There are two ways of examining the students at the end of the course. If the number of students is limited, a piece of coursework on a self-chosen subject is the most favourable examination method. But with large groups this is too laborious. At such occasions a 'sit-in' (i.e formal) examination is the only way of examining the students.

A case especially designed for formal examination purposes needs to be constructed in such a way that both a proper and an improper solution is plausible when the background theory is not thoroughly trained and internalised.

The cases the students choose themselves for a piece of coursework also need to meet several criteria. It is important that the central dilemma of the case is a true dilemma for the student. In other words, the student does not know in advance what the right solution is. Furthermore, the case has to be complex but realistic.

Table 1. Motivational, analytical and examination phase.

Motivation	Analytical	Examination (Sit-in)	Examination (Take-home)
100-200 words; two hidden perspectives	500 words; multilevel case	100-200 words; example	2000 words; case is presented by the student in a piece of coursework
self discovery; exploring	application of theory; learning skills	testing knowledge and skills with a focus on knowledge	testing knowledge and skills with a focus on skills
group	group	individual (in large group)	individual (in a small group)

4. Educational goals

During a course in bioethics, students are trained in ethical reasoning and problem solving. Ethical problem solving requires a different approach from the one professionals in the natural sciences are used to. Natural scientists have a tendency to focus on objectively or scientifically proven facts. In ethics - often because of the absence of hard facts or opposing interpretations of the relevance of 'objective facts' - emphasis is put on consistent argumentation and philosophical theories and principles. For instance, the empathy of many people for small lambs is a mere 'emotion' from a scientific, agricultural point of view and therefore of little value. From a psychological or sociological point of view this kind of human behaviour towards small lambs is an observed and hard fact and therefore highly relevant.

Instead of starting a debate between the 'hard' and 'soft' sciences, it is more constructive to collect all the arguments that are found relevant by the various parties. Some of these arguments are hard facts in the classical natural science sense (body weight, level of stress hormones), other arguments are less objective (level of pain, level of welfare) and some depend strongly on theoretical presuppositions (level of well-being, expectation of positive or negative claims in the future). When the students face such a mixture of different types of arguments, they may panic and reject the whole ethical endeavour altogether. What they are about to learn is that one can use logic and common-sense rationality to systematize the problem.

A fruitful approach is the use of 'if...then...' constructs to highlight specific lines of argument: '*If* this action really induces pain in these animals *then* we have ..... , but *if* this is not the case *then* financial arguments carry more weight.' The ethicist's 'tool box' consists of argumentation (distinction between fallacies and logically consistent arguments), ethical theories and principles (virtue ethics, consequentialism, deontology) and several tools and techniques to apply theoretical knowledge to practical problems. Cases used in bioethical training need to be constructed in such a way that they describe a realistic bioethical issue and in addition serve as an 'eye opener' for a specific element of the ethical tool box.

In the process of selecting cases for teaching purposes there are two possibilities. Either one chooses an existing case (examples are discussed in the following paragraph) or one designs a new case. If one chooses to design a new case, the main question is: 'What is the educational goal to be reached?'. An educational case consists of a description of a bioethical issue and additional questions. By adding questions one forces students to look at the case with focussed attention. A good source for raw material are newspapers and weekly magazines. Bioethical issues are very often discussed in newspapers, so it should not be hard to find an interesting case. However, formulating the right questions often proves to be the real challenge for teachers.

The central question is usually this: 'What is the dilemma of this case?'. What are the possible solutions to the problem and why is this a dilemma? What are the arguments for and against the different solutions? And what are the underlying (ethical) principles of these arguments?

If different points of view are discussed in one and the same article, such an article can be used to train students to look at an ethical issue from different perspectives. Possible questions to ask are: Who are the different actors involved in this case? What is the proper level of analysis? Who has to make a decision? As a take-home assignment students can be asked to prepare a line of argumentation from a single, preferably not their own, point of view.

When the main objective is to train students to value available data, facts and principles, it is recommended to search for articles on issues that involve complex scientific facts which lead to strong public debate. Good examples are cloning, animal experimentation and gene therapy. Scientists tend to disqualify arguments put forward by the public as 'science fiction' or ignorance and therefore declare them irrelevant. By stimulating students to analyse the argumentative structure of the debate they are forced to look at the ethical principles behind the objections made by the public. The principle of reflective equilibrium can serve as a tool for this exercise. The basic questions are the following: 'What are the relevant facts put forward by the different parties?', 'What are the moral intuitions of the different parties?' and 'To what principles do the different parties (consciously or unconsciously) refer?'.

To summarize, some general and some specific goals can be distinguished in bioethical training. When designing or choosing a case for teaching purposes one should take into consideration which teaching goal is to be achieved. A schematic overview of general teaching goals and teaching tools and approaches is given below.

5. Different types of cases: illustrative procedures and games

Although the general procedure of a case exercise follows three successive steps, the method needs not be limited to a round-table discussion.

Table 2. Schematic overview of general teaching goals and teaching tools and approaches.

General goals	Teaching tools and approaches
Motivation and sensitisation for the hidden theoretical structures behind a bioethical case	Be provocative, choose a case close to the experience of the students.
Cooperation	Work in small groups, demand one voice as a solution.
Participation	Work in small groups, ask directly for individual opinions.

Specific Goals	Teaching tools and approaches
Basic knowledge of argumentation theory: separate the strategy used in argumentation, from the content of the argument	Select persuasive articles and let the students find the informal fallacies of the strategies used in persuasion.
The ability to separate opinion from argument	Select a case which invokes a strong moral response. Before discussing and analysing the case, ask students for their first (non-reflective) moral intuitions. After analysing the case ask again for opinions and solutions. Ask if, how and why opinions are changed.
Disinterested perspective Impartial perspective	Let the students change roles. Let students argue in favour of the role opposite to the one they would choose intuitively.
The ability to find hidden presuppositions	Make a list of the arguments for and against each position and presuppositions.
The ability to value the use of available data	Select a case that involves complex scientific and moral facts and principles.
The ability to connect the right principles to the case	Formulate supporting questions to name the ethical principles. Alternatively, ask for the case to be reframed as a set of competing principles.
The ability to recognise the proper analytical level	Make the case personal. Let there be a 'John' who is put in a dilemma (the level of personal ethics against loyalty towards the level of the company or the state).

5.1 General procedure of the case exercise

1. All cases start with a concise description of the case. This can be a description on paper, but a video or a lecture by an invited speaker will also do. The description contains relevant facts, an introduction of relevant actors and a clear definition of the dilemma. In more advanced groups, however, a hidden dilemma offers a good opportunity to test the students' ability to recognise dilemmas.

2. Secondly, this question is asked in some form: 'What would you do if you were that person?'. This person can operate at different levels (micro level, meso level or macro level): a scientist, a laboratory director, a member of parliament, a school teacher and so on.

The level dictates to what extent he or she emphasises his or her personal morality (micro level) or where more generally accepted ethical principles are the main frame of reference, as at the meso and macro levels.

3. Now go and find the answer. Make a sound and morally balanced decision. Choose between the different options and actions on the basis of defendable grounds.

5.2 Variety in methods: some illustrations of exercises and games

Various exercises and games can be organised to facilitate effective (in terms of time and quality) analyses by means of interactive discussions.

Some basic teaching settings are discussed below.

5.2.1 Classroom distribution

One corner of the classroom represents one ethical position regarding the case and the opposite corner represents the other solution. All students have to position themselves over the diagonal. The next step for the students is to ask their neighbours whether they are on the right spot. This will give rise to discussions on subtle differences in opinion between the students. The teacher has to be aware here of dominant or popular students who might blur the analytical exercise.

This game is very useful in phase one. All students have to participate and initial reluctance to speak in front of the others is challenged in a playful way.

5.2.2 Write and pass

All participants of the group (5-10) have to read the case description. All students are given a blank sheet of paper with a provocative statement about the case written at the top of the paper. The students have to write their own argument in favour or against this statement and then pass the page to their neighbour on the right-hand side. The neighbour, who has just passed on his own remark to his right-side neighbour, reads the response from his left-hand neighbour and adds an additional reason to this position. So whatever his or her own opinion is, he or she is forced to take a different position and has to focus on the argument that is presented by the left-hand neighbour. Again, they pass the page to the neighbour on the right-hand side. They read the new sheet with the two answers and then answer the next question: 'List one assumption made in the argument described above'. This question challenges the participants to find hidden presuppositions or counter arguments, even when they agree intuitively with the constructed argument of that opposite position.

Finally, the first participant is asked to read out the content of the sheet in front of him or her and to comment upon it. Then the other participants are asked whether they follow the same line of argument. Some participants may indicate that they depart from the same starting point and base their opinion upon a different kind of assumption or counter argument. After this analysis the right-hand neighbour is asked to present his or her sheet, and so on. After the third person, the group can be asked whether there are any participants with a completely different line of argument, and so on.

This exercise is suitable for large groups. Depending on the questions it can be used in the motivational and analytical phases. The example below is typical for the motivational phase to make an inventory of relevant pros and cons.

Food-producing animals are selectively bred to increase their productivity. This production goal may conflict with the animals' health and well-being.

1. Do you think it is morally wrong to breed animals that cannot keep their health and fitness? Answer YES or NO and provide one reason in defence of your answer.
    *   Pass this paper to the person on your right

2. Write down one good reason that supports the claim stated in (1).
    *   Pass this paper to the person on your right

3. List one assumption made in this argument.
    *   Pass this paper back to the person on your left

Figure 1. Write and pass exercise.

5.2.3 Dilemma game

A case is presented on a sheet and the main actor in the case has to make a decision. Four possible solutions (a, b, c and d) are presented under the case (see example below). Each participant reads the description of the case and its solutions carefully and makes up his or her mind for one of the offered solutions.

The teacher asks one of the participants to defend his or her choice in front of the group. Then the group members are asked to show their approval or disapproval regarding the choice and the argumentation by holding their thumb up (I agree) or down (I disagree). The teacher invites one of the members who disagreed to defend his or her choice ('I have chosen option 'c' instead because...'). Then the teacher asks whether someone has chosen one of the other two options and, if so, on the basis of what arguments.

Finally, the teacher announces the number of points each option receives. There are at most three points in the A category and three in the B category. Sometimes an option receives a mix (two A points and one B points). The teacher does not reveal the meaning of A or B points for the first 10 cases, but challenges the participants to hypothesize what the point might mean. Some students build a large number of

Figure 2. Dilemma game 1: Stress in pigs

Pigs are raised in boring, intensive farming stables. This is the reason why they tend to panic when they are brought into new environments. In particular, when they are transported to the slaughterhouse the situation can be too stressful for them. Sometimes this stress leads to a loss of meat quality and even to sudden death. A new breeding program is proposed to select pigs that can stand more stress. You, as an employee of the Minister of Agriculture, are asked to give your advice.

1. Yes, this is a good idea. Although we use animals for meat production, we ought to respect them. Stimulating a better stress response, is a good thing to do.
2. Incredible! Intensive farming is horrible! Now someone proposes to breed them in such a way that they fit this practice better. This shows disrespect for nature.
3. There are economic interests at stake. This option goes for better quality of meat. One should not ignore that we have a large interest in a good competitive position on the meat market.
4. Although the pigs will be better off, intensive farming will be encouraged again. This is not what we want. We have to look for alternative practices instead. This program does not encourage the search for alternatives, so we have to advise negatively.

Teacher: attribution of points

1. B 3 points   (positive deontologic: be respectful..)
2. B 3 points   (negative deontologic: .. shows disrespect..)
3. A 3 points   (positive consequentialistic: .. competitive position..)
4. A 3 points   (negative consequentialistic: ...wrong direction of policy..)

A points, others show a heavy accumulation of B points and some score equally on both.

The options are constructed in such a way that for the A category two answers possess a positive and a negative argument in a consequentialistic (utilitarian) framework and the two other options of the B category are arguments (positive and negative) in a deontological framework (virtues, norms, principles). This is a good game to help students realize that their ethical solutions are prescribed by some hidden types of argument: looking at consequences or emphasizing the rightness or wrongness of the act itself (deontology).

The same game can be applied to help students learn other presuppositions such as: anthropocentric or zoocentric perspectives; private or public perspectives; self-interest or disinterest and so on.

This approach is quite productive in the motivational phase because it helps students to explore their own hidden structures of argument. When these are recognized, further theoretical teaching on theories and principles can build on this experience. The approach is suitable for groups of around five to ten persons.

5.2.4 A typical discussion and development of the case

Maria: All the options have some point... but I think that option three fits best with my position. It is unrealistic to be romantic about animals. Farming is business.
Teacher: Who agrees (thumbs up) or disagrees (thumbs down) with Maria?
Teacher: John, you put your thumb down. What is your choice and could you explain why you favour this one?
John: I think that Maria's position is quite remote from an ethical position. Our world would deteriorate to materialism if everyone argues like her. My choice is number two. Farmers are not only money-makers. They have chosen for a way of life close to animals.
Teacher: Who has different positions from Maria and John? Could you explain?

(students exchange arguments)

Teacher: Does anybody now know better what to choose?... Maria you stick to three? Then you receive three points in A... John you get three points in B. Let us turn to the next case.

(after some cases)

Teacher: Does anyone have any idea what A and B stand for?

David: I think it has to do with money. When you lose or gain something you get a B mark?

Teacher: hmmm...and what about A marks?

David: I don't know, to me they all appear unrealistic positions.

Christine: No, no it has to do with politics. The A marks, look at the pig case, stand for the green parties.

Peter: That does not make sense. I am not green at all. As a veterinarian I chose very pragmatically for option one in the pig case and yet I also received an A mark.

Teacher: Difficult? You are on the right track, but you haven't reached the right distinction between A and B marks yet. Let's try another case and look carefully at yourself. What kind of arguments do you value? And what is the difference with the arguments of your colleagues.?

....(after some cases)...

Maria: I only get points in A and John in B? I think it is because he is a boy (laughter).

John: I get the feeling that Maria is more principled. I am more pragmatic.

Christine: Yeah, that is true. Let us try another case to test it!

Teacher: OK, You discovered the essential difference between the two types of possible answers. The A type of approach is called 'consequentialistic', that is to say that your first and main analysis of the case is focussed on the consequences. In some situations this is in favour of animals and in other situations it is in favour of the farmers. So it is not necessarily the same as the difference between animal-friendly or business-friendly. The B type of approach is called 'deontologic'. Your first and main analysis of the case is focussed on the 'act' itself. Irrespective of the consequences, there is something right or wrong in the intention of the act. Intensive farming is considered wrong in itself, regardless of the possibility that it will produce more meat for the global society.

So, besides all kinds of scientific and hard facts to underpin your arguments, you appear to have a more or less strong habit to think consequentialistically (Maria) or deontologically (John). The other group members don't show such a strong predisposition for either approach. They adapt their approach to the case. In some cases they feel a consequentialistic approach is more suitable and in other cases they opt for a more deontological view.

For a professional ethical analysis, it is important to free yourself from these unconscious biases towards one of the approaches and to try instead to analyse every case from at least these two different angles. Your final choice should also contain an argument why you favour, say, a consequentialistic approach over a deontological approach.

5.2.5 Simulation (change of roles)

Play the ethical review committee by assigning roles to the participants. The roles must be defined in ethical terms, which implies that this approach is more suitable for the analytical phase when the students have obtained enough insight in the ethical frameworks. Select an anonymized but realistic proposal for the simulation and assign the roles: scientists (freedom of research, anthropocentric, consequentialistic); animal caretakers (zoocentric, promotion of well-being); ethicists (encourage balanced solutions, compromises critical regarding the value and relevance of facts versus estimations).

In order to structure the simulated discussion, the chairman can introduce an ethical decision model (see chapter on reflective equilibrium) which serves as an agenda for the 'meeting'.

Again, in the light of the general goal to stimulate impartial analysis, it might be better to assign roles to persons in such way that they have to defend positions they would probably not have chosen themselves. Complicated cases can be best handled by giving couples one specific role, so that they can cooperate and analyse the case in advance.

Large groups can be split into smaller groups which deal with different proposals or cases. While one group is carrying out their simulation exercise another group can act as the critical public. After the simulation the teacher invites them to give their own comments on the simulation. Such comments can be about the quality of the arguments (suggestive, straightforward), the attitudes and emotions of the members, the validity of the decision and so on. Similarly, one can provide groups with the same case, but ask them to apply different ethical decision models.

Although simulations are fun, the educational issue only emerges during the evaluation of the simulation. It is important that in a round table discussion the teacher addresses the formulated arguments, the theories and the principles that were put forward in the simulation.

T. de Cock Buning and E. ter Gast

Acknowledgement

The authors wish to thank all the students who contributed to the presented case methodology in the education of applied ethics by their participation and responses.

The ethical matrix: a framework for teaching ethics to bioscience students

Ben Mepham

Centre for Applied Bioethics, University of Nottingham, Sutton Bonington Campus, Loughborough LE12 5RD, United Kingdom

Abstract

The chapter describes an approach used to introducing bioethical theory to applied biology undergraduates over a number of years. It employs a framework for ethical analysis, the ethical matrix, which is designed to assist students in:

- Identifying ethical concerns
- Examining information and opinions relevant to those concerns
- Making considered ethical judgements on these concerns, and
- Through class discussion, comparing these judgements with the views of others.

In the form in which it used here, the matrix applies the three ethical principles of respect for wellbeing, autonomy and justice to the interests of defined groups who might be affected by the application of a novel agricultural biotechnology. To illustrate the use of the matrix, ethical issues raised by the employment in dairying of the milk yield-boosting hormone, bovine somatotrophin (BST), are assessed by examining claimed impacts of its use on these three principles as they affect dairy farmers, consumers, treated dairy cows and the ecosystem. The example chosen highlights critical concerns, since BST is employed commercially in the USA but is banned in the EU. It is stressed that in order to reach sound ethical decisions, in addition to a suitable framework, it is also necessary that assessments be performed by competent moral judges, whose attributes are defined.

Although illustrated here in relation to an agricultural biotechnology, in principle, the Matrix can be applied to any field of human endeavour in which choices involving a range of interests have to be made.

1. Introduction

There are many different approaches to addressing ethical issues, but for bioscience students one that appeals to *principles* is generally helpful, in that a systematic, rational approach accords well with their scientific training. The aim is to stimulate authentic ethical deliberation, without suggesting that ethics is either facile or too complicated. The following approach has been adopted in introducing bioethical theory to applied biology undergraduates over a number of years.

2. The theoretical background

Western ethical traditions are often said to stem from two quite distinct modes of reasoning, both of which however are widely acknowledged, and thus contribute to the reasoning most people apply to ethical questions. The *utilitarian* approach, most famously articulated by Bentham and Mill (18/19th centuries) aims to provide justification for actions by maximising *pleasure* over *pain*, (or, in other formulations, *benefit* or *preference* over *cost* or *harm*). This consequentialist approach often appeals to a scientific outlook because it appears to depend on a *hedonic calculus*. While it is obviously incumbent on us to behave responsibly, alert to the consequences of our actions, utilitarianism has several logical and logistical limitations. For example, there are difficulties in deciding who (or *what*, if we refer to animals in that way) is to count in the cost/benefit analysis, and over how long a period of time the costs and benefits are to be aggregated.

The *deontological* approach advanced by Kant (18th century) stresses the importance of rights and duties, irrespective of consequences. Categorical imperatives are the bedrock principles of this approach, which Kant grounded in reason not emotion. Most people do recognise the force of some absolute principles (e.g. 'murder is always wrong') but difficulties arise when respect for different categorical imperatives is in conflict. If we are duty bound both to tell the truth and to protect others from harm it is likely that we shall sometimes be unable to respect both duties.

The ancient Greeks placed emphasis on virtues, of which *justice* remains critical in modern democratic society. The late John Rawls, the US political theorist, saw *justice as fairness* as the basis of modern democratic society (Rawls, 1972), and it can be seen as the third pillar of a principled approach to ethics. In fact, all three approaches may be said to contribute to the *common morality*, which forms the basis of the implicit (or rarely articulated) norms informing ethical reasoning in Western society.

In the 1970s, the biomedical ethicists, Beauchamp and Childress (1994), devised an approach to resolving ethical issues in modern medicine by appeal to four *prima facie* principles derived from the common morality. Of course, it may not be possible to respect all the principles fully, but the approach seeks to ensure that in addressing ethical issues in medicine due attention is paid to a range of questions, in ways which demonstrate consistency and are explicit.

3. An Ethical Matrix

Mepham (1996) has adapted Beauchamp and Childress's approach to ethical issues arising from modern agricultural and food biotechnology, which (unlike the simplest types of medical issue) usually entail consideration of impacts on several interest groups, e.g. consumers, farmers, farm animals, and the ecosystem (animal and plants in the environment). By applying three principles, viz. respect for wellbeing (utilitarianism), autonomy (Kantianism), justice (Rawlsianism), derived from Beauchamp and Childress's four principles, to the different interest groups, a table (Matrix) is produced, which facilitates ethical deliberation and analysis. Figure 1 illustrates *one* form of the Matrix.

Some of the specifications of the principle may appear more problematical than others, but generally the approach has received support from those who have employed it in workshops on specific issues.

The Matrix is a conceptual tool. It aims to facilitate ethical reasoning, rational debate and transparency, and to identify areas of agreement and disagreement. In some

Respect for:	Wellbeing	Autonomy	Justice
People in the agricultural industry	Satisfactory income and working conditions, AW	Appropriate freedom of Action, AA	Fair trade laws and practices, AJ
Citizens	Food safety and acceptability, Quality of life CW	Democratic, informed choice e.g. of food, CA	Availability of affordable food, CJ
Farm animals	Animal welfare, FW	Behavioural freedom, FA	Intrinsic value, FJ
The ecosystem	Conservation, EW	Biodiversity, EA	Sustainability, EJ

Figure 1. An Ethical Matrix showing, in twelve individual cells, the interpretation of respect for the principles of wellbeing, autonomy and justice in terms appropriate to the interests of people working in the agricultural industry, citizens, farm animals and the ecosystem, respectively.

formulations, the ways in which a principle (e.g. animal welfare or biodiversity) is respected or infringed by a prospective practice (e.g. a biotechnology) can be 'scored' (by assigning it, say, +1 or -2). But since the different principles are likely to have different *weights*, no simple calculus of ethical acceptability is possible. In other formulations, the Matrix merely serves to structure analysis, and is not used to 'score'.

4. Aims and limitations

It is important to appreciate the aims and limitations of the Matrix.
- The impacts defined for each of the separate cells usually depend on rigorous examination of objective (often, but not invariably, scientific) data
- The duties described are *prima facie* duties: circumstances will frequently arise when there are conflicts between different duties and compromises will have to be made
- Construction of the Matrix is in principle ethically neutral, i.e. it is an *analytical* tool
- By contrast, ethical *evaluation* or *judgement* requires a weighing or ranking of the different impacts, so that e.g. an animal rightist might consider *any* exploitation of animals inadmissible, while a more utilitarian view might accept that substantial human benefits outweigh minor harms inflicted on animals
- The Matrix records ethical impacts in one set of circumstances (e.g. the prospective introduction of a technology) with another set of circumstances (usually, the *status quo*). Hence, the impacts recorded are *relative* to a pre-existing condition, which itself might be far from ethically acceptable by reference to some other actual or possible condition
- While it might guide individual ethical judgements, the principal aim of the Matrix is to facilitate rational decision-making by articulating the ethical dimensions of any issue in a manner which is transparent and broadly comprehensible.

It must be appreciated that the types of evidence considered in the different cells of the Matrix are of variable nature. Some might be based on numerical data, others on predictions of future consequences, which could be highly speculative. In some cases, those assessing ethical impacts might place emphasis on immediate effects while others might discount short-term consequences and prefer to place emphasis on future developments. In all cases, the degree of trust assigned to those presenting the relevant information is certain to influence the ethical assessments made.

5. Application of the Ethical Matrix to the case of bovine somatotrophin

To exemplify the issues, the Matrix is used here to analyse ethical impacts of the use of bovine somatotrophin (BST) in dairying. This genetically engineered protein hormone is used in the USA and some other countries to increase milk yield in dairy cattle, but in 1999 it was banned for commercial use in the EU, after many years of being subject to a moratorium.

Responses to fortnightly subcutaneous injection of BST are claimed to increase yield by 10-15% (2 - 7 litres of milk per day). Benefits of BST use are primarily economic - fewer cows being required to produce a given quantity of milk, but it is also claimed to have some beneficial environmental impacts since production of certain undesirable products, such as manure and methane gas, might be reduced.

In the following, where there is perceived *respect for a principle* this is indicated by + (listed first) whereas perceived *infringement of a principle* is indicated by ø. The symbol? indicates an uncertain outcome. Abbreviations (e.g. *FW*) refer to cells of the Matrix (see Figure 1), but some of the interest groups are defined more narrowly. Thus the following application of the Ethical Matrix to BST considers impacts on the dairy farmers, consumers of dairy products, the treated animals and the ecosystem.

In view of space limitations, the analysis is summary in the extreme, but it serves to illustrate the way in which the Matrix might be used. Moreover, the selection of interest groups is to a degree subjective, even they have been chosen to conform to the common morality. Some might consider the list too restrictive: e.g. it excludes economic impacts on society as a whole - although these are included indirectly in consumer issues, since milk consumption is almost universal in Western countries.

5.1 Dairy farmers

Wellbeing (AW)

? Official predictions (US Government, 1994) envisaged that financial returns for farmers adopting BST would be increased by $3 per cow in 1999. However, Tauer and Knoblauch (1997), reporting on the economic effects of BST use on New York dairy farms stated that net farm income increased only by a mean of $27 per cow, which *"was not significantly different from zero"*. Moreover, a more recent economic analysis (Tauer, 2001) indicated that at least half the farmers using the product did so at a loss, but that the complexity of farm systems and inter-year variability precluded identification of the 'winners' and 'losers'.

ø Official predictions (US Government, 1994) envisaged that financial returns for farmers not adopting BST would be decreased by $84 per cow in 1999.

Autonomy (AA)

+ Farmers are allowed to use BST in several countries: most prominently in the USA

ø Farmers are not allowed to use BST in other countries, notably the EU

ø In countries such as the USA some farmers may have adopted BST, against their real wishes, out of perceived economic necessity (i.e. unwilling recruits to the 'technological treadmill'). For example, in a survey of dairy farmers in the United Kingdom, Millar *et al.* (1999) reported that 79% did not consider BST 'ethically acceptable'. It has been argued that each new technological advance both limits freedom of choice and leaves the industry more dependent on external inputs and the commercial imperatives of biotechnology companies.

Justice (AJ)

+ It is claimed that because BST is applied on an individual cow basis it is scale-neutral, so that benefits are available even to small farmers (Bauman, 1992)

ø It is claimed that the effectiveness of BST use depends on 'a high level of management skills' (Bauman, 1992), which are, however, defined by the scale of the milk yield response observed (a circular, and hence questionable, argument). However, there is considerable variation in responses to BST injection (see *AW*) and in some cases no significant yield increase is observed. Its economic use might thus depend on sufficiently large herds (to allow for poor responders), adequate feed (crucial to sustain the response) and recourse to veterinary treatment to manage increased illness. Such conditions are unlikely to apply for most livestock farmers in less developed countries, whose ability to compete with dairy products produced using BST might thus be adversely affected. The principle of *respect for justice* thus has international implications.

5.2 Consumers

Food safety (CW)

While BST, itself, is unlikely to exert any adverse biological effects in humans because it is substantially different chemically from human ST, BST administration increases the concentration in milk of a substance with undisputed biological activity in humans, viz. insulin-like growth factor-1, also known as IGF-1 (EU, 1999b).

+ One official report claims *"rbST can be used without any appreciable health risks to consumers"* because *inter alia* IGF-1 is degraded in the gut. (Joint FAO/WHO, 1998)

ø However, according to a later EU report: *"clear evidence is provided that orally ingested IGF-1 reaches the receptor sites in the gut in its biologically active form".* Moreover, *"The diverse biological effects attributable to the intrinsic activity of IGF-1, exerting a broad variety of metabolic responses through endocrine, paracrine and autocrine mechanisms, make the definition of an in vivo quantitative dose-effect relationship virtually impossible."* (EU, 1999b)

ø The EU report claimed there is a need to evaluate *"the possible contribution of life span exposure to IGF-1 and related proteins, present in milk from BST treated cows, to gut pathophysiology, particularly of infants, and to gut associated cancers"* and draws attention to *"an association between circulating IGF-1 levels and an increased risk of breast and prostate cancer"*(EU, 1999b).

ø Data on the amount of IGF-1 and of a more active form (so-called 'truncated IGF-1') in milk of BST-treated cows were described as *'incomplete'*"(EU, 1999b).

ø Increased use of antimicrobial substances (to treat BST-associated mastitis) might lead to the selection of resistant bacteria "(EU, 1999b).

ø According to a survey in the EU, milk consumption would decline by 11% if BST use were to be legalised EU Commission, 1993). This raises a different type of public health concern, in that milk is an important source of dietary nutrients (particularly calcium) and any significant reduction in its consumption might have adverse effects on public health. For example, inadequate calcium intake increases the risk of osteoporosis

Respect for wellbeing also includes attitudes encompassed by *ethical acceptability.*

ø   In a United Kingdom consumer survey (Millar *et al.*, 1999), carried out prior to the announcement of the EU ban on BST use, 65.4% considered that the use of BST was not 'ethically acceptable.'

Autonomy (CA)

Respect for their *autonomy* would allow consumers to chose whether or not to purchase the dairy products that result from BST use e.g. by requiring such products to be labelled.

+   Negative labelling (that the product is derived from cows not treated with synthetic BST) is allowed in the USA, at the expense of the non-BST user. The wording must include the statement: *The FDA has said no significant difference has been shown and no test can now distinguish between milk from rBGH* (i.e. BST) *treated and untreated cows".*

ø   Milk from cows treated with BST in the USA is not labelled.

Justice (CJ)

There appears to be no evidence that BST use in USA has benefited consumers through reduced prices. There thus seems to be no impact on affordability.

5.3 Dairy Cows

Welfare (FW)

Although there is an extensive body of scientific literature on BST's physiological effect in increasing milk yields, the number of studies which have specifically considered its impacts on animal welfare is relatively small. For current purposes, an EU report (EU, 1999a) is used as an authoritative source of reference. The authors of the report expressed the opinion that "*animal welfare does not appear to have been an issue in the decision making process on BST in the USA*" i.e. as performed by the Food and Drug Administration, and concluded that an adequately wide range of studies on welfare indicators in animals receiving BST had not been performed, making accurate assessment of risks impossible. Despite this, the authors were able to draw a number of conclusions.

+ It has been claimed that in some cases the reduction in herd size required to meet a given milk yield target may allow stockpersons to give increased time to individual cow welfare.

+ In the particular circumstances in which BST injections given in late lactation allowed economic returns from extending the normal lactation, thus obviating the necessity for annual calving, welfare might be improved by reducing the total lifetime stress associated with pregnancy and parturition (van Amburgh *et al.*, 1997). (This is not, however, widely practised.)

ø Use of BST increases the risk of clinical mastitis, a painful disease resulting from inflammation of the udder. The magnitude of the increase in incidence of this condition following BST use has been variously recorded as: 15-45%; 23%; 25%; 42%; and 79% and the duration of treatment for mastitis was longer than normal in cows receiving injections of BST (EU, 1999a).

ø Increased incidences of foot and leg disorders associated with long term administration of BST have been reported, e.g. in the largest study, the number of multiparous cows with foot disorders and the number of days affected were both more than doubled (EU, 1999a).

ø Reproductive capability is reduced by BST, with some studies showing that the pregnancy rate (i.e. the number of inseminated animals which become pregnant) fell from 90% to 63% in primiparous cows and from 82% to 73% in multiparous cows. Rates of multiple births, which may reduce welfare, were substantially increased by BST (EU, 1999a).

ø BST-treated cows often have reduced body condition at the end of the lactation period and experience increased periods of being 'off feed'. This results from the increased demands on both the energy reserves of the body and the digestive capacity of the gut (EU, 1999a).

ø A number of other conditions are associated with BST use e.g. increased incidences of bloat, indigestion and diarrhoea; reduced ability to cope with raised environmental temperatures; and an increased culling rate in multiparous cows (EU, 1999a).

While the occurrence of increased levels of morbidity in BST-treated cows is generally acknowledged (21 identifiable health risks are listed on the Monsanto Posilac 'package insert'), dispute has arisen as to whether the morbidity is attributable to the BST *per se* or to its effect in increasing yield. Some (e.g. White *et al.*, 1994) argue that because increased morbidity may also be associated with increased milk yields achieved by other means, it is inaccurate to blame BST. Others (e.g. Willeburg, 1994) consider this argument spurious. However, it clearly does not apply to the following:

ø Adverse injection site reactions occur in BST-treated animals, with severe reactions in at least 4% of cows (EU, 1999a). Although injections made at the tail head (as recommended) cause less swelling than those behind the shoulder, because there is less subcutaneous space to accommodate inflammation at this site, the resulting pressure and pain may be greater. The procedures involved in administering the injection may be stressful in some cases.

Behavioural freedom (FA)

ø Increased incidences of lameness reduce mobility (see *FW*).

ø The need to supply the cows with greater amounts of concentrate feed (to fully exploit the effect of BST) tends to favour regimes in which cows are kept indoors and deprived of opportunities to graze. This also has important implications for *AW*, since permanent housing is associated with increased lameness and mastitis.

Intrinsic value (FJ)

ø The concept of *respect for animals' intrinsic value* implies that animals should receive fair treatment in accordance with their perceived rights as sentient beings. (For a more detailed description of this principle and the related concept of *telos*, see Mepham, 2000a). In such terms, the enforced alteration of physiological and behavioural norms (reflected, for example, in reduced reproductive fertility) may be seen as an infringement of the standards of treatment to which animals 'under human care' are entitled.

5.4 The ecosystem

The living environment is represented in the Ethical Matrix as the ecosystem (i.e. wildlife in the form of flora and fauna). The ethical principles are translated as respect for *conservation (EW)* of the living environment, *biodiversity (EA)* and *sustainability (EJ)*, which are considered here collectively.

It is difficult to define these ethical impacts with precision because most effects are likely to be secondary to BST treatment and depend principally on effects on herd sizes and locations, feeding and housing regimes, and alternative forms of land use if grazing is reduced. Thus, in theory, BST use might lead to some effects which respect the ethical principles and to others which infringe them. Much will depend on management practices, predictions of which will doubtless be based on evidence derived from analogous previous developments.

+ Some claim that because BST increases yield per cow, the total number of cows needed to supply milk requirements will be reduced, and this could free up land for more environmentally friendly purposes.

+ Reducing cow numbers could also reduce environmental pollution caused by dairy farming, e.g. on a global scale fewer dairy cows would produce less methane (a greenhouse gas) (Johnson *et al.,* 1992).

ø Conversely, increased intensification, considered by some to be a likely result of widespread BST use, often leads to more point-source pollution, e.g. from slurry and silage effluents.

ø Moreover, if increased feeding of concentrates (to more fully exploit the effect of BST) leads to reduced grazing (in extreme cases zero grazing) land not being used to graze animals may be used for increased monoculture, risking a decrease in biodiversity.

5.5 Ethical evaluations of BST use

It is clear that two opposing outcomes are possible using such an analysis. Those approving the ethical acceptability of BST use would probably cite the economic benefit to the manufacturers of BST, to the economies of countries in which it is manufactured, to some of the farmers who use it, and, were prices to fall, to consumers of dairy products. Moreover, if its use led to reduced cow numbers it might result in marginally reduced emissions of methane. This case also rests on perceptions that the welfare of treated cows is not affected significantly (or that increased morbidity can be effectively treated) and that there are no risks to human safety, so that labelling is unnecessary. Job losses in the dairy industry are not seen as an ethical issue, being merely a feature of market economies.

The ethical case of those wishing to ban (or at least limit) BST use would probably focus on respects in which it appears to infringe several commonly accepted ethical principles. They would point to authoritative reports which suggest that use of BST substantially increases the risk of pain and disease in dairy cows, and that it might present a risk to human safety through ingestion of increased IGF1 in milk. Moreover, they might consider that BST use will compromise the autonomy of farmers; undermine consumer autonomy because milk products from BST-treated cattle are not labelled; jeopardise public health if a widespread rejection of dairy products followed the licensing of BST in the EU; and increase local pollution as a consequence of the intensification of dairying.

6. Competent moral judges

The Matrix *per se* does not adjudicate on ethical evaluations, both because the principles allow room for the exercise of judgement with respect to specific cells of the Matrix and because individual judgements also depend on the 'ethical weight' attached to different cells

Clearly, however, making ethical judgements, particularly when they have such widespread impacts, should not be a matter of expressing personal prejudices. For example, it would make a mockery of the procedure if those with a vested commercial interest were allowed to have a deciding influence on whether any particular technology were to be approved.

Rawls (1951 p. 177) claimed that ethical decision-making depends on the existence of *"a class of competent moral judges"*, who should have the following characteristics: normal intelligence; reasonable knowledge of world affairs; a capacity to *reason*, i.e. see both sides of a question, making allowance for personal bias; and an imaginative appreciation of other people's predicaments. Certain constraints were attached to the judgement procedure viz. the judge must be immune from, and have no vested interest in, the judgement; which must only be arrived at after a careful enquiry into the facts of the case, be delivered with appropriate certitude and be stable. More recently in the United Kingdom, the Nolan Committee on Standards in Public Life made similar recommendations, which form an essential element of the selection procedure for government advisory committees (British Council, 1999).

However, it is a feature of democratic societies, particularly modern multicultural societies, that the normal outcome of the exercise of human reason is *"a plurality of reasonable yet incompatible doctrines"* (Rawls, 1993). Hence, consensus on a moral orthodoxy is probably an unrealistic, if not a dangerous, objective. A sounder aim might be that of devising a social contract which benefits from social cooperation despite the differences of opinion between the contractors. Thus, the role of ethical theory in this process is not to determine *the right* policies but to act as a means of assessing whether specific proposed policies are ethically acceptable.

7. Worldviews

In conclusion, it is worth considering briefly the justifications people advance to support their ethical judgements. An important factor appears to be the different 'worldviews' people bring to these judgements. For present purposes, in very general terms, worldviews may be characterised as:
- Anthropocentric: placing value chiefly in humans and human achievements, and seeing the natural world as a resource
- Ecocentric: placing value in the natural world, of which humans are only one species, whose activities ought not to unduly disturb the ecosystem
- Ambicentric (a neologism): a moderated anthropocentrism, which considers that human interests demand precedence, but also recognising that people have qualified duties to the ecosystem.

It would, of course, be possible to multiply the list indefinitely by adding other qualifications, such as the belief that animal life is intrinsically no less valuable than human life (a view which would probably be supported by most vegans).

According to such a characterisation, respect for the different principles due to BST use might be represented as in Figure 2, which seeks to provide a rationale for the different positions adopted by the USA and the EU. The judgements are shown as skewed to the 'disallow' option because this is the clear opinion indicated by an opinion poll in the United Kingdom (Millar *et al.*, 1999).

Worldview	AW	AA	AJ	CW	CA	CJ	FW	FA	FJ	EW	EA	EJ	Judgement
Anthropocentric	+	+	+	0	Ø	+	Ø	Ø	0	+	0	0	Allow
Ambicentric	+	0	+	Ø	Ø	0	Ø	Ø	Ø	0	0	Ø	Disallow
Ecocentric	Ø	Ø	Ø	Ø	Ø	Ø	Ø	Ø	Ø	Ø	Ø	Ø	Disallow

Figure 2. The influence of worldviews.
The impact of worldview on assessment of the impacts of BST use on the ethical principles defined in Figure 1. The symbols +, Ø and 0 represent 'respect for the principle', 'infringement of the principle', and 'no effect', respectively. Note that these are the author's speculations, and are not research data.

This chapter is based on an analysis originally published in Mepham (2000b), listed below, which provides a fuller account of how ethical analysis relates to public policy.

8. Conclusions

This chapter has described a framework for ethical analysis which is designed to assist students in:
- Identifying ethical concerns
- Examining information and opinions relevant to those concerns
- Making considered ethical judgements on these concerns, and
- Through class discussion, comparing these judgements with the views of others

Although illustrated here in relation to an agricultural biotechnology, in principle, the Matrix can be applied to any field of human endeavour in which choices involving a range of interests have to be made.

References

Bauman, D.E., 1992. Bovine somatotropin: review of an emerging animal technology. Journal of Dairy Science 75, p. 3432-3451.

Beauchamp, T.L. and Childress, J.F., 1994. Principles of Biomedical Ethics. 4th edition, New York and Oxford: Oxford University Press.

British Council (1999) Governance and law: ethics in public life and corporate governance. (London: British Council)

EU Commission, 1993. Concerning bovine somatotrophin (B.S.T) COM(93) 331 final.

EU Consumer Policy and Health Protection Directorate, 1999a. Report on Animal Welfare Aspects of the Use of Bovine Somatotrophin.

EU Consumer Policy and Health Protection Directorate, 1999b. Report on Public Health Aspects of the Use of Bovine Somatotrophin.

Johnson, D.E., Ward, G.M. and Torrent, J., 1992. The environmental impact of bovine somatotropin use in dairy cattle. Journal of Environmental Quality 21, p. 157-162.

Joint FAO/WHO Expert Committee on Food Additives (1998) Rome. WHO Food Additive Series 41, p. 125-146.

Mepham, T.B., 1996b. Ethical analysis of food biotechnologies: an evaluative framework. In T.B. Mepham (ed.) Food Ethics (London: Routledge). p. 101-119.

Mepham, T.B., 2000a. A framework for the ethical analysis of novel foods: the Ethical Matrix. Journal of Agricultural and Food Ethics 12, p. 165-176.

Mepham, T.B., 2000b. The role of food ethics in food policy. Proceedings of the Nutrition Society 59, p. 609-618.

Millar, K.M., Tomkins, S.M. and Mepham, T.B., 1999. Ethical attitudes of consumers and farmers to the use of two technologies: bovine somatotrophin and automated milking systems. Proceedings of the British Societry for Animal Science, p. 195.

Rawls, J., 1951. Outline of a decision procedure for ethics. The Philosophical Review 60, p. 177-197.

Rawls, J., 1972. A Theory of Justice, Oxford: Oxford University Press.

Rawls, J., 1993. Political Liberalism, Oxford: Oxford University Press.

Tauer, L.W., 2001. The estimated profit impact of recombinant bovine somatotropin in New York dairy farms for the years 1994 through 1997. AgBioForum 4, p. 115-123.

Tauer, L.W. and Knoblauch, W.A., 1997. The empirical impact of recombinant bovine somatotropin in New York dairy farms. Journal of Dairy Science 80, p. 1092-1097.

US Government, 1994. Use of bovine somatotropin in the United States: its potential effects. Washington DC: US Government Printing Office.

Van Amburgh, M.E., Galton, D.M., Bauman, D.E. and Everett, R.W., 1997. Management and economics of extended calving intervals with use of bovine somatotropin. Livestock Production Science 50, p. 15-28.

White, T.C., Madsen, K.S., Hintz, R.L., Sorbet, R.H., Collier, R.J., Hard, D.L., Hartnell, G.F., Samuels, W.A., Kerchove, G. and Adriaens, F., 1994. Clinical mastitis in cows treatred with sometribove (recombinant bovine somatotropin) and its relationship to milk yield. Journal of Dairy Science 77, p. 2249-2260.

Willeburg, P., 1994. An internal perspective on bovine somatotropin and clinical mastitis. Journal of the American Veterinary Medical Association 205, p. 538-541

Winkler, E.R., 1993. From Kantianism to Contextualism: the rise and fall of the paradigm theory in bioethics. In Applied Ethics E R Winkler and J R Coombs (eds), Oxford, Blackwell, p. 343-365.

The use of the reflective equilibrium method of moral reasoning in teaching animal and veterinary ethics

Bart Rutgers[1] and Robert Heeger[2]
[1]*Department of Animal, Scence & Society, Faculty of Veterinary Medicine, Utrecht University, P.O. Box 80168, 3508 TD Utrecht, The Netherlands,* [2]*The Ethics Institute, Utrecht University, P.O. Box 80103, 3508 TC Utrecht, The Netherlands*

Abstract

Veterinarians are often faced with ethical issues regarding the human-animal relationship. Moral decision making requires understanding of moral reasoning. In seeking answers to moral questions, the reflective equilibrium method of moral reasoning appears to be a suitable approach. This method is used in teaching veterinary and animal ethics to veterinary students at the Faculty of Veterinary Medicine of Utrecht University. Here an outline is given of both the course on animal ethics and the reflective equilibrium method. With this method of moral reasoning we strive for coherence of moral judgements, moral principles and background thinking. Such reflection enables us to give a justification of our moral judgement about an ethical issue. In order to show how this method works, it will be applied to the issue of the castration of pigs. Decision making according to the reflective equilibrium method bears a strong resemblance to the way veterinarians try to solve clinical problems. Therefore, it should not be difficult for veterinarians who are experienced in clinical thinking to make themselves familiar with the way of moral thinking along the lines of the reflective equilibrium method.

1. Introduction

In all branches of veterinary practice - in companion animal, sport animal and farm animal practice - veterinarians are confronted with ethical issues relating to the human-animal relationship. The resulting professional activities take place within a triangular relationship between animal owners, their animal(s) and the veterinarian. The veterinarian and the owner decide by mutual agreement how to treat the animal. Situations arise within veterinary practice that confront the veterinarian with moral problems: he or she feels uncertain about what is morally right to do in a particular instance.

Moral problems in veterinary practice arise mainly on two levels: (1) on the level of the relationship between veterinarian and the owner, and (2) on the level of the relationship between the veterinary profession and society. Moral problems on the first level may be characterised as conflicts between the interests of animal owners and the interests of animals. The veterinarian ought to take into account both human and animal interests. Sometimes it is quite difficult to decide whose interests - human or animal - should take precedence.

Veterinarians do not only have to deal with animal owners. Society - including ordinary citizens, consumers, government, and animal protectionists - looks critically at how veterinarians do their job and how they deal with their professional responsibilities. Moreover, the veterinary profession has been increasingly required to take a clear stance on particular ethical issues. In this field of tension of interests and expectations, individual veterinarians have to find their way. The problems are often so complex that it is not easy to determine a course of action or policy that is morally right.

Moral decision-making requires insight into moral reasoning. In seeking answers to moral questions, the reflective equilibrium method of moral reasoning appears to be a suitable approach. This method is used in teaching veterinary and animal ethics to veterinary students at the Faculty of Veterinary Medicine of Utrecht University (the Netherlands). We will start by providing an outline of this course in animal and veterinary ethics. Next, we go into the reflective equilibrium method of moral reasoning. We begin with explaining what moral reasoning according to this method means. Subsequently, we give an indication of how we use this method in decision making about a concrete moral problem. In order to show how this method works, it will be applied to the ethical issue of the castration of pigs. Surgical castration of pigs without prior anaesthesia is a common practice in many countries in Europe and in the USA. For many years this practice has been a subject for debate, since surgical castration without anaesthesia is a painful procedure that adversely affects the health and welfare of the pig. For this reason, within the EU there is a strong call for banning surgical castration of pigs.

2. Course in animal and veterinary ethics

One of the subjects of the veterinary curriculum at the Utrecht Faculty of Veterinary Medicine is 'Veterinary Medicine and Society'. In this course, which is taught during the fourth year of the curriculum, particular attention is given to the ethical and legal aspects of the use of farm animals, companion animals and laboratory animals, to the ethical and legal aspects of the veterinary profession and veterinary practice, and to the legal aspects of veterinary public health care. More specifically, the subject Veterinary Medicine and Society consists of three parts: (1) animal ethics and veterinary ethics, (2) veterinary law and (3) forensic veterinary medicine.

The course in ethics consists of two parts: animal ethics and veterinary ethics. The most important teaching goal of *animal ethics* is to train students in practical moral reasoning and moral decision making. Moral reasoning requires knowledge of ethical theories, norms, values and moral principles. Our approach is case-oriented, that is to say, the students in working groups discuss different types of ethical issues regarding the human-animal relationship. Some examples are: performing surgery on animals for non-therapeutic purposes, the use of growth promoters in farm animals, reproduction technology, biotechnology, the killing of animals, and the use of laboratory animals.

The course in *veterinary ethics* focuses on professional responsibility. The teaching goals are that the students understand what is meant by professional responsibility and that they gain insight into professional norms, values and attitudes, and the different ethical codes of practice.

3. The reflective equilibrium method of moral reasoning

In instructing our students in moral reasoning, we use a version of the reflective equilibrium method (DeGrazia, 1996, Sayre-McCord, 1996, Van der Burg and Van Willigenburg, 1998). Let us first sketch this version of the method and then give an impression of how we use it in reasoning about a concrete moral problem.

3.1 Coherence and justification

With a reflective equilibrium method we strive for *coherence* among our moral beliefs. Coherence means not only that our moral beliefs are logically consistent. It means also that they enjoy argumentative support: justifying reasons are offered for a moral judgement, so that the judgement, the reasons, and their implications can be tested for consistency with other judgements that we affirm. Coherence means that our beliefs hang together, and that it is clear how they hang together. For example, if we believe that moral norms of how to treat animals differ from moral norms of how to treat humans, we should explain this division.

According to a reflective equilibrium method, to render our moral beliefs coherent is the best we can do in order to achieve a set of *justified* moral beliefs. A reflective equilibrium method is a coherence method of justification. To justify one's beliefs means to show them to be right. A coherence method says that the various beliefs one holds can be shown to be right only if they cohere well with the other things one believes. Whether one's beliefs are justified is a matter of how well they hang together.

If we follow this method, we do not need to establish one or more general moral principles by which all correct moral judgements would be justified and from which all of them could be derived. In reflective equilibrium thinking, no set of moral norms is privileged such that all justification would have to be based, ultimately, on those norms. An emphasis is placed on the fact that we have moral beliefs of differing degrees of generality. For example, we make particular judgements, that is judgements about a particular case or action ("The veterinarian has done right by revealing the cost of treatment in advance"), or we make fairly general judgements, say judgements about a set of actions or a practice ("Keeping layer hens in battery cages is wrong"), but we also hold very general principles, such as "We ought not to inflict suffering on animals". According to a reflective equilibrium method justification occurs at all levels of generality.

3.2 Interplay between judgements and general principles

The statement that justification occurs at all levels of generality requires some explanation. It means in the first place that justification is not a one-way traffic, from the general to the particular, but can proceed in both directions. We can justify, for instance, particular moral judgements by appealing to one or more general principles, which *support* them. But we can just as well justify a general principle by showing that it is *in keeping with*, or even *accounting for*, many particular moral judgements. In both cases, the particular judgements and the general principles cohere.

But what is to be done if they do not cohere? Suppose that, while striving for coherence among moral beliefs, we discover that there are frictions or even clashes between the beliefs that we in fact are holding. In order to alleviate or eliminate such clashes, we can try to render our beliefs coherent by *revising* at least one of the clashing beliefs. The above statement on justification also concerns this endeavour. It says that judgements at any level of generality can be used to revise judgements on any other level. So, if two judgements conflict and if one of them is a particular judgement while the other is a general principle, then one should not always revise the particular judgement leaving the principle unchanged. Even a principle may be in need of revision. Two examples can illustrate these assertions. First, a member of an animal experiment committee may initially make the judgement that in the case at hand it is morally permissible to keep mice one by one in isolation, because this would be useful for gaining the wanted data. But she may also hold the general principle that we ought not to inflict suffering on animals, and she may realize that mice are social animals, which suffer from being kept in isolation. Then the principle may suggest that she should revise her particular judgement. But, secondly, revision may also go in the opposite direction, from a particular judgement about a case to a general principle. For example, confronted with the urgent need of proving the effectiveness of a new vaccine, we might feel compelled to revise the principle of not inflicting suffering to allow exceptions in this sort of case.

3.3 Intuitive judgements

Searching for justified moral beliefs, one should pay attention to the fact that some moral beliefs are expressed in *intuitive* judgements. These are judgements made simply because they seem correct, whether or not they are supported by arguments. A reflective equilibrium method takes intuitive judgements seriously. It pays heed to their plausibility. A particular moral judgement or moral principle is the more plausible the more it is believable in its own right. It can at least have some initial credibility. One example is the principle that cruelty to animals is wrong. Yet, a reflective equilibrium method guards against excessive reliance on intuitive judgements. The main reason is that even intuitive judgements are not privileged, thus not immune to being revised. An intuitive judgement may be in need of revision, for instance, if under the pressures of reflection it turns out to be insufficiently subtle or excessively parochial.

3.4 Background thinking

There is one more feature of a reflective equilibrium method that should be mentioned. Reflective equilibrium thinking is not confined to an interplay between moral judgements, be they particular or fairly general, and general moral principles.

We should also strive for coherence of our moral beliefs with whatever else we know or reasonably believe. The latter is often called background theories or *background thinking*. The main reason for including background thinking is that it can help to protect against bias. We must be on our guard against bias because moral judgements must be justifiable to others. This requires that our judgements be backed by reasons that are recognized as good reasons not just by ourselves but by persons generally. Background thinking can help to identify and root out bias-infected assumptions in moral reasoning. For example, attention to findings about the mental life of animals may affect our moral judgement that consuming meat from factory farms is, for the most part, unproblematic.

In summary, in order to achieve a set of justified moral beliefs we should try to render our moral beliefs coherent. If the beliefs we are holding do not cohere, we should try to revise them so that they support each other. We may begin with particular or fairly general moral judgements that have some initial credibility, and then look for moral principles that account for those judgements. If we find such principles, then the judgements are supported by the principles. If many judgements are accounted for by the principles but some are not, then there are two possibilities. Mostly, in view of their support, these recalcitrant judgements seem less credible than the principles and may be revised or dropped. But sometimes particular or fairly general moral judgements may require a modification of a moral principle since they express an insight that needs to be better accounted for. After all, a principle finds support from its ability to account for particular judgements. Finally, both judgements and principles should cohere with our background thinking. Striving for reflective equilibrium is striving for a coherent set of beliefs.

4. Reasoning about a concrete moral problem

Our version of the reflective equilibrium method is intended to serve a practical purpose: it should provide guidance in concrete moral decision-making. We therefore try to practise the method by confronting our students with cases and asking them to strive for answers. In order to teach the students both to discern some important components of reflective equilibrium thinking and to structure their reasoning we make use of a step-by-step design (Van Willigenburg *et al.*, 1998). This design starts with the description of a case and subsequently puts three groups of questions to the students:

1. Characterization
- What is the moral question?
- Which possibilities for action are open at first sight?

2. Analysis
- Who are involved in this moral issue?
- Which arguments are relevant for answering the moral question?

3. Weighing
- What is the importance of these arguments for this case?
- Which possible action is preferable on the basis of weighing the arguments?

To give an impression of how this works, here is one example.

4.1 Case: The castration of pigs

In many countries, male pigs are usually surgically castrated. This is done because the meat of uncastrated boars, when heated, gives off an unpleasant smell, "boar taint". Consumers do not want pork with this smell. In most European countries, meat from uncastrated boars is not marketable. This is true in particular for Germany to where the greater part of, for example, Dutch pork is exported. For financial reasons, pig farmers usually castrate their piglets in the first week of life and that without prior anaesthesia. The EU Directive regarding the protection of pigs allows the castration of boars without anaesthesia up to the age of one week, whereas surgical castration of pigs after the seventh day of life may only be performed under anaesthesia and hence by a veterinarian.[100] However, the Royal Netherlands Veterinary Association (1999) has evaluated both surgical castration and a recently proposed alternative, immunocastration, and states that castration of pigs should be rejected, and that alternative methods for the prevention of boar taint should be optimized.

4.1.1 Characterization

Two questions are to be answered: What is the central moral question in this case? Which possibilities for action are open at first sight?

In order to take up the first question, let us begin by asking whether the castration of boars is a moral problem. This question is directed to everyone. But let us make it come alive by means of an example: Is the castration of boars a moral problem for a veterinary surgeon? She may deplore the current castration practice but this practice is the pig industry's concern. Under existing legislation the government entrusts the surgical castration of male piglets to pig farmers. From a legal point of

---

[100] Commission Directive 2001/93/EC of 9 November 2001 amending Directive 91/630/ECC laying down minimum standards for the protection of pigs.

view, veterinary surgeons cannot be held responsible for the current practice with regard to the castration of pigs. Yet, the surgical castration of boars does concern the veterinary surgeon morally. She thinks it a grave operation on the pigs. Her professional responsibility is to care for the well-being of animals, and for all that she is forced to condone the suffering of many animals. At the same time she realizes that, on her own, she would hardly be able to alter the prevailing practice. So, what is the *moral question* for this veterinary surgeon? Probably this: Ought she to make the effort to bring about a policy of her professional group aimed at doing away with the castration of pigs?

Suppose this is the central moral question with regard to the case. Then the second question occurs: Which *possibilities for action* are open at first sight? One answer may be that the veterinary surgeon could explain to the pig farmer that castration without anaesthesia is a grave operation on the pigs, and that the farmer should let a veterinary surgeon perform the castration under anaesthesia. Another answer may be that the veterinary surgeon could advocate that surgical castration without anaesthesia be abolished and replaced by immunocastration, a sterilization technique by immunological means.[101] A third answer may be that the veterinary surgeon could join her colleagues from the Royal Netherlands Veterinary Association and plead with the pig industry and the government both for the abolition of castration in the medium to long term and for efforts to optimize available methods for the prevention of boar taint. These are three possible actions open at first sight.

4.1.2 Analysis

Under this heading two questions are raised: Who are involved in this moral issue? Which arguments are relevant for answering the moral question?

Ought the veterinary surgeon to make the effort to bring about a policy of her professional group aimed at doing away with the castration of pigs? In order to give a considered answer to this question we must try to find out which arguments are relevant for answering it. If we go in search of such arguments, it is advisable to guard against bias or selective perception. A suitable means is to clarify *who are involved* in the moral issue, and then to have as much regard as possible for the perspectives of all those concerned. In our case, apart from the veterinary surgeon herself and her like-minded colleagues, those involved are at least the pig farmers

---

[101] Immunocastration, by which castration is achieved by immunological means, is a recently developed technique. The technique is based on the use of antibodies against gonadotropin-releasing hormone (GnRH). The neutralisation of this hormone by antibodies suppresses the development and function of the testes, and thereby inhibits the development of boar taint. GnRH vaccines are currently being developed.

for whom the current practice is of economic interest, the whole pig industry which benefits by the protection of the pork market at home and abroad, and - last but not least - the animals, even though they cannot stand up for themselves.

Keeping in mind all those concerned we are in a better position to get hold of *arguments relevant* for answering the central moral question. Let us begin with the pig farmers and the pig industry. Taking their perspective into account we can bring forward the fairly general moral judgement that the current policy of castration is right. The current practice is in their economic interest. This would seem to mean, among others things, that the current practice is important for pig farmers and other workers in the pig industry. Therefore, it is likely that we, in support of the moral judgement, can appeal to the general moral principle of beneficence. This principle says that we ought to do and to promote good. The good to be promoted includes the well-being of humans.

If we pay attention to the perspective of the animals, then another fairly general moral judgement appears to be relevant: It is wrong to castrate piglets without anaesthesia. Surgical castration without anaesthesia is detrimental to the welfare of the animals because it is a painful intervention that causes suffering. For the moral judgement we could find support in the general moral principle of non-maleficence which states that we ought not to inflict harm on living beings, be they humans or animals. One could also argue for the propriety of the moral judgement and for the applicability of the general principle by appealing to background thinking. It has been scientifically determined, using physiological and ethological parameters, that surgical castration without anaesthesia is a painful intervention. The view that higher animals, such as pigs, can experience pain is common sense and goes together with the so-called analogy postulate (Stafleu *et al.*, 1992). This postulate says that the similarity in anatomy (structure of the pain system), physiology (pain perception), and behaviour (expression of pain) between humans and higher animals makes it reasonable to believe that the sensation of pain is analogous in humans and higher animals. In evaluating pig castration, the Federation of Veterinarians of Europe (FVE) has recognized that castration without anaesthesia causes suffering to the animals. Therefore, the FVE has stated that the castration without prior anaesthesia should be banned.[102]

Attempting to do justice to the perspective of the animals we could also consider the fairly general moral judgement that immunocastration is right. Immunocastration can be carried out by means of injection of an immunovaccine.

---

[102] FVE Position Paper on Pig Castration (Report FVE/01/083), adopted by the FVE General Assembly on 17 November 2001.

The pain caused by the injection is less than the pain of surgical castration without anaesthesia. We could therefore think that the moral judgement about immunocastration is supported by the general moral principle of non-maleficence.

From the perspective of the animals there is still another fairly general moral judgement that may be relevant: All castration of pigs for the sake of protecting the pork market or production is wrong. This judgement concerns both surgical castration and immunocastration. It would seem to be relevant, if we do not from the start restrict judgements of animal ethics to matters of suffering and contentment. The moral judgement could be supported by a general moral principle of respect for animal integrity (Rutgers and Heeger, 1999, Bovenkerk *et al.*, 2002). It is not easy to delimitate the content of this principle in precise terms. But we may perhaps indicate what it typically contains: We ought not to impair the biological intactness of an animal, we ought not to infringe those functions and operations of an animal that a member of the species is normally capable of performing. Examples are growth, reproduction, motion, and social capacities.

Our search has resulted in *four arguments* that are relevant to the central question. The arguments are four moral judgements supported by moral principles. The moral judgements say (1) that the current policy of castration is right, (2) that it is wrong to castrate piglets without anaesthesia, (3) that immunocastration is right, and (4) that all castration of pigs for the sake of protecting the pork market or production is wrong. There are collisions between the moral judgements. But even the supporting principles collide with each other: the principle of beneficence collides with the principle that we ought not to inflict harm on living beings, or attack their well-being, and it collides with the principle that we ought to respect animal integrity.

4.1.3 Weighing

Again, two questions are to be answered: What is the importance of the arguments for this case? Which possible action is preferable on the basis of weighing the arguments?

If we try to make up our mind about the *importance* of the arguments, we should look out for a pitfall. The conflict of arguments is also a conflict of principles. But this does not mean that we could cope with the conflict of arguments by reducing it to a conflict of principles. The principles mentioned do not collide as such. We could subscribe to all of them without contradicting ourselves. Rather the principles collide with each other in the case we are dealing with. So, in order to resolve conflicts between them, we must reflect on them in the context of the concrete situation, not isolated from the moral judgements, but together with them. We must ask what weight they have in the concrete situation, and whether or not in this situation they differ in importance.

Let us begin by considering the moral judgement that the current practice of castration without anaesthesia is right. This judgement is supported by the principle of beneficence. The principle captures something important for pig farmers and other workers in the pig industry: their experienced well-being ought to be promoted. Their economic earnings do not fall from heaven like manna. The principle of beneficence does justice to the interest of pig farmers and other workers in the pig industry.

However, if we follow the principle of beneficence along these lines, we come into collision with the principle of non-maleficence which says that we ought not to inflict harm on animals. This principle provides an objection to the current castration policy. It points out that we have moral duties not only towards humans but also towards animals, and that these duties include that we abstain from inflicting suffering on animals. Since castrating pigs without anaesthesia entails inflicting suffering on them, this intervention is not morally permissible.

We could try to resolve this conflict by dropping the judgement that castration without anaesthesia is morally right, and instead subscribing to the moral judgement that it is wrong to castrate piglets without anaesthesia. In order to meet the requirements of the principle of non-maleficence we could opt for the moral judgement that surgical castration under anaesthesia is right. This judgement is more in line with the principle because anaesthesia leads to a reduction of suffering. But if we consider adopting the judgement as a guide for a general policy, then we get into trouble. To castrate all male piglets under anaesthesia is not feasible. The animals would have to be anaesthetized by veterinary surgeons. Calling in veterinary surgeons would have enormous logistic and financial consequences, given the very large numbers of animals involved (in the Netherlands over 10 million yearly). This background thinking about feasibility would seem to prevent us from endorsing the judgement as a guide for a general policy.

But we could perhaps contemplate the moral judgement that immunocastration is right. This judgement might be compatible with the principle of non-maleficence since the pain caused by the injection of the vaccine for immunocastration is less than the pain of surgical castration without anaesthesia. However, some background thinking may make us hesitate. Until now little attention has been paid to the effects of immunocastration on the health and welfare of pigs. There is insufficient information available to be able to determine whether immunocastration really is less detrimental to the welfare (health and behaviour) of animals than surgical castration. So, we do not know whether the moral judgement that immunocastration is morally right actually meets the principle that we ought not to inflict harm on animals. Still, we may say that it meets the principle in one important respect: immunocastration avoids the suffering caused by surgical castration without anaesthesia.

But there is one more argument that has to be considered. Both surgical castration and immunocastration collide with the moral judgement that all castration of pigs for the sake of protecting the pork market or production is wrong. This judgement is supported by the principle of respect for animal integrity. According to this principle we ought not to impair the biological intactness of an animal. We ought not to infringe those functions and operations of an animal that a member of the species can normally perform. By surgical castration we take away something that belongs to the animal. This is also true for immunocastration. By neutralising the gonadotropine-releasing hormone we suppress the development and functioning of the testes. We interfere with the biologically intact system of the animal. We infringe the functions and operations that are characteristic of it. We detract from the "pigness" of the pig.

What can we say about the weight of these different arguments? The moral judgement that surgical castration without anaesthesia is right can be supported by the general principle of beneficence which implies that the well-being of pig farmers and other workers in the pig industry ought to be promoted. However, this is incompatible with the principle of non-maleficence which supports the moral judgement that surgical castration without anaesthesia is wrong. If we hold the principle of non-maleficence, we should consider the suffering of millions of boars more weighty than the contribution of the current practice to the well-being of pig farmers and other workers in the pig industry. Since surgical castration under anaesthesia is not feasible for a general policy, we may contemplate the moral judgement that immunocastration is right. In one important respect, this judgement is supported by the principle of non-maleficence: immunocastration avoids the suffering caused by surgical castration without anaesthesia. But in another respect, it collides with the moral judgement that castration, be it surgical or immunological, is wrong. This moral judgement is supported by a principle of respect for animal integrity which is not confined to matters of suffering but also captures the normal functions and operations of an animal that should be left intact.

How can we cope with this conflict? To be sure, the principle of respect for animal integrity does not say that we ought to refrain from any intervention in the life of animals. What it does demand is that we abstain from impairing their characteristic functions and operations unless there are valid and sufficiently weighty reasons that justify our interference. Protecting the pork market and production is an important aim. It is, however, not a valid reason of sufficient weight to justify the castration of boars, be this surgical or immunological. We may refer to some background thinking here. We could stand our ground by using other means than castration. There are realistic, practicable alternatives for the prevention of boar taint: we could slaughter boars before they are sexually mature, and we could develop the already available methods for the detection of boar taint on the slaughter line. If the slaughterhouses would be ready to process younger boars, and if the pig industry would make an effort to optimize adequate detection methods, then we could abolish castration in the medium to long term.

If this weighing of the arguments merits our assent, then we can also answer the question as to which possible action is *preferable*. We should plead with the pig industry and the government both for the abolition of castration in the medium to long term and for efforts to optimize available methods for the prevention of boar taint.

5. Moral thinking and clinical thinking

Moral decision making according to the reflective equilibrium method is a process of balancing moral judgements, moral principles and background theories. Reflecting on an ethical issue along the lines of the reflective equilibrium method enables us to give a justification of our moral judgement about an ethical issue. It provides us with a structuring of moral justification that closely resembles our day-to-day reasoning processes when trying to reach a moral decision. This makes the method attractive to use.

On closer inspection, decision making according to the reflective equilibrium method bears a strong resemblance to the way veterinarians try to solve clinical problems. Clinical decision-making is a process of balancing intuitive clinical judgements (the 'clinical eye'), biological principles and background theories. For example, a dog owner consults a veterinary surgeon because his dog is diseased. From the story of the owner about the troubles of the dog and the clinical symptoms the veterinarian gets an idea of what is going on with the dog. To be sure about the diagnosis she wants to test whether her initial intuition ('clinical eye') is right. For that purpose she examines the dog more closely. She may use different diagnostic tools, such as a blood test, bacteriological research or X-ray diagnostics. Then the veterinarian has to interpret the results of the diagnostic tests in order to determine their significance for the diseased dog. The final diagnosis can be made by a continuous interplay between the veterinarian's initial intuition about what is wrong with the dog and her general knowledge of diseases. In this process, biological principles and background theories play an important role. The knowledge of a disease as a biological phenomenon is constructed by the interaction of several biomedical disciplines, such as epidemiology, diagnostics, clinical trials and therapy development. A coherent structure of knowledge of the disease can be obtained by the interaction of arguments (facts and theories) of each of these biomedical disciplines (Nederbragt ,2000). The initial diagnosis of the veterinary surgeon can only be justified when there is coherence between her clinical intuition and the scientific knowledge of the disease.

It can be concluded that clinical thinking and moral reasoning show a strong methodological resemblance. Therefore, it should not be difficult for a veterinarian who is experienced in clinical thinking to make herself familiar with the way of moral thinking along the lines of the reflective equilibrium method.

References

Bovenkerk, B., Brom, F.W.A. and van den Bergh, B.J., 2002. 'Brave New Birds. The Use of 'Animal Integrity' in Animal Ethics', Hasting Centre Report 32 (1), p. 16-22.

DeGrazia, D., 1996. Taking Animals Seriously: Mental Life and Moral Status, Cambridge: Cambridge University Press, Ch. 2.

Nederbragt, H., 2000. 'The biomedical disciplines and the structure of biomedical and clinical knowledge', Theoretical Medicine 21, p. 553-566.

Royal Netherlands Veterinary Association, 1999. 'Castration of pigs: to do or not to do?' (in Dutch), Tijdschrift voor Diergeneeskunde 124 (19), p. 584-586.

Rutgers, L.J.E. and Heeger, F.R., 1999. 'Inherent worth and respect for animal integrity', in M. Dol, M. Fentener van Vlissingen, S. Kasanmoentalib, M. Visser and H. Zwart (eds.), Recognizing the Intrinsic Value of Animals: Beyond Animal Welfare, Assen: Van Gorcum, p. 41-51.

Sayre-McCord, G., 1996. 'Coherentist Epistemology and Moral Theory, in W. Sinnot-Armstrong and M. Timmons (eds.), Moral Knowledge? New Readings in Moral Epistemology, Oxford, Oxford University Press, p. 137-189.

Stafleu, F.R., Rivas, E., Rivas, T., Vorstenbosch, J., Heeger, F.R., and Beynen, A.C., 1992. 'The use of analogous reasoning for assessing discomfort in laboratory animals', Animal Welfare 1, p. 77-84.

Van den Burg, W. and Van Willigenburg, T. (eds.), 1998. Reflective Equilibrium, Dordrecht, Kluwer Academic Publishers.

Van Willigenburg, T., Van den Beld, A.,. Heeger, F.R. and Verweij, M.F., 1998. Ethiek in praktijk, Assen, Van Gorcum, Ch. 7.

Biographies

Arnaud Aubert holds a PhD in Behavioural Neurosciences from the University of Bordeaux, France. His research involves the study of the behavioural implications of immune activation as well as the analysis of emotional processes and how they are related to cognition. Since 1998, he is lecturer at the University of Tours, France, and initiated the introduction of several teachings in animal ethics issues for students in Biology, Physiology and Psychology. He is currently the supervisor of the Department of Behavioural Sciences at the University of Tours.

Xavier Boivin is ethologist at the French National Institute of Agricultural Research (INRA). He is working at the Herbivores Research Unit of the Research Centre of Clermont-Ferrand/Theix. He holds its PhD from the University of Rennes (France). Since 1989, he studied farm animal's behaviour. He has been responsible for research programs on farm ungulates' responses to humans and handling in order to improve animal welfare, farmer's comfort and safety. He published many scientific papers on this topic. He is very experienced with beef cattle but also worked with sheep, goats and horses in different countries.

Donald M. Broom (MA, PhD, ScD) has been Professor of Animal Welfare at Cambridge University since 1986. His group have developed concepts and methods of scientific assessment of animal welfare, publishing over 500 papers on welfare in relation to housing and transport, behaviour problems of pets, attitudes to animals and ethics of animal usage. He served on UK and Council of Europe committees and has been Chairman or Vice Chairman of EU Scientific Committees on Animal Welfare since 1990. Amongst his seven books are *Stress and Animal Welfare* (Broom and Johnson 1993, Kluwer) and *The Evolution of Morality and Religion* (2003, CUP).

Stine B. Christiansen is a DVM from The Royal Veterinary and Agricultural University (RVAU) in Copenhagen, Denmark (1993) and holds an MSc in Applied Animal Behaviour and Animal Welfare from the University of Edinburgh, Scotland (1995). Currently she is doing a PhD at the RVAU. Since 1998 she has been involved in research projects concerning animal ethics and has served as the scientific secretary of the Danish Animal Ethics Council. She has been part of the development of and is now part of the teaching team for the "Introductory Philosophy Course for Veterinary Students" at the RVAU.

Tjard de Cock Buning trained and graduated in Biophysics (University of Leiden) and Philosophy (University of Amsterdam), and accomplished his PhD research in sensory neurophysiology. From 1986 he is appointed on the chair for Lab Animal Issues (Utrecht University), dealing with ethics, history and alternatives in animal experimentation. In 1986 he joint, as the first ethicist in the Netherlands, the Animal Care and Use Committee of the Leiden University, and initiated the first training and education workshop for members of such committees. He developed ethical decision models for ethical review committees on animal experimentation and wrote various articles on this topic. He is involved in methodological projects that focus on societal dialogue in relation to life science innovations.

Trine Dich holds an MSc in Geography (1989) from University of Copenhagen, Denmark. From 1989 to 1998 she has been teaching Geography at high school and university (B.Sc.) level and writing textbooks. In 1999 she was employed as an Assistant Professor at the Royal Veterinary and Agricultural University in Copenhagen to teach and develop courses for agricultural science and veterinary students in problem based project work and since 2001 also in bioethics and philosophy of science.

Sandra Edwards graduated with a BA (Hons) in Natural Sciences from Cambridge University and then completed a PhD in farm animal behaviour at the University of Reading. She has worked for more than 25 years in applied research on farm livestock for different UK organisations. From her position as Head of the Animal Management and Health Department of the Scottish Agricultural College, she moved in 1998 to become Reader in Animal Science in the Department of Agriculture at the University of Aberdeen. In 2000, she was appointed to the Chair of Agriculture at the University of Newcastle, where her research interests focus particularly on livestock production systems and their implications for animal welfare.

Jaume Fatjó studied at the Autonomous University of Barcelona from which he received a degree in Veterinary Medicine in 1993. In 1994 he spent some months as a visiting veterinarian at the Animal Behaviour Clinic (New York State College of Veterinary Medicine). Since 1995 he is responsible for the Animal Behaviour Clinic at the Barcelona School of Veterinary Medicine. In addition to clinical activity, since 2004 he is an associate professor of ethology and animal welfare at the same University. His research is focused on aggressive behaviour in dogs.

Gustavo C. Gandini holds a DVM degree from Parma University, Italy. He focused his research interests in animal genetics and farm animal genetic resources conservation and management. Presently he is associate professor at the Faculty of Veterinary Medicine, University of Milan. Since 2002 he is Chairman of the Working Group on Animal Genetic Resources of the European Association for Animal Production. He is involved in different nature conservation and animal protection programmes and organisations.

Alison J. Hanlon holds a PhD on animal stress from the University of Aberdeen, Scotland, after an MSc (by research) from University College Dublin, Ireland. In 1995, Dr Hanlon was appointed Lecturer in Animal Behaviour and Welfare at the Faculty of Veterinary Medicine, University College Dublin. She has developed teaching programmes on animal behaviour and welfare, using innovative strategies such as problem-based learning. In 2001, nominated by her students, she was the first member of the Faculty to receive the '*President's Teaching Award*'. She is currently involved in the development of e-learning software on animal ethics and a pedagogic study on approaches to learning and learning styles.

Tina Hansen was educated at the University of Roskilde (MA in Geography and International Studies, 1991). From 1992 to 1998 she was a Teaching Assistant and PhD student at the Department of Geography and International Studies at the University of Roskilde. Since January 1999 she is Assistant Professor at the Royal Veterinary and Agricultural University in Denmark. Since 2001 she has been involved in developing courses and teaching in bioethics and philosophy of science.

Robert Heeger is emeritus professor of ethics at the theology and philosophy departments of Utrecht University and at present working at the Ethics Institute of this university. He received his doctor's degree from Uppsala University, Sweden (1975), where he was lecturer in ethics. 1991-1995 he was president of Societas Ethica: European Society for Research in Ethics, 1993-2002 he was a member of the Dutch Council for Animal Matters, and 1994-2003 he was president of the Netherlands School for Research in Practical Philosophy. Since 1986 he has been teaching animal and veterinary ethics at the Faculty of Veterinary Medicine, Utrecht University.

John Hodges (PhD) was responsible for animal breeding and genetic resources in the Food and Agriculture Organization of the UN from 1982-90. Previously he was Professor of Animal Genetics at the University of British Columbia, Canada and earlier taught at Cambridge University. He is the former Head of Production Division of the Milk Marketing Board of England and Wales. He has an Honours Bachelor degree in Agriculture (Reading), Masters in Animal Production (Cambridge), Doctorate in Animal Genetics (Reading) and a degree in Business Administration (Harvard, USA). He now writes and speaks on genetics, ethics, biotechnology, agriculture and food.

Pernille Kaltoft is a Senior Scientist within Environmental Sociology at the National Environmental Research Institute in Denmark since 2000. From 1998 to 2000 she was Assistant Professor at the Royal Veterinary and Agricultural University (KVL) in Copenhagen in the Department of Natural Resources and Economics. She is educated as an engineer from the Technical University of Denmark and holds a PhD (on relations between ethics, view of nature, knowledge and practices in organic farming) from the same university. Since 1999 she has taught ethics for agronomy students and students in biotechnology at KVL, starting as a pioneer activity, in collaboration with Professor Peter Sandøe.

Linda J. Keeling has a PhD in Zoology from Edinburgh University and is now Professor of Animal Welfare at the Swedish University of Agricultural Sciences. Her current research interests are assessment of animal welfare including the development of techniques to assess the emotional states of animals, behavioural problems in farm animals and human-animal interactions. She is responsible for education in animal welfare and legislation for veterinary students and agriculture students.

Ute Knierim obtained her Veterinarian and Doctoral Degree from the School of Veterinary Medicine Hannover, and her MSc in Applied Animal Behaviour and Animal Welfare from the University of Edinburgh. Three years of work at the Animal Welfare Section of the German Federal Ministry of Agriculture provided her with experience in animal welfare legislation and implementation. As a lecturer and researcher at the School of Veterinary Medicine Hannover, she covered the area of Animal Welfare and Applied Ethology. Currently she is Professor for Farm Animal Behaviour and Husbandry at the Faculty of Organic Agricultural Sciences of the University of Kassel, continuing to teach and research welfare and behaviour issues in farm animals with special emphasis on cattle and poultry.

Tadeusz Kuczynski, Professor at the University of Zielona Gora, Poland, is currently the Dean of Faculty of Civil and Environmental Engineering. He specialises in sustainable management of natural resources, sustainable rural area planning and development, agricultural pollutants emission, environmental control in animal housing, animal welfare on farm and in transportation. For last 11 years he has been a Board member of the International Organisation of Agricultural Engineering. Since 1993 he actively participates in two UNECE TFEIP working groups: Ammonia Expert Group and the Agriculture and Nature Panel. He is proposed for chairing the Working Group on Ethics in the SOCRATES Thematic Network: Redefining the curricula for the multifunctional rural environment - agriculture, forestry and the rural society (MRENet).

Pierre Le Neindre, first trained as an agricultural engineer, received his PhD from the University of Rennes on animal ethology. He works in the French National Institute of the Agricultural Research (INRA), were he develops studies on the social environment of the ruminants and on the effect of this social environment (mother young relationships, man animal relationships) on the adaptation of the animals and on their welfare. He has been for more than ten years member of the European scientific committee on animal health and animal welfare. Former president of the International Society of Applied Ethology (1997-1998), he is currently president of the INRA research centre of Tours and vice-president of the scientific committee of the European Food Safety Authority.

Monica Libell holds a PhD in history of ideas and science at Lund University. Part of her doctoral work was conducted at UC Berkeley in California. She is currently working as a lecturer and researcher at the department of cultural sciences at Lund University. Her research involves animal ethical issues, and she is engaged in work on medical bioethics and agricultural ethics. She is currently also working on a book on cultural and historical meanings of "human" and "animal" and linking them to concepts of "dehumanization" and "anthropomorphism."

Stefan Mann holds a PhD in Agriculture (1997) from Hohenheim University and one in Economics (2004) from the University of Greifswald. After working in Germany's federal administration for five years, he taught agricultural policy from 1999-2002 at Rostock University where he started his specialisation on welfare economics. Since 2002, he heads the Forecasting group at the Swiss Federal Research Station of Agricultural Economics. Although he does research on diverse fields of agricultural policy, his main publications focus on contradictions between individual willingness to pay and utility, i.e. the concept of merit goods.

Xavier Manteca Vilanova holds a Master's degree in Applied Animal Behaviour and Animal Welfare from the University of Edinburgh and a PhD in physiology from the Autonomous University of Barcelona. Currently, he is associate professor at the School of Veterinary Science in Barcelona, where he teaches animal behaviour and animal welfare. His main research interests are farm animal welfare -particularly during transport and slaughter-, feeding and social behaviour of pigs and ruminants and behavioural problems of companion animals. He is member of the Panel on Animal Health and Animal Welfare of the European Food Safety Authority and chairman of the Institutional Animal Care and Use Committee of the Autonomous University of Barcelona.

Michel Marie holds a Master degree in animal physiology (Claude Bernard University, Lyon) and a PhD in animal endocrinology (P. & M. Curie University, Paris). He taught animal reproduction and production in the Hassan II Agronomic and Veterinary Institute (Rabat, Morocco), and now in the National School of Agronomy and Food Industries (INPL, Nancy), were he developed a curriculum in bioethics. His research interests are in farm sustainability and animal ethics assessment. He co-ordinated the "Aristoteles" group devoted to animal bioethics teaching in the Socrates AFANet Thematic Network, and is the chairman of the Ethics Working Group of the European Association for Animal Production.

Ben Mepham is Special Professor of Applied Bioethics at the University of Nottingham. Graduating in physiology from University College London in 1961, he completed a PhD in biochemistry at the ARC Institute of Animal Physiology, Cambridge before moving to the University of Nottingham, where he was first lecturer and then reader in physiology. Since 1993 he has been Director of the Centre for Applied Bioethics in the School of Biosciences. He was Executive Director of the Food Ethics Council (1998-2003), and, from 2000, a founder member of the UK Government's Biotechnology Commission (AEBC) and the European Society for Agricultural and Food Ethics. Oxford University Press published his latest book, 'Bioethics: an introduction for the biosciences,' in 2005.

Vincent Pompe has a Master degree in biology of the University of Leiden and a Master in philosophy of the University of Utrecht. He is a senior lecturer Animal Ethics at the Van Hall Institute in Leeuwarden, since 1992. The development of ethical oriented education on animal issues is his main task. He has special interests in socio-psychological dimensions of animal use and philosophy of animal welfare. Additionally, he is linked to the University of Groningen as a member of the Institutional Animal Care and Use Committee and as a lecturer ethics of the post-doctoral course on Laboratory Animal Science.

Michael Reiss is Professor of Science Education at the Institute of Education, University of London and Head of its School of Mathematics, Science and Technology. He is Chief Executive of Science Learning Centre London, Honorary Visiting Professor at the University of York, Docent at the University of Helsinki, Director of the Salters-Nuffield Advanced Biology Project and a member of the Farm Animal Welfare Council. After a BA in Natural Science and a PhD in animal behaviour he did post-doctoral research before qualifying as a teacher (PGCE). He taught in schools before taking up academic posts at the University of Cambridge and Homerton College, Cambridge. He has an MBA and is a Priest in the Church of England.

Laurent Rollet holds a PhD in philosophy and history of science from the University of Nancy 2 (France). In 1999 and 2000 he was teaching and research assistant at the National Institute of Applied Sciences in Lyon. Since 2001, he is in charge of ethics and epistemology teaching at the Lorraine National Polytechnics Institute and at the University Henri Poincaré - Nancy 1. As a researcher, he is a member of the Henri Poincaré Archives and he is in charge of the edition of Henri Poincaré's private correspondence.

Bart L.J.E. Rutgers received his degree in Veterinary Medicine from the Utrecht University, The Netherlands in 1978. Until 1989 he worked as a surgeon at the Department of Large Animal Surgery of the Faculty of Veterinary Medicine in Utrecht. Given his interest in animal ethics, he obtained an appointment at the Department of Animals, Science & Society of the same faculty in 1989. Since that time he is concerned with teaching and research in animal and veterinary ethics. In 1993 he was given the PhD degree in this field. Since 1989 he has been a member of the Ethical Committee of The Royal Dutch Veterinary Medical Association.

Peter Sandøe was educated at the Universities of Copenhagen (MA in philosophy) and of Oxford (D.Phil. in philosophy). From 1985 to 1997 he was appointed at the Department of Philosophy of the University of Copenhagen were he headed the Bioethics Research Group. Since 1992, he has served as Chairman of the Danish Animal Ethics Council. He is now Professor in Bioethics at the Royal Veterinary and Agricultural University in Copenhagen, and director of the Centre for Bioethics and Risk Assessment. As from August 2000 he has been president of the European Society for Agricultural and Food Ethics. The major part of his research has been within bioethics with particular emphasis on ethical issues related to animals, biotechnology and food production.

Benjamin Taubald studied theology in Vienna and Muenster and received a Dr.theol. in Foundational Theology in 2000. Since 1997, he works at the Institute for Social Ethics at the University of Vienna. His fields of research include borderline questions of theology and philosophy, political theology, and media ethics. Since 2001, he co-operates with the Institute for Animal Husbandry and Animal Protection at the University of Veterinary Medicine, Vienna.

Ellen ter Gast holds a Master degree in both medical biology (University of Amsterdam) and philosophy (Free University). Currently she is undertaking a postgraduate study at Ph.D level on the ethics of genetic modification of laboratory mice at the Radboud University in Nijmegen. Since 2001 she is a lecturer on bioethics, philosophy of science and communication of science at both the Free University (till 2002) and the Radboud Universtity. From 2000 to 2004 she was the chairman of the animal experimentation committees of the Dutch Cancer Institute and Sanguin.

Antonio Velarde studied Veterinary Medicine at the Autonomous University of Barcelona (UAB); where he received his BVSc degree in 1995, a Master degree in Animal Production in 1998, and the PhD degree in Animal Welfare and Meat Science in 2000. He is currently working in the Meat Research Centre of the Catalan Institute for Food and Agricultural Research and Technology (IRTA), leading different projects on farm animal behaviour and welfare. His research involves the effect of pre-slaughter handling on animal welfare and meat quality in pigs, cattle and sheep. He has been involved with the working groups established by the European Food Safety Authority on welfare aspects of animal stunning and killing methods, and welfare aspects of castration in piglets.

Henk Verhoog studied biology at the University of Amsterdam. From 1968 till 1999 he was lecturer on social and ethical aspects of biology at the Institute of Theoretical Biology of Leiden University. His Philosophical dissertation was about 'Science and the social responsibility of natural scientists'. He published on the relationship between science and ethics, and the philosophical aspects of the human-animal (human-nature) relationship. He has been member of state advisory boards on the ethics of animal experimentation, the genetic modification of animals, and the release of GMOs in the environment. Now is associate of the Louis Bolk Institute, an independent research centre on organic agriculture, nutrition and health care, with interest for the values and principles underlying organic agriculture.

Caroline Vieuille ended studies focused on animal ethology by a doctoral degree in Biology of organisms (Mention Biology of behaviour, 1991), while being responsible for a 500 grouped-sows production unit (within the framework of a 'CIFRE' research training). Still highly motivated by animal behaviour and by human-animal relationship but not much by economical management, she answered INRA researcher missions on animal welfare, working freelance until today. She has mainly been working on sows welfare's evaluation and piglets' crushing. Since 1994, she has also been half-time associated teacher at Tours' University, teaching ethology, animal welfare and ethics.

Eberhard von Borell has a degree in Agricultural Biology and a PhD in Life Sciences from the University of Hohenheim in Germany. After his PhD he spent in total 6 years in Canada (University of Guelph) and the United States where he held a faculty position in Animal Behaviour and Stress Physiology at Iowa State University. In 1994, he became a Professor in Animal Husbandry and Ecology at the Martin-Luther-University Halle-Wittenberg in Germany. His main expertise is in animal welfare, animal housing and management assessment and behavioural physiology. Currently he serves as the President of the Commission of Management and Health of the European Association for Animal Production.

Anne Vonesch obtained a diploma of protestant theology in Basel, Switzerland, in 1972, and of general medicine in Strasbourg in 1981. This became the background of voluntary involvement with associations in various issues such as river ecology and landscape, air pollution, transport, road safety, and human rights. Since 1995 she concentrates on farm animal welfare, including various contacts on a national and European level. Insisting on dialogue with farmers, she works on regional animal production issues with the Alsatian Consumers Chamber, and she represents Alsace Nature in the agricultural domain.

Index – Animal bioethics: cases, situations

Animal Bioethics: Principles and Teaching Methods

Index – Teaching objectives, strategies, methods

Index – Philosophical concepts, schools, authors